AIDS Testing

Gerald Schochetman
J. Richard George
Editors

AIDS Testing

A Comprehensive Guide to Technical, Medical, Social, Legal, and Management Issues

Second Edition

Foreword by Walter R. Dowdle

With 59 Illustrations

Springer-Verlag
New York Berlin Heidelberg London Paris
Tokyo Hong Kong Barcelona Budapest

Gerald Schochetman, Ph.D.
Chief
Laboratory Investigations Branch
Division of HIV/AIDS
National Center for Infectious Diseases
Centers for Disease Control and
Prevention
Atlanta, GA 30333
USA

J. Richard George, Ph.D.
Chief
Developmental Technology Section
Laboratory Investigations Branch
Division of HIV/AIDS
National Center for Infectious Diseases
Centers for Disease Control and
Prevention
Atlanta, GA 30333
USA

Library of Congress Cataloging-in-Publication Data
AIDS testing: a comprehensive guide to technical, medical, social, legal, and management issues / [edited by] G. Schochetman, J.R. George.—2nd ed.
p. cm.
Includes bibliographical references and index.
ISBN 0-387-94291-2.—ISBN 3-540-94291-2
1. AIDS (Disease)—Diagnosis. 2. Medical screening.
I. Schochetman, Gerald. II. George, J. Richard.
[DNLM: 1. Acquired Immunodeficiency Syndrome—diagnosis. 2. HIV Infections—diagnosis. 3. Diagnosis, Laboratory, 4. Ethics, Medical. WD 308 A288468 1994]
RC607.A26A359 1994
616.97′92075—dc20
DNLM/DLC
for Library of Congress 94-7986
CIP

Printed on acid-free paper.

Production coordinated by Chernow Editorial Services, Inc., and managed by Theresa Kornak; manufacturing supervised by Gail Simon.
Typeset by Best-set Typesetter Ltd., Hong Kong.
Printed and bound by R.R. Donnelley and Sons, Harrisonburg, Virginia.
Printed in the United States of America.

9 8 7 6 5 4 3 2 1

ISBN 0-387-94291-2 Springer-Verlag New York Berlin Heidelberg
ISBN 3-540-94291-2 Springer-Verlag Berlin Heidelberg New York

Foreword

During the two years since the publication of the first edition of this book, the global spread of human immunodeficiency virus/acquired immunodeficiency syndrome (HIV/AIDS) has continued. HIV was estimated by the World Health Organization (WHO) in 1993 to have infected at least 13 million individuals worldwide, with 1 million infected in the United States. HIV/AIDS in the United States has become the leading cause of death among men 25 to 44 years of age and the fifth leading cause of death among women of the same age group. Prevention of HIV infection remains a global challenge.

Testing for HIV is the cornerstone for surveillance and prevention programs and for the provision of appropriate medical care for those who are infected. Such testing is equally essential to the search for effective antivirus drugs and vaccines.

This second edition of *AIDS Testing* incorporates the most current thinking on test methodology and interpretation, some of which has changed considerably over the past two years. This edition also has been expanded to include a section consisting of six chapters on test applications and a section consisting of four chapters on management issues. This edition, like the first, describes in clear terms all the complex elements of testing, including applications, scientific principles, quality assurance, safety, and medical, ethical, and legal considerations.

For laboratory workers and epidemiologists involved in surveillance and prevention, this book provides a framework for program direction and quality data. For health care providers who must understand and apply testing for detection, counseling, and caring of HIV-infected persons, this book provides a comprehensive overview of the spectrum of issues related to HIV testing. Tests are discussed in terms of the information they provide and the application of that information for the benefit of the patient. For persons involved in HIV/AIDS research and for all laboratory workers this book serves as a handy reference for the design and safe implementation of quality HIV tests.

Since the first edition, notable advances have been made in the use of the polymerase chain reaction (PCR) in a variety of research and clinical settings. The chapter on nucleic acid amplification techniques (e.g., PCR) in this edition reflects these more recent demonstrations of test utility.

Like PCR, other laboratory tests for HIV/AIDS and associated host responses will continue to evolve as our knowledge expands and new applications are required. In whatever direction test technology may evolve, the basic scientific principles described in this book for performance of HIV/AIDS tests of high sensitivity, specificity, and quality will remain.

Walter R. Dowdle
Deputy Director
Centers for Disease Control and Prevention

Contents

Contributors

David M. Bell, M.D., Hospital Infections Program, National Center for Infectious Diseases, Centers for Disease Control and Prevention, Atlanta, GA 30333, USA

Elizabeth A. Bolyard, R.N., M.P.H., Division of HIV/AIDS, National Center for Infectious Diseases, Centers for Disease Control and Prevention, Atlanta, GA 30333, USA

Barbara H. Bowman, Ph.D., Roche Molecular Systems Inc., Alameda, CA 94501, USA

Michael P. Busch, M.D., University of California, San Francisco, and Department of Scientific Services, Irwin Memorial Blood Center, San Francisco, CA 94118, USA

Denise M. Cardo, M.D., Hospital Infections Program, National Center for Infectious Diseases, Centers for Disease Control and Prevention, Atlanta, GA 30333, USA

Kenneth G. Castro, M.D., Division of Tuberculosis Elimination, National Center for Prevention Services, National Center for Infectious Diseases, Centers for Disease Control and Prevention, Atlanta, GA 30333, USA

Carol A. Ciesielski, M.D., Surveillance Branch, Division of HIV/AIDS, Centers for Disease Control and Prevention, Atlanta, GA 30333, USA

Lynda S. Doll, Ph.D., Division of HIV/AIDS, National Center for Infectious Diseases, Centers for Disease Control and Prevention, Atlanta, GA 30333, USA

Jay S. Epstein, M.D., FDA/CBER/OBRR, Office of Blood Research and Review, Center for Biologics Evaluation and Research, U.S. Food and Drug Administration, Rockville, MD 20852, USA

J. Richard George, Ph.D., Chief, Developmental Technology Section, Laboratory Investigations Branch, Division of HIV/AIDS, National Center for Infectious Diseases, Centers for Disease Control and Prevention, Atlanta, GA 30333, USA

Nancy J. Haley, Ph.D., Director, Insurance Testing Laboratory, Metropolitan Life Insurance Company, Elmsford, NY 10523, USA

Thomas L. Hearn, M.S., Public Health Program Practice Office, Centers for Disease Control and Prevention, Atlanta, GA 30333, USA

Walid Heneine, Ph.D., Division of Viral and Rickettsial Diseases, National Center for Infectious Diseases, Centers for Disease Control and Prevention, Atlanta, GA 30333, USA

Priscilla Holman, National AIDS Information and Education Program, Centers for Disease Control and Prevention, Atlanta, GA 30333, USA

C. Robert Horsburgh, Jr. M.D., Division of HIV/AIDS, National Center for Infectious Diseases, Centers for Disease Control and Prevention, Atlanta, GA 30333, USA

Ann N. James, Ph.D., J.D., Jenkens & Gilchrist, Houston, TX 77002, USA

Robert S. Janssen, M.D., Division of HIV/AIDS, National Center for Infectious Diseases, Centers for Disease Control and Prevention, Atlanta, GA 30333, USA

Jonathan E. Kaplan, M.D., Division of HIV/AIDS, National Center for Infectious Diseases, Centers for Disease Control and Prevention, Atlanta, GA 30333, USA

Meaghan B. Kennedy, M.P.H., Division of HIV/AIDS, National Center for Infectious Diseases, Centers for Disease Control and Prevention, Atlanta, GA 30333, USA

Rima F. Khabbaz, M.D., Division of Viral and Rickettsial Diseases, National Center for Infectious Diseases, Centers for Disease Control and Prevention, Atlanta, GA 30333, USA

Fred Kroger, Director, National AIDS Information and Education Program, Centers for Disease Control and Prevention, Atlanta, GA 30333, USA

Alan L. Landay, Ph.D., Director of Clinical Immunology, Office of Consolidated Laboratory Services, Rush-Presbyterian St. Luke's Medical Center, Chicago, IL 60612, USA

Alison C. Mawle, Ph.D., Division of Viral and Rickettsial Diseases, National Center for Infectious Diseases, Centers for Disease Control and Prevention, Atlanta, GA 30333, USA

Bruce J. McCreedy, Ph.D., CMB Division, Roche Biomedical Laboratories, Research Triangle, NC 27709, USA

J. Steven McDougal, M.D., Chief, Immunology Branch, Division of HIV/AIDS, National Center for Infectious Diseases, Centers for Disease Control and Prevention, Atlanta, GA 30333, USA

Janet K.A. Nicholson, Ph.D., Division of HIV/AIDS, National Center for Infectious Diseases, Centers for Disease Control and Prevention, Atlanta, GA 30333, USA

Jacquelyn A. Polder, B.S.N., M.P.H., Hospital Infections Program, National Center for Infectious Diseases, Centers for Disease Control and Prevention, Atlanta, GA 30333, USA

Mark A. Rayfield, Ph.D., Laboratory Investigations Branch, Division of HIV/AIDS, National Center for Infectious Diseases, Centers for Disease Control and Prevention, Atlanta, GA 30333, USA

Barry S. Reed, M.D., J.D., Medical Department, Metropolitan Life Insurance Company, New York, NY 10010, USA

Jonathan Y. Richmond, Ph.D., Director, Office of Health and Safety, Centers for Disease Control and Prevention, Atlanta, GA 30333, USA

Martha F. Rogers, M.D., Chief, Epidemiology Branch, Division of HIV/AIDS, National Center for Infectious Diseases, Centers for Disease Control and Prevention, Atlanta, GA 30333, USA

Gerald Schochetman, Ph.D., Chief, Laboratory Investigations Branch, Division of HIV/AIDS, National Center for Infectious Diseases, Centers for Disease Control and Prevention, Atlanta, GA 30333, USA

R.J. Simonds, M.D., Epidemiology Branch, Division of HIV/AIDS, National Center for Infectious Diseases, Centers for Disease Control and Prevention, Atlanta, GA 30333, USA

John J. Sninsky, Ph.D., Roche Molecular Systems, Inc., Alameda, CA 94501, USA

John W. Ward, M.D., Chief, Surveillance Branch, Division of HIV/AIDS, National Center for Infectious Diseases, Centers for Disease Control and Prevention, Atlanta, GA 30333, USA

Thomas J. White, Ph.D., Vice President, Roche Molecular Systems, Inc., Alameda, CA 94501, USA

1
Testing for Human Retrovirus Infections: Medical Indications and Ethical Considerations

JOHN W. WARD

In 1985 the first tests became available to identify antibody to the retrovirus subsequently called the human immunodeficiency virus (HIV). Since that time HIV antibody testing has become the cornerstone of public health efforts to stop the epidemic of HIV infection. The tests were used initially to screen donated blood and plasma, and their use for this purpose dramatically decreased the rate of transfusion-associated (TA) HIV transmission.[1,2] Soon thereafter, testing linked with counseling was recommended as a method to encourage safer sexual and drug use practices[3,4] and to aid in the diagnosis of some conditions associated with HIV infection, including those in the acquired immunodeficiency syndrome (AIDS) surveillance case definition.[5]

The major benefits of HIV testing are counseling to promote the behavior change necessary to reduce HIV transmission and the referral of HIV-seropositive persons for medical evaluation and treatment. As part of counseling, information is provided regarding the modes of HIV transmission and how the risk of transmission can be reduced. This information frequently results in these individuals changing high-risk sexual or drug use behaviors to prevent HIV infection in themselves and others.[6–10] Women at risk for infection receive information regarding perinatal transmission of HIV, and women found to be HIV-seropositive may elect to avoid pregnancy. Although counseling can be performed in the absence of HIV testing, knowledge of HIV seropositivity may be an additional motivator for behavior change.[6–8]

The development and use of medical therapies for HIV infection and associated illnesses have strengthened the medical indications for testing. During the summer of 1989 the use of zidovudine was found to delay the progression to AIDS in HIV-infected persons.[11] Although it is still uncertain when zidovudine therapy should be initiated for optimal effect on survival, persons with AIDS-defining conditions who receive zidovudine survive longer than do those with AIDS who do not receive this treatment.[12,13] Other prophylactic and therapeutic regimens, such as trimethoprim/sulfamethoxazole and aerosolized pentamidine for *Pneumocystis carinii* pneumonia, chemotherapy for tuberculosis (TB), and pneumococcal vaccination, may delay or prevent HIV-associated conditions.[14–16] Of the approximately 1 million HIV-infected persons in the United States, an estimated 460,000 to 675,000 have sufficient immune deficiency to warrant medical therapy according to the most recent treatment guidelines.[17]

As the uses of the tests have expanded, there has been a concomitant increase in concerns regarding the indications for testing, and the possible adverse consequences for persons testing positive for HIV infection.[18–21] There was initial concern about the reliability of the tests and the meaning of a positive test result. Subsequent studies of test performance have demonstrated the HIV antibody test to be highly sensitive and specific.[22–24] Studies consistently have shown that HIV serologic assays typically detect HIV antibodies within 3 to 12 weeks after infection, and seroconversion beyond 6 months after infection is uncommon.[25] A positive screening antibody assay followed by a positive Western blot or other more specific serologic assay almost invariably indicates an HIV infection.[26,27] However, false-positive test results remain a possibility and may cause persons to receive incorrect test information.[28] The laboratory reporting of test results to the clinician may also be erroneous or vague.[29,30] Although these instances of poor reporting are uncommon, such instances may result in unnecessary anxiety and changes in career and family planning.[31]

Protection from discrimination is particularly important with HIV testing, as the populations at greatest risk for HIV infection—homosexual and bisexual men, injecting drug users (IDUs), and racial and ethnic minorities—are already prone to discrimination by society. Persons known to be infected with HIV have experienced loss of employment, housing, and health or life insurance.[32–35] Health care workers may also refuse to care for HIV-seropositive persons.[32] In addition, HIV-seropositive persons are asked to inform their sexual partners, which may result in the destruction of marriages and other personal relationships.[35]

To maximize the benefits and reduce the risk of HIV antibody testing, the U.S. Public Health Service and other authorities have reviewed the medical and ethical issues regarding HIV testing and have recommended certain practices and procedures.[4,33,35–37] To benefit the person tested, testing should have the proper medical indications and should be linked with counseling and referral for appropriate evaluations and treatment. To increase the safety of the test procedure, HIV testing is recommended to be voluntary except for situations such as blood and tissue donation, with informed consent, and with procedures to protect the confidentiality of test results.[36,37]

This chapter primarily addresses testing for HIV antibody, the most common screening test for HIV infection. The efforts to maximize benefits of testing and minimize risks for persons consenting to be tested, however, apply to other tests for HIV infection and for other retrovirus testing as well. Recommendations for counseling individuals infected with the human T lymphotropic virus (HTLV) types I and II have been reported elsewhere.[38]

Indications for Testing

Not everyone requires HIV antibody testing. Knowledge of the behaviors that place persons at risk for HIV infection, the clinical conditions associated with HIV infection, and the rate of HIV infection or AIDS in the community or health care setting is required to properly target HIV counseling to those most in need of these services. Testing persons who are not at risk for HIV infection unnecessarily lowers the predictive value of a positive test and needlessly adds to the cost of public health programs and medical care. Those who should be offered HIV counseling and testing are discussed below.

Individuals with Behavioral Risks for HIV Infection

Persons at risk for HIV infection include homosexual/bisexual men, present or past IDUs, male or female prostitutes, sexual partners of infected persons or persons at increased risk, all persons with hemophilia who received clotting factor concentrate before 1985, and newborn infants of mothers with risk for HIV infection.[4,39] To identify risks associated with HIV transmission, physicians and other health care providers should interview patients regarding their sexual and drug use practices. HIV counseling and testing should be recommended for persons who self-report risks or for whom risks for HIV infection have been identified.

Individuals with Clinical Conditions Associated with HIV Infection

The test for HIV is a useful diagnostic tool for evaluating patients with clinical or laboratory evidence of HIV infection and is a necessary procedure before therapy specific for HIV infection is initiated. The Centers for Disease Control and Prevention (CDC) allows the reporting of AIDS cases based on a diagnosis of certain conditions indicative of underlying immunosuppression if persons are positive for HIV antibody.[5,40] Moreover, some other medical conditions are found more often among persons infected with HIV, including pneumonia caused by *Streptococcus pneumoniae* or *Hemophilus influenzae*. In addition, *Mycobacterium tuberculosis* infection is recognized as an HIV-associated condition, and pulmonary TB was added to the AIDS surveillance case definition on January 1, 1993. All persons with TB should be offered counseling and testing for HIV infection.[16,41]

Recipients of Blood and Blood Products Between 1978 and 1985

Persons who received transfusions of blood or blood components during the period 1978–1985 are at the greatest risk of TA HIV transmission. Physicians should obtain a transfusion history to determine the risk of TA HIV infection. An estimated 12,000 persons were infected with HIV before HIV testing of blood donors was initiated.[41] The risk of TA transmission is greatest for persons who received relatively large numbers of units of blood in areas of countries with high rates of AIDS and HIV infection.[42]

Individuals with Other Sexually Transmitted Diseases

All persons seeking treatment for sexually transmitted diseases (STDs) should be routinely offered HIV counseling and testing.[37] HIV testing is particularly important for those with other STDs because (1) they are practicing behaviors that may also transmit HIV; (2) the presence of some STDs facilitates the transmission of HIV; and (3) persons with STDs may not otherwise recognize their risk for HIV infection. STD clinic patients may have appreciable rates of HIV infection.[43,44] During 1991–1992, studies of 348,758 specimens from patients attending 112 STD clinics in 46 metropolitan areas of the United States found a median HIV seroprevalence of 1.6% (range 0.1–25.1%; the median HIV seroprevalence was higher for male heterosexual attendees who did not report IDU (0.9%, range 0–6.7%) than for female attendees who did not report IDU (0.6%, range 0–6.6%).[44]

Individuals Receiving Treatment for IDU

Injecting drug use is associated with practices that place persons at high risk for HIV transmission. The sharing of needles and other equipment among users facilitates exposure to the blood of HIV-infected persons. In addition, the higher rates of STDs among HIV-infected IDUs than among uninfected users suggest that unsafe sexual practices (e.g., exchange of sex for drugs or money) increases the HIV infection rates in this population. Accordingly, persons in drug treatment programs have high rates of HIV infection. During 1991–1992 surveys in 35 cities observed a median HIV seroprevalence rate of 7.5% (range 1–51%) among IDUs entering drug treatment programs. Although the rate of HIV infection was higher in the Atlantic coast states than in the rest of the United States, all IDUs can benefit from counseling and testing to promote safer behavior and reduce HIV transmission. Previous prevention efforts targeted to IDUs have been successful in decreasing high risk needle-sharing practices but have produced less improvement in reducing the frequency of high risk sexual activity.[6,45]

Women of Reproductive Age

Women at risk for HIV infection should be offered counseling and testing services. Women at risk include those who have used intravenous drugs, engaged in prostitution, received a transfusion of blood during the period 1978–1985, or had sexual partners who are infected with HIV or who are at risk for HIV infection: Women who live in communities where there is a known or suspected high prevalence of HIV infection are also at high risk.[4]

Although the overall HIV seroprevalence among women is low, women benefit from testing because they are more likely than men to be unaware of their risk for HIV infection.[46] Offering HIV testing to women in areas with high HIV seroprevalence is particularly important because many women may not report a risk for HIV infection during an interview.[47] During 1991–1992, a CDC survey of 254,828 serologic specimens from women of reproductive age attending 144 clinics in 39 metropolitan areas in the United States and Puerto Rico indicated the median rate of HIV infection to be 0.2% (range 0–2.3%). The median HIV seroprevalence rate among African-American women was 0.4% (range 0–8.2%) and was greater than the median seroprevalence of 0% among white (range 0–8.2%) and Hispanic women (range 0–1.5%) who attended the clinics surveyed.[48]

In addition, the availability of HIV testing and counseling for women is a prevention strategy for reducing the risk of perinatal HIV transmission. The offering of HIV counseling and testing allows women of childbearing age to determine their serologic status and to become aware of the risk of perinatal HIV transmission. These women may then decide whether they wish to delay becoming pregnant or to continue current pregnancies.

Children Born to Mothers with HIV Infection or Increased Risk for HIV Infection

Children born to infected mothers should be evaluated as early as possible after delivery for laboratory and clinical evidence of HIV infection so appropriate prophylactic and therapeutic interventions can begin. During

1991–1992 the CDC conducted a study of anonymous residual blood samples collected from newborns for routine metabolic screening in 42 states in the United States to detect the prevalence of maternal HIV antibody.[44,49] Based on data from 35 states, there were approximately 7000 annual births to HIV-infected women during 1991–1992, for an estimated annual national HIV infection prevalence of 1.7 per 1000 child-bearing women.[50] Assuming a perinatal transmission rate of 20% to 30%, an estimated 1400 to 2100 infants born in the United States during 1992 were HIV-infected.

Persons with Tuberculosis

All persons with tubercular infection (TB) must be assessed for HIV infection.[16,41] *Mycobacterium tuberculosis* infection is frequently found among HIV-infected persons. Among 20 TB treatment clinics surveyed during 1988–1989 in the United States, HIV seroprevalence ranged from 0% to 46% with a median rate of 3.4%.[51] The highest rates were found in the Northeast and along the Atlantic Coast and among persons with extrapulmonary disease. More than 100,000 persons in the United States are estimated to be co-infected with HIV and *M. tuberculosis*.[52] In San Francisco and Seattle, the rates of HIV infection among persons with TB are 29% and 23%, respectively.[51,53] Persons co-infected with HIV and *M. tuberculosis* have a much greater likelihood of developing clinical TB and may be more difficult to diagnose than non-HIV-infected persons.[54,55] In a retrospective study of IDUs with documented positive tuberculin skin tests, the observed incidence of active TB was 7.9 per 100 person-years for 49 HIV-infected persons compared with no cases among 62 HIV-seronegative persons. Knowledge of a patient's HIV serostatus is also important for proper medical management of TB because longer courses of therapy and prophylaxis are recommended for HIV-infected patients with TB.[56,57] Persons with HIV infection also appear to be at increased risk for active infection with drug-resistant strains of *M. tuberculosis*. Persons with positive tuberculin skin test reactions should be interviewed regarding their risk for HIV infection, and HIV counseling and testing should be strongly encouraged if risks are identified. Persons with TB or positive skin test reactions should be given or referred for the appropriate therapy.[18,41]

Patients in Acute Health Care Settings in Some Areas

The rate of HIV seroprevalence among patients in acute health care facilities in some areas is high. Studies have described HIV seroprevalence rates ranging from 0.3% to 6.0% among various patient populations.[47,58–62] In anonymous, unlinked serologic surveys conducted by the CDC, 0.2%

to 8.9% of persons receiving care in emergency departments and 0.1% to 7.8% of persons admitted to acute-care hospitals were HIV antibody-positive.[63,64] In two studies that obtained data on previous HIV testing, most of the HIV-seropositive patients were unaware of their HIV infection before hospital admission.[58,61] During 1989–1991, a survey of persons admitted to 20 U.S. hospitals observed an HIV seroprevalence of 4.7% (range 0.2–14.2%).[65] Based on this survey, an estimated 225,000 HIV-infected persons were hospitalized in 1990, with an estimated 72% admitted for conditions other than HIV or AIDS. Therefore testing patients on the basis of acknowledged risk behaviors or clinical signs and symptoms does not recognize all HIV-infected patients.

Based on these and other data, all hospitals and their associated providers should consider whether HIV counseling and testing services should be routinely offered to all or selected patients who receive care in their facilities.[36] Routine, voluntary HIV counseling and testing programs for patients 15 to 54 years of age are recommended for hospitals and associated clinics with a high rate of HIV infection. A high rate of HIV infection is defined as an estimated HIV seroprevalence rate of at least 1% or an AIDS diagnosis rate of more than 1.0 per 1000 discharges. Guidelines for the proper delivery of HIV counseling and testing services in acute health care settings are available.[36]

Individuals with Occupational Exposures Placing Them at Risk for HIV Infection

Occupational exposures that place a worker at risk of HIV infection include percutaneous injuries and contact of mucous membranes or skin (especially when the skin is chapped, abraded, or afflicted with dermatitis or the contact is prolonged or involves an extensive area) with blood and other body fluids to which universal precautions apply.[66] Data from several prospective studies among health care workers indicate that the average risk of seroconversion after a needlestick injury with HIV-infected blood is approximately 0.3%.[67–69] After an exposure, if the source patient has HIV infection or refuses testing, the worker should be evaluated clinically and serologically for evidence of HIV infection as soon as possible after the exposure and, if seronegative, should be retested for a minimum of 6 months after exposure (e.g., 6 weeks, 12 weeks, and 6 months after exposure) to determine if HIV infection has occurred.

Guidelines for HIV Counseling and Testing

Proper HIV counseling and testing procedures are important to maximize the benefits and reduce the risks of HIV testing. More individuals seek

HIV testing if they are confident about the testing program and can recognize the benefits of testing. The following recommendations were developed to standardize the methods of providing HIV counseling and testing services for persons in the acute clinical care setting.

Laboratory Testing and Reporting

Laboratories that perform HIV testing should be familiar with HIV test procedures and should use standardized criteria for test interpretation.[29] The HIV counseling process begins with reporting the test results. Therefore laboratory reports should contain clear language describing what tests were performed and their interpretation.[30] HIV test results should be recorded on the medical record, and every effort should be made to ensure the confidentiality of test results. A careful review of record-keeping practices and how medical information is maintained and distributed can help to identify ways to stop unnecessary disclosure of patient information. Except for reporting test results to local and state health departments where required, test results should not be distributed to persons or institutions outside the hospital or testing center without consent of the patient.

Informed Consent

Testing for HIV infection should be voluntary, and persons should be able to refuse testing. Mandatory HIV testing is not recommended as a method to prevent HIV transmission. The public health uses of mandatory testing are primarily limited to the screening of organ, plasma, blood, and other tissue donors. The testing of patients to prevent the risk of occupational HIV transmission has not been shown to be effective and is not a substitute for universal precautions.[64,69] In one study the frequency of parenteral or cutaneous contact was not influenced by a health care worker's knowledge of the patient's HIV seropositivity or behavioral risks for HIV infection.[69]

Prevention efforts continue to depend on persons receiving prevention messages and voluntarily changing high risk behaviors. Persons must seek or agree to HIV testing voluntarily to initiate the process of voluntary behavioral change. Specific documented informed consent is required for HIV testing.[35,37] Other than for immediate medical care, HIV counseling and testing should be offered in settings that allow persons to make an informed, voluntary decision regarding HIV testing. Before testing, the patient should be informed why the tests are being performed, the benefits and risks of testing, and how the test results are protected from disclosure; the patient should also be given an option to obtain more information before testing.

HIV-Related Counseling Services

The HIV-related counseling services should be provided to individuals before and after testing. HIV counseling should not be viewed as a one-time event conducted during a single patient encounter but, rather, as an on-going process to facilitate the necessary behavioral change to reduce HIV transmission risks. Therefore counselors should view all clinical encounters with patients as opportunities to provide and reinforce HIV prevention messages. Persons who conduct counseling sessions should have a good working knowledge of HIV infection and should be able to explain ways to modify unsafe sexual and drug use practices. The counseling message should be tailored for the individual patient and should be structured to the cultural background, sexual identity, developmental level, and language abilities of the patient. For persons found HIV-seropositive, counselors should be able to provide or arrange for appropriate medical evaluation and, if necessary, notification of sexual or needle-sharing partners.

Before testing counseling should consist in information about the procedures of the testing program, the risks for HIV infection, and the meaning of either a positive or negative test.[34] Pretest counseling should also include an assessment of the patient's risk for HIV infection and ways to reduce that risk. Persons who report IDU should be advised to discontinue their drug use or, if they cannot do so, to avoid sharing needles and other injection equipment. If sharing of needles and other equipment continues, the equipment should be washed with bleach before each use. Persons who are sexually active should be informed that sexual abstinence or avoiding sexual partners who are not known to be free of HIV are optimal ways to eliminate the risk of sexual transmission of HIV. Otherwise, condoms should be used during sexual contact. The risk of perinatal transmission should be discussed with women and with men who have sexual contact with women.

After testing, patients should be informed of their test results in a confidential manner. HIV-seronegative persons should be informed that a negative test result does not indicate protection from acquiring HIV infection and that continued high risk sexual or drug use behaviors could result in HIV infection. These persons should also be advised that HIV antibody tests may not detect infection that occurred during the several weeks or months immediately before testing. Persons concerned about a recent exposure should be advised to seek repeated testing at least 6 months after the exposure.

All HIV-seropositive patients should be counseled by persons who are able to interpret the test results correctly and who can discuss the medical, social, and psychological implications of HIV infection.[36] Reassessing a patient's risk for infection is also helpful for focusing risk reduction messages and evaluating the validity of test results. Repeat HIV testing may be necessary to verify the test result for HIV-seropositive persons without identified risk factors. Accurate counseling helps infected persons avoid behaviors that may transmit HIV and helps them to cope with the test information. Counseling sessions should convey hope and cautious optimism. It is not known if all HIV-infected persons develop AIDS, and therapeutic approaches are increasingly more effective in delaying or preventing HIV-related illnesses.

Partner Notification

Sexual and needle-sharing partners of HIV-infected persons are at risk for HIV. Persons who are HIV antibody-positive should be instructed how to notify their partners and to refer them for counseling and testing. If the HIV-infected persons are reluctant to notify partners directly, counselors should offer to inform the partners or to refer the names of the partners to the local health department.[34,36] As a last resort, health care providers may need to exercise their legal and ethical duty to inform known sexual and needle-sharing partners of HIV-infected persons who refuse to inform these persons.[31,32,34,42] When others notify partners of the infected patient, the name of the patient should not be used unless consent is given.

Medical Evaluation

Persons found to be HIV-seropositive require a medical evaluation, including immunologic monitoring, screening for other STDs and TB, prophylaxis against certain opportunistic illnesses, vaccinations, antiretrovirus therapy, and other preventive and therapeutic services.[36] Physicians, clinics, or institutions that offer testing should be able to provide a medical evaluation or have an effective referral mechanism.

Conclusion

The HIV counseling and testing must be based on sound medical indications and have goals that benefit the health of the public and the health of the individual. The benefits of testing are greatest when persons are tested by a reliable and experienced laboratory in a supportive atmosphere with proper consent and confidentiality procedures and then receive effective prevention messages and appropriate medical referral services. Access to HIV testing programs with these characteristics can encourage more persons to seek HIV testing services and reduce the rate of HIV transmission and HIV-related illnesses.

References

1. CDC. Provisional Public Health Service inter-agency recommendations for screening donated blood and plasma for antibody to the virus causing acquired immunodeficiency syndrome. MMWR 1985;34:1–5
2. Ward J, Holmberg SD, Allen JR, et al. Human immunodeficiency virus (HIV) transmission by blood transfusions screened negative for HIV antibody. N Engl J Med 1988;318:473–478
3. CDC. Additional recommendations to reduce sexual and drug abuse related transmission of human T-lymphotropic virus type III/lymphadenopathy-associated virus. MMWR 1986;35:152–155
4. CDC. Public health service guidelines for counseling and antibody testing to prevent HIV infection and AIDS. MMWR 1987;36:509–515
5. CDC. Revision of the CDC surveillance case definition for acquired immunodeficiency syndrome. MMWR 1987;36:1–15S
6. DesJarlais DC, Friedman SR. The psychology of preventing AIDS among intravenous drug users: a social learning conceptualization. Am Psychol 1988;43:865–870
7. Godfried JR, Van Griensven MS, Ernest MM, et al. Impact of HIV antibody testing on changes in sexual behavior among homosexual men in The Netherlands. Am J Public Health 1988;78:1575–1577
8. McCusker J, Stoddard AM, Mayer KH, et al. Effect of HIV antibody test knowledge on subsequent sex behaviors in a cohort of homosexual men. Am J Public Health 1988:78462–78467
9. McCusker J, Stoddard AM, Zapka JG, Zorn M, Mayer KH. Predictors of AIDS preventive behavior among homosexually active men: a longitudinal study. AIDS 1989;3:443–338
10. Higgins DL, Lavotti, O'Reilly KR, et al. Evidence for the effect of HIV antibody counseling and testing on risk behavior. JAMA 1991;266:2419–2429
11. Volberding PA, Lagakos SW, Koch MA, et al. Zidovudine in asymptomatic human immunodeficiency virus infection. N Engl J Med 1990;322:941–949
12. Abowker J-P, Swart AM. Preliminary analysis of the Concorde trial. Lancet 1993;341:889–890
13. Ragni MV, Kingsley LA, Zhou SJ. The effect of antiviral therapy on the natural history of human immunodeficiency virus infection in a cohort of hemophiliacs. J Acquir Immune Defic Syndr 1992;5:120–126
14. CDC. Guidelines for prophylaxis against Pneumocystis carinii pneumonia for persons infected with human immunodeficiency virus disease. MMWR 1992;41(RR-4)
15. CDC. General recommendations on immunization. MMWR 1989;38:207–227
16. CDC. Screening for tuberculosis and tuberculous infection in high-risk populations and the use of preventive therapy for tuberculous infection in the United States: recommendations of the Advisory Committee for Elimination of Tuberculosis. MMWR 1990;39:RR-8
17. CDC. Estimates of HIV prevalence and projected AIDS cases: summary of a workshop, October 31–November 1, 1989. MMWR 1990;39:110–112
18. Henry K, Maki M, Crossley K. Analysis of the use of HIV antibody testing in a Minnesota hospital. JAMA 1986;259:229–232

19. Sherer R. Physician use of the HIV antibody test: the need for consent, counseling, confidentiality, and caution. JAMA 1988;259:264–265
20. Dickens BM. Legal rights and duties in the AIDS epidemic. Science 1988; 239:580
21. Walters L. Ethical issues in the prevention and treatment of HIV infection and AIDS. Science 1988;239:597
22. CDC. Update: serologic testing for antibody to human immunodeficiency virus. MMWR 1988;36:833–840, 845
23. Weiss SH, Goedart JJ, Sarngadharan MB, et al. Screening test for HTL-III (AIDS) agent antibodies; specificity, sensitivity, and applications. JAMA 1985;253:221–225
24. Ward J, Grindon AJ, Feorino PM, et al. Laboratory and epidemiologic evaluation of an enzyme immunoassay for antibodies to HTLV-III. JAMA 1986;256:357–361
25. Horsburg CR J, Ou CY, Jason J, et al. Duration of human immunodeficiency virus infection before detection of antibody. Lancet 1989;2:637–640
26. Pan LZ, Sheppard HW, Winkelstein W, et al. Lack of detection of human immunodeficiency virus in persistently seronegative homosexual men with high or medium risks for infection. J Infect Dis 1991;164:962–964
27. Ou CY, Kwok S, Mitchell SW, et al. DNA amplification for direct detection of HIV-1 in DNA of peripheral blood mononuclear cells. Science 1988;239: 295–297
28. Saag MS, Britz J. Asymptomatic blood donor with a false positive HTLV-III Western blot. N Engl J Med 1986;314:118
29. CDC. Interpretation and use of the Western blot assay for serodiagnosis of human immunodeficiency virus type 1 infections. MMWR 1989;38(suppl 7S)
30. Benenson AS, Peddercord M, Hofherr LK, et al. Reporting the results of human immunodeficiency virus testing. JAMA 1989;262:3435–3438
31. Lo B, Steinbrook RL, Cooke M, et al. Voluntary screening for human immunodeficiency virus (HIV) infection. Ann Intern Med 1989;110:727–733
32. Gostin LO. The AIDS litigation project; a national review of court and human rights commission decision. Part 2. Discrimination. JAMA 1990;263: 2086–2093
33. Landesman SH, DeHovitz JA. HIV diagnostic testing: ethical, legal, and clinical considerations. Sex Transm Dis 1990;81:975–983
34. Bayer R. Editorial review: ethical and social policy issues raised by HIV screening: the epidemic evolves and so do the challenges. AIDS 1989;3: 119–124
35. Levine C, Bayer R. The ethics of screening for early intervention in HIV disease. Am J Public Health 1989;79:1661–1667
36. CDC. Recommendations for HIV testing services for inpatients and outpatients in acute-care hospital setting; and technical guidance on HIV counseling. MMWR 1993;42(no. RR-2)
37. CDC. 1989 Sexually transmitted diseases treatment guidelines. MMWR 1989; 38:S-8
38. CDC. Recommendations for counseling persons infected with human T-lymphotrophic virus, types I and II: recommendations on prophylaxis and therapy for disseminated *Mycobaterium avium* complex; four adults and adolescents infected with human immunodeficiency virus. MMWR 1993; 42(no. RR-9):1–13

39. Pizzo PA, Eddy J, Faloon J. Acquired immune deficiency syndrome in children: current problems and therapeutic considerations. Am J Med 1988; 85:195–202
40. CDC. 1993 Revised classification system for HIV infection and expanded surveillance case definition for AIDS among adolescents and adults. MMWR 1992;41(no. RR-17)
41. CDC. Tuberculosis and human immunodeficiency virus: recommendations of the Advisory Committee for the Elimination of Tuberculosis (ACET). MMWR 1989;3:236–238, 243–250
42. CDC. Human immunodeficiency virus infection in transfusion recipients and their family members. MMWR 1987;36:137–140
43. McCray E, Onorato IM, et al. Sentinel surveillance of human immunodeficiency virus infection in sexually transmitted disease clinics in the United States. Sex Transm Dis 1992;19:235–241
44. Centers for Disease Control and Prevention (CDC). National HIV Serosurveillance Summary—Results Through 1992. Publication HIV/NCID/11-93/036. Atlanta: US Department of Health and Human Services, Public Health Service, 1993
45. Guydish JR, Abramowitz A, Woods W, et al. Changes in needle sharing behavior among intravenous drug users: San Francisco, 1986–88. Am J Public Health 1990;2:253–271
46. Ward J, Kleinman SH, Douglas DK, Grindon AJ, Holmberg SD. Epidemiologic characteristics of blood donors with antibody to human immunodeficiency virus. Transfusion 1988;28:298–301
47. Lindsay MK, Peterson HB, Feng TI, et al. Routine antepartum human immunodeficiency virus infection screening in an inner city population. Obstet Gynecol 1989;74:289–294
48. Sweeney PA, Onorato IM, Allen DM, et al. Sentinel surveillance of human immunodeficiency virus infection in women seeking reproductive health services in the United States, 1988–1990. Obstet Gynecol 1992;79:503–510
49. Gwinn M, Pappaioanou M, George R, et al. Prevalence of HIV infection in childbearing women in the United States: surveillance using newborn blood samples. JAMA 1991;265:1704–1708
50. Davis S, Gwinn M, Wasser S, et al. HIV prevalence among US childbearing women, 1989–1992 [abstract 27]. In: Program and Abstracts of the First National Conference on Human Retroviruses and Related Infections, Washington, DC, December 12–16, 1993:60
51. Onorato IM, McCray E, Field Services Branch. Prevalence of human immunodeficiency virus infection among patients attending tuberculosis clinics in the United States. J Infect Dis 1992;165:87–92
52. Raviglione MC, Narain JP, Kochi A. HIV-associated tuberculosis during the AIDS era. JAMA 1993;269:2865–2868
53. Chaisson RE, Theuer CP, Schecter GF, Hopewell PC. HIV infection in patients with tuberculosis [abstract]. In: Abstracts, IV International Conference on AIDS, Stockholm, Book 2, 1988:313
54. Selwyn PA, Hartel D, Lewis VA, et al. A prospective study of the risk of tuberculosis among intravenous drug users with human immunodeficiency virus infection. N Engl J Med 1989;320:545–550
55. Selwyn PA, Sckell BM, Alcabes P, et al. High risk of active tuberculosis in HIV infected drug users with cutaneous anergy. JAMA 1992;268:504–509

56. CDC. Tuberculosis and human immunodeficiency virus infection: recommendations of the Advisory Committee for the Elimination of Tuberculosis (ACET). MMWR 1989;38:236–238, 243–250
57. CDC. Initial therapy for tuberculosis in the era of multidrug resistance: recommendations of the Advisory Council for the Elimination of Tuberculosis. MMWR 1993;42(no. RR-7):1–8
58. Kelen GD, DiGiovanna T, Bisson L, et al. Human immunodeficiency virus infection in emergency department patients. JAMA 1989;262:516–522
59. Soderstrom CA, Furth PA, Glasser D, et al. HIV infection rates in a trauma center treating predominantly rural blunt trauma victims. J Trauma 1989; 29:1526–1530
60. Lewandowski C, Ognjan A, Rivers E, et al. HIV-1 and HTLV-1 seroprevalence in critically ill resuscitated emergency department patients. In: Abstracts for the V International Conference on AIDS. Montreal: Health and Welfare Canada, 1989:142
61. Gordin FM, Gibert C, Hawley HP, Willoughby A. Prevalence of human immunodeficiency virus and hepatitis B virus in unselected hospital admissions: implications for mandatory testing and universal precautions. J Infect Dis 1990;161:14–17
62. Risi GF, Gaumer RH, Weeks S, Leeter JK, Sanders CV. Human immunodeficiency virus: risk of exposure among health care workers at a southern urban hospital. South Med J 1989;82:1079–1082
63. St. Louis ME, Rauch KJ, Petersen LR, et al. Seroprevalence rates of human immunodeficiency virus infection at sentinel hospitals in the United States. N Engl J Med 1990;323:213–218
64. Marcus R, Culver DH, Bell DM, et al. Risk of human immunodeficiency virus infection among emergency department workers. Am J Med 1993;94: 363–370
65. Janssen RS, St. Louis ME, Satten GA, et al. HIV infection among patients in US acute-care hospitals: strategies for the counseling and testing of hospital patients. N Engl J Med 1992;327:445–452
66. Recommendations for prevention of HIV transmission in health-care settings. MMWR 1987;36(suppl 2S):1S–18S
67. Tokars JI, Marcus R, Culver DH, et al. Surveillance of HIV infection and zidovudine use among health care workers after occupational exposure to HIV-infected blood. Ann Intern Med 1993;118:913–919
68. Henderson DK, Fahey BJ, Willy M, et al. Risk for occupational transmission of human immunodeficiency virus type 1 (HIV-1) associated with clinical exposures: a prospective evaluation. Ann Intern Med
69. Tokars JI, Bell DM, Culver DH, et al. Percutaneous injuries during surgical procedures. JAMA 1992;267:2899–2904

2 Biology of Human Immunodeficiency Viruses

GERALD SCHOCHETMAN

The etiologic agent of acquired immunodeficiency syndrome (AIDS), the human immunodeficiency virus type 1 (HIV-1), was first isolated during 1983–1984 from patients with AIDS-related complex (ARC) and AIDS.[1–3] HIV-1 is the predominant AIDS virus and is found worldwide, primarily in Central Africa, Europe, and North and South America. A second AIDS virus, HIV-2,[4–6] closely related to the simian immunodeficiency virus (SIV), was discovered in 1986 and was shown to be endemic in parts of West Africa with limited spread in Western Europe. HIV-2 is only now beginning to appear in the Americas, mainly in the United States, Canada, and Brazil. Infection with either HIV-1 or HIV-2 results in a number of biologic and pathologic changes leading to a spectrum of immune dysfunctions, neurologic disorders, enteropathy, and AIDS.

Retrovirus Family

The human immunodeficiency viruses are members of the retrovirus family of viruses.[7] The retroviruses are so called because at the beginning of their life cycle they reverse the usual flow of genetic information in a cell. In all living organisms and in many other viruses, genetic information is stored as DNA and later transcribed into RNA. By contrast, retroviruses store their genetic information as RNA and contain a unique enzyme, reverse transcriptase (RT), which catalyzes the reverse transcription of the RNA genome (its entire complement of genes) into a DNA copy. The resulting DNA, termed the provirus, can be perceived by the host cell as its own and is integrated into its chromosomal DNA, where the provirus can remain dormant for weeks, months, or even years without being expressed (i.e., remain latent). The integrated state is responsible for the persistent nature of retrovirus infections.

The retrovirus family is composed of three subfamilies: oncoviruses, spumaviruses, and lentiviruses (Table 2.1, Fig. 2.1). The oncoviruses, or cancer-causing viruses, are found to be transmitted horizontally by host-to-host contact and vertically as integrated viruses in germ cells. When integrated into the host DNA, oncoviruses efficiently transform the host cells that have a tumor producing potential. The lentiviruses and spumaviruses (spuma, for foamy) do not cause cancer and do not integrate into the host's germ cell lines (i.e., no genetic transmission through the maternal or paternal germ line). Both the lentiviruses and the spumaviruses produce a persistent lifelong infection of the host cells. Of the two, however, only the lentiviruses have been identified as causes of human and animal diseases. Classification of HIV as a lentivirus is based on its fine structure, biologic properties, and protein and nucleic acid sequence homology (Table 2.2). As with other lentiviruses, mature extra-cellular particles of HIV are characteristically 90 to 130 nm in diameter and have a double-membraned envelope surrounding an electron-dense cylindrically shaped core.

Genetic Variation of HIV

It is clear now that extensive genetic variation is a characteristic feature of both HIV-1 and HIV-2.[8] This genetic variation can result in biologically

TABLE 2.1. Subfamilies of retroviruses.

Subfamily	Examples
Oncoviruses: associated with the activation of certain cell genes leading to tumor development	
Type A	Mouse intracisternal type A
Type B	Mouse mammary tumor virus
Type C	Murine leukemia virus Human T cell leukemia virus types I and II Feline leukemia virus Bovine leukemia virus
Type D	Mason-Pfizer virus SAIDS virus
Spumavirus: readily isolated from humans and other primates, but have not been associated with any specific disease	
	Simian foamy virus Human foamy virus
Lentiviruses: produce acute cytocidal infection followed by a slowly developing multisystem disease	
	Visna maedi virus Caprine arthritis encephalitis virus Equine infectious anemia virus Feline immunodeficiency virus Bovine immunodeficiency virus Simian immunodeficiency virus Human immunodeficiency virus type 1 (HIV-1) and type 2 (HIV-2)

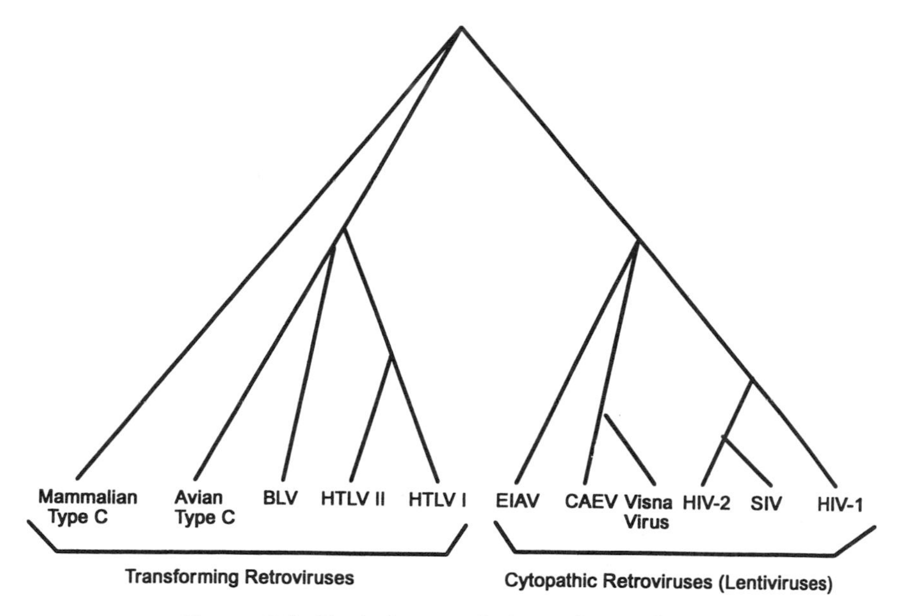

FIGURE 2.1. Evolutionary relations of retroviruses.

altered viruses by modifying the structure, function, and immunogenicity of the various viral gene products. It is also clear that no two HIV variants are alike; and even within a single person HIV exists as a swarm of microvariants that are highly related to, but genetically distinct from, each other. These intraperson genetic variants have been termed "quasispecies." The inordinately high genetic diversity of HIV-1 confers on the virus a high degree of versatility to respond to in vivo selection

TABLE 2.2. HIV characteristics resembling those of lentiviruses.

Biologic characteristics
Persistent or latent infection
Cytopathic effects (syncytia formation) on selected cells
Capable of infecting macrophages
Associated with immune suppression
Long incubation period
Central nervous system involvement
Affects hematopoietic system
Molecular biologic characteristics
Similar genomic organization
Morphology of virus (cone nucleoid)
Accumulation of unintegrated proviral DNA
Polymorphism, particularly in the envelope gene
Primer binding site ($tRNA_{lys}$)

pressures, which can lead to the rapid development of immunologic escape mutants and drug-resistant mutants. Because HIV-1 has been shown to replicate continuously throughout all stages of the infection cycle (see below), it is likely that genetic variants and escape mutants are also produced continuously in the infected individual. This property of the virus could pose a major problem for the development of effective therapies and vaccines.

The V3 region of the envelope gp120 protein, because of its functional importance (it represents the principal epitope to which virus-neutralizing antibodies are directed), has spawned an international effort to obtain sequences from HIV-1s worldwide. Based on the phylogenetic analysis of genetic sequences of the V3 loop, these global HIV-1 variants have been grouped into six subtypes: A through F (Los Alamos). The variants included in these subtypes originate from a wide range of geographic regions. The subtype A viruses were found primarily in persons from Western and Central Africa. The subtype B viruses were found mostly in persons from North and South America, Europe, and Asia; they represent the most extensively studied of the HIV-1 variants because of the early availability. The subtype C viruses were found in Southern and Central Africa, India, and Malaysia. The subtype D viruses have been found in Central Africa and the subtype E viruses in Thailand and Central Africa. Viruses from the subtype F group have now been identified in Brazil and Romania.

A seventh HIV-1 subtype has recently been reported and has been designated subtype O[8,8a,8b] (see Chapter 5 for additional information). The subtype O group comprises the newly reported ANT70 and MVP5180 viruses from the Cameroon in addition to the chimpanzee viruses, CPZ_{gab} and CPZ_{ant} from Gabon in Africa. These viral strains represent the most divergent of the HIV-1s that have been reported with overall genetic distances of 65% to HIV-1 and 56% to HIV-2. The env gene of these viruses exhibit genetic distances to HIV-1 and HIV-2 of 53% and 49%, respectively. Additonal HIV-1 variants have reportedly been detected from countries such as Gabon, Zaire, Zambia, Taiwan, Finland and Russia, and are tentatively classified as subtypes G, H, and I. It is anticipated that additional variants of HIV will be detected as the effort to study HIV globally continues.

Origins of HIV-1 and HIV-2

The genetic similarity between HIV-1 and HIV-2[9,10] is significantly less (40–50% nucleotide identity) than is found among different HIV-1 isolates (>85% nucleotide identity). However, serologic cross-reactive antibodies, mainly to the *gag* and *pol* proteins, from patients infected with either virus can be detected using commercially available serologic tests.

TABLE 2.3. Primate lentiviruses.

Virus	Host
HIV-1	Human
HIV-2	Human
SIV_{CPZ}	Chimpanzee
SIV_{SM}	Sooty mangabey
SIV_{MAC}	Rhesus macaque
SIV_{MNE}	Pig-tailed macaque
$SIV_{STM_{ver}}$	Stump-tailed macaque
$SIV_{AGM_{gri}}$	Vervet monkey
$SIV_{AGM_{sab}}$	Grivet monkey
$SIV_{AGM_{tan}}$	Sabaeus monkey
SIV_{AGM}	Tantalus monkey
SIV_{MND}	Mandrill
SIV_{SYK}	Sykes' monkey

In addition to HIV-1 and HIV-2, other lentiviruses have been discovered in a wide variety of nonhuman primates[11–13] (Table 2.3). The nonhuman primate lentiviruses are collectively known as simian immunodeficiency viruses (SIVs). To date, natural SIV infections not been shown to cause disease in the infected animal. There is now mounting evidence for a simian origin of HIV. All of the known primate lentiviruses appear to be more closely related to each other than they are to lentiviruses of nonprimate origin (e.g., cats, cows, and sheep). HIV-1 and HIV-2 appear to be closely related to the primate lentiviruses isolated from chimpanzees and sooty mangabey monkeys. Both HIV-1 and HIV-2 are believed to have arisen through relatively recent primate lentivirus infections of humans from nonhuman primates. The geographic coincidence of HIV-2 and the highly related SIVs that naturally infect sooty mangabeys implicates this species as the immediate source of the human virus. It is not yet clear whether chimpanzees or another, as yet unidentified primate species represent the natural host of the HIV-1 lineage.

Structure of HIV

The structure of HIV resembles that of all retroviruses but particularly that of the lentiviruses (Fig. 2.2). HIV has a cylindrical eccentric nucleoid, or core. The nucleoid contains the HIV genome, which is diploid (i.e., is composed of two identical single-stranded RNAs). Encoded in the RNA genome are the entire complement of genes of the virus. These genes code (contain the genetic information) for the structural proteins that are used to assemble the virus particles and the regulatory proteins involved in the regulation of viral gene expression (Table 2.4, Fig. 2.3).

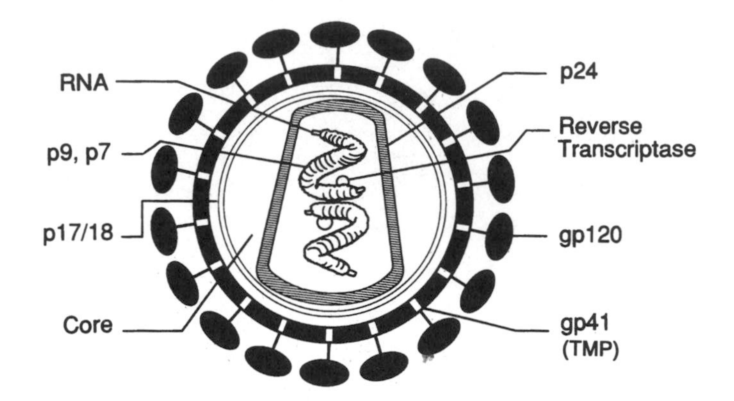

FIGURE 2.2. Morphologic structure of HIV-1 shown in cross section.

The HIV RNA genome is associated with a basic nucleic acid binding protein p9 and the RT (Fig. 2.2). The core or capsid antigen p24 encloses the nucleoid components, completing the nucleocapsid structure. The matrix antigen p17 encircles the viral core and lines the inner surface of the envelope of the virus. The surface of HIV manifests external knob-like structures formed by the envelope glycoprotein (sugar-containing protein) gp120. The transmembrane protein (TMP) gp41 spans the viral membrane and has both external and internal domains. The TMP anchors the external gp120 to the viral envelope. The membrane lipid bilayer is derived from the host cell plasma membrane.

The envelope gene (*env*) of HIV codes for a precursor polyprotein p85, which is glycosylated to form the *env* precursor protein gp160 (Fig. 2.4). The gp160 is then cleaved by a cellular protease to form the surface virion extracellular envelope glycoprotein gp120 and the TMP gp41.[14] The gp120 protein contains specific amino acid domains responsible for binding the virus to the CD4 virus receptor protein on the surface of the cell. The gp41 protein contains stretches of amino acids that play an important role in syncytia formation. Fusion, mediated by the gp41 TMP, may also play a role in penetration of the viral core into the interior of the cell. Another domain of the TMP spans the membrane and anchors the envelope gp120 protein onto the infected cell or virus particle.

The *gag* gene encodes a 55-kDa polyprotein precursor molecule (Pr55*gag*) that is cleaved into the major structural proteins of the virus capsid (p17 and p25) and core structures (p15, which is further cleaved to p7 and p9) surrounding the virion RNA. During synthesis of the Pr55*gag* polyprotein the fatty acid myristate is attached to the amino-terminal end (p17) of the precursor protein. Myristoylation of Pr55*gag* appears to be required for production of infectious virions.[15] The p9 protein contains multiple cysteine residues that probably help to form specific secondary structure involved in nucleic acid binding. The function of the proline-rich virion core protein p9 remains unclear.

A second polyprotein precursor, the Pr180*gag-pol*, contains the *gag*, protease, polymerase, and integrase gene products translated from the same genomic messenger RNA as the *gag* polyprotein Pr55*gag*. The *pol* part of the message actually codes for three proteins that are cleaved from the larger precursor molecule (Pr180*gag-pol*). They include the protease, RT, and integrase. The viral protease is involved in processing Pr180*gag-pol* and Pr55*gag*.[16] Cleavage of the *gag-pol* precursor by a functional protease releases RT and integrase.

TABLE 2.4. Summary of HIV proteins.

Gene product	Description	Localization
env (envelope)		Virion
gp160	Precursor of *env* glycoprotein	
gp120	Outer *env* glycoprotein	
gp41	Transmembrane glycoprotein	
gag (core)		Virion
p55	Precursor of *gag* proteins	
p24	*gag* Protein (capsid structural protein)	
p17	*gag* Protein (matrix protein)	
p15, p9, p7	*gag* Proteins	
pol (polymerase)		Virion
p66	Reverse transcriptase of *pol* gene	
p51	Reverse transcriptase of *pol* gene	
p31	Endonuclease of *pol* gene	
p15	Protease	
p11	Integrase	
tat (transactivator)		Nucleolus/nucleus
p14	Transactivator of viral RNA synthesis	
rev (regulator)		Nucleolus/nucleus
p19/20	Regulates viral mRNA expression	
vif (infectivity factor)		Golgi?
p23	Increases virus infectivity	
vpr		Virion
p18	Assists in virus replication	
nef		Cytoplasm/plasma membrane
p27	Pleiotropic, including suppression of virus	
vpu (only in HIV-1)		Membrane
p16	Involved in virus release	
vpx (only in HIV-2)		Virion
p14	Involved in viral infectivity	

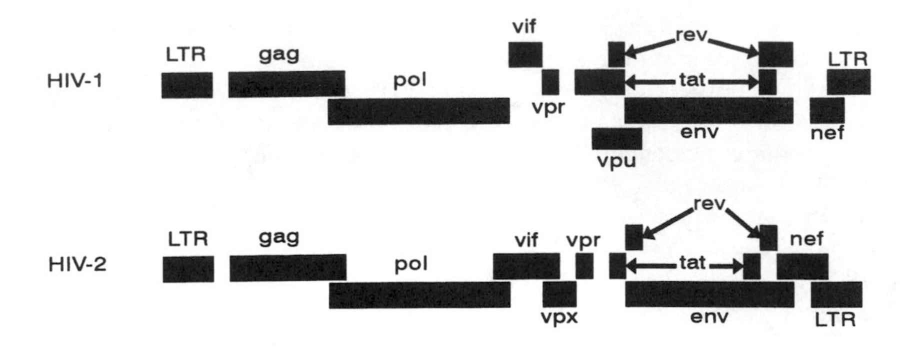

GENE	FUNCTION
GAG	CORE PROTEINS
POL	RT/RNase H (found inside core)
PROTEASE	POSTTRANSLATIONAL PROCESSING OF VIRAL PROTEINS
INTEGRASE	INTEGRATION VIRAL DNA
ENV	ENVELOPE PROTEINS
TAT	UPREGULATES VIRAL RNA EXPRESSION
REV	REGULATES VIRAL mRNA EXPRESSION
VIF	INCREASES VIRAL INFECTIVITY
VPR	INVOLVED IN VIRAL REPLICATION
VPU (HIV-1 ONLY)	INVOLVED IN VIRAL RELEASE
VPX (HIV-2 ONLY)	INVOLVED IN VIRAL INFECTIVITY
NEF	PLEIOTROPIC

FIGURE 2.3. Structure of the genomes of HIV-1 and HIV-2 with a listing of their individual viral genes and functions.

Life Cycle of HIV

The first step in the infection of a susceptible cell is attachment of the HIV to the cell through a specific interaction of the viral envelope glycoprotein gp120 with the cell-surface-associated CD4 protein (Fig. 2.5). The TMP gp41 noncovalently interacts with the gp120 and is involved in virus–cell and cell–cell fusion events. Jointly these two HIV surface proteins are responsible for virion binding of CD4-bearing cells and for syncytia formation between infected and uninfected cells. The binding step to the CD4 receptor protein could potentially be inhibited by anti-gp120 virus-neutralizing antibodies (Fig. 2.5). After the virus binds to the CD4 protein of the cell, the cell is penetrated by the virus capsid via fusion of the viral and cellular membranes. Infection seems to be species-specific, as murine cells with the human CD4 antigen bind but do not take up HIV.

After the virus is internalized it is partially uncoated, thereby activating the viral RT, which converts the single-stranded viral RNA into double-stranded linear DNA. This step can be inhibited by compounds such as azidothymidine (zidovudine), which are RT inhibitors that prematurely terminate the growing DNA chain, preventing formation of the HIV provirus (Fig. 2.5). The DNA product is then transported from the cytoplasm of the cell to the nucleus. It is believed that the linear, not a circular, form becomes the integrated provirus.[17] The integration of the linear viral DNA into the host's chromosomes is mediated by the virus-coded integrase.

The HIV gene expression involves synthesis of viral messenger RNA (mRNA) transcripts using the host cell's RNA polymerase II and other

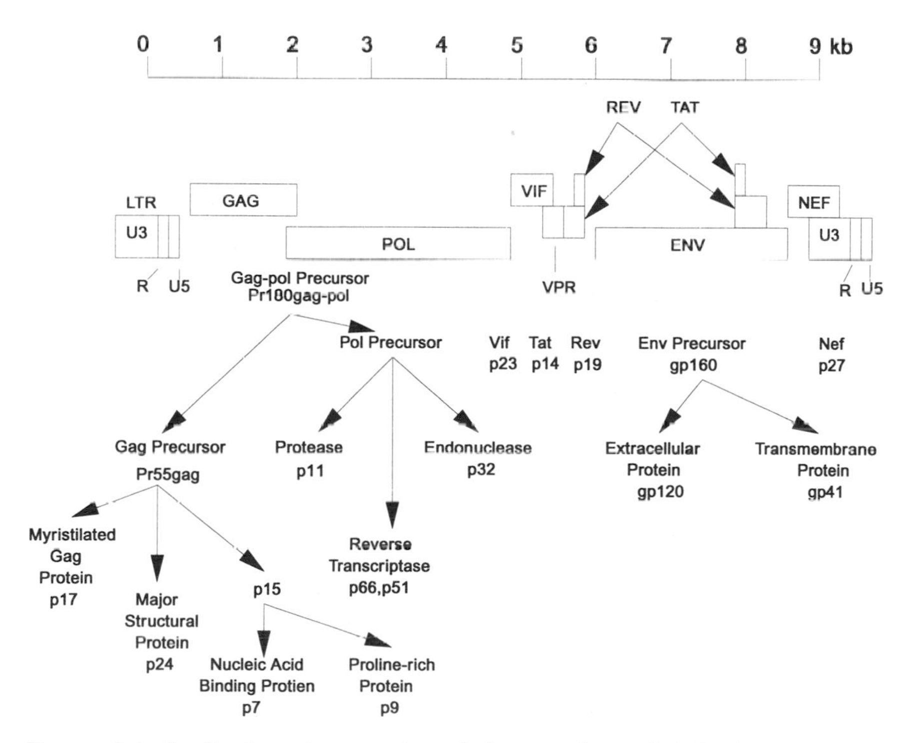

FIGURE 2.4. Synthesis and processing of the proteins coded for by the HIV-1 genome.

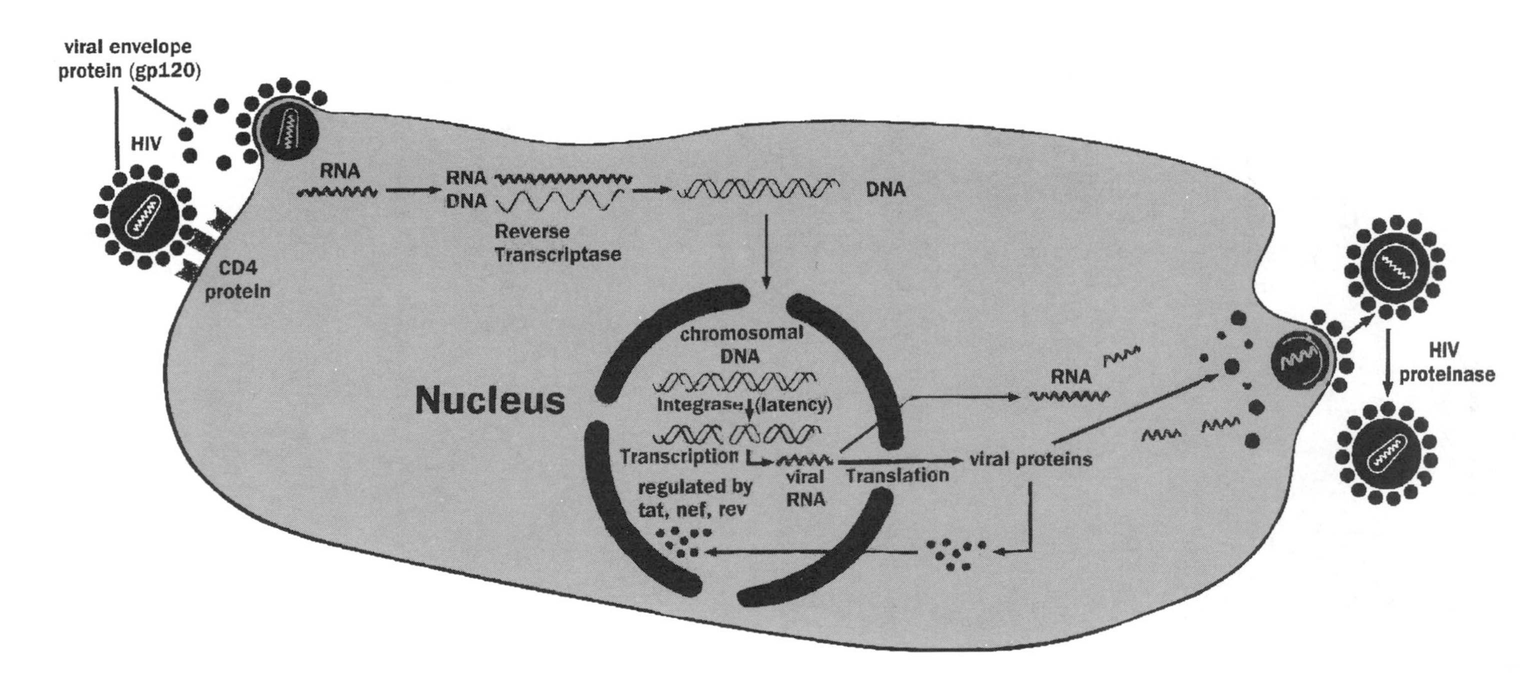

FIGURE 2.5. Life cycle of the virus. See text for details.

cellular and viral transcriptional factors. The mRNAs for the *gag* and *gag-pol* precursors and for the viral RNA present in the mature virions represent full-length transcripts of the viral genome. A smaller, singly spliced transcript codes for the envelope components. The regulatory proteins of HIV (*nef*, *rev*, *tat*, *vif*, *vpr*) are synthesized via translation from multiply spliced transcripts that are not found in the viral particle, although they do play a significant role in virus expression.[18] The control of HIV replication is regulated through interplay of the regulatory gene products and specific sequences present in the viral RNA or provirus.[19] The proportion of the different mRNAs transported from the nucleus to the cytoplasm for translation into viral proteins is determined by the regulatory protein encoded by the HIV *rev* gene.

Assembly of the newly synthesized virion occurs in the cytoplasm and involves formation of a dimer of identical viral RNA molecules in a complex containing *gag* and *pol* gene products. The synthesis and assembly of the virion components could be potentially inhibited by compounds such as interferon (Fig. 2.5). The resulting core structure buds from the cellular membrane, where it acquires its envelope glycoprotein coat (consisting of gp120 and gp41) and a cellular lipid bilayer. Cleavage of the internal core proteins by the viral protease occurs during this process, yielding extracellular infectious progeny HIVs. The extracellular virus could potentially be attacked and destroyed at this stage by sensitized T cell antibodies (Fig. 2.5). Infectious HIV can undergo cell-to-cell spread, independent of virus release. It can occur through the fusion or syncytia formation of an infected cell with an uninfected cell.

The mechanism for HIV latency is still unclear. It is known, however, that the level of HIV replication is strikingly affected by activation of resting lymphocytes. Resting T cells are nonpermissive for the replication of HIV-1, despite efficient binding of the virus to the CD4 receptor molecule displayed at the surface of the cell. The transition from a latent to a productive infection has been suggested to occur in response to T cell mitogens, phorbol esters, calcium ionophores, and gene products of other viruses, such as HTLV-I, herpes simplex viruses, cytomegalovirus, adenovirus, hepatitis B virus, and human herpes virus 6. T cell activation is important for virus penetration. Once the host cell is activated, inducible host cell transcription (the process of conversion of DNA to RNA) factors stimulate a low level of early HIV-1 gene expression. The first mRNA molecules that reach the cytoplasm from the nucleus are composed entirely of multiply spliced messages that code for the HIV-1 regulatory proteins. Once the *tat* protein is produced, it elicits a strong positive effect on viral transcription. The *tat* protein acts by increasing the rate of transcription of all the gene sequences linked to the HIV long terminal repeat (LTR).

In Vivo HIV Infectivity

An HIV infection of susceptible cells in a person occurs via interaction of the external membrane glycoprotein gp120 on the virus and the cell surface receptor CD4. The primary CD4-bearing cells in HIV-infected persons are the T-helper lymphocytes, selected marrow progenitor cells, and monocytes and macrophages. Additionally, a variety of nonhematopoietic cells expressing low levels of CD4 on their surface are also susceptible to HIV infection.[20] They include the epidermal Langerhans cells, follicular dendritic cells of lymph nodes, and certain cells of the central nervous system. Interestingly, infection of glial cell lines, colorectal cells, and fetal brain cells that do not appear to express detectable surface CD4 suggests that another mechanism may also be responsible for virus infection.[21]

After primary infection with HIV, rapid virus replication and an early burst of viremia are often evident. During this early period, large numbers of viral particles spread throughout the body, seeding themselves in the lymphoid organs as lymphocytes retraffic to the lymph nodes.[22] Also during this period an estimated 50% to 75% of infected persons develop an initial infection characterized by flu-like symptoms, high levels of HIV in the peripheral circulation, high levels of p24 antigenemia (Fig. 2.6), and a significant drop in the number of circulating $CD4^+$ T cells.[23,24] Generally, an HIV-specific immune response is usually seen within 4 to 6 weeks of the onset of symptoms, accompanied by a dramatic decline in plasma viremia.[25] As the acute syndrome resolves, levels of $CD4^+$ T cells

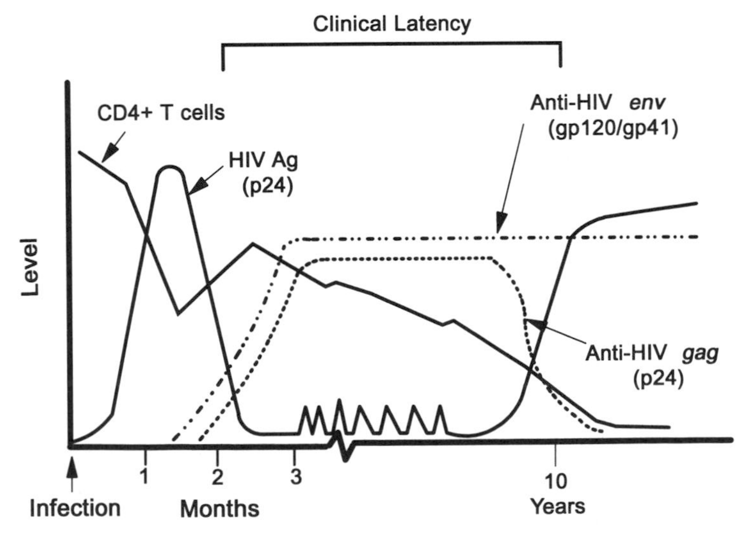

FIGURE 2.6. Temporal serologic relation between HIV-1 antigen (p24) production and detection of HIV-specific antibodies.

may rebound to 80% to 90% of their original level (Fig. 2.6), although many patients stabilize to a $CD4^+$ T cell level that is moderately or even markedly depressed. With most infections, viral antigen becomes undetectable when antibodies to p24 emerge, although antigen may appear later in the course prior to the development of AIDS. Transient IgM and persistent IgG and IgA antibodies to a broad range of virus-encoded proteins can be detected. Although there have been reports of delayed antibody response to HIV-1 infection, 95% or more of infected persons develop detectable antibodies to HIV-1 within 3 to 6 months.[26] The interval between infection and seroconversion or detection of antibodies has been called the "window period." This period has received much attention because of implications for the safety of the blood supply and the possibility of allowing HIV-1-infected seronegative blood to be used for transfusion. Studies of more than 1 million blood donors in the United States, West Germany, and Austria have failed to demonstrate a single instance of an antigen-positive, antibody-negative collection, presumable because antigenemia is short-lived in relation to the time between infection and donation. This result implies that routine use of a screening test for HIV antigen (at least in a low prevalence population, e.g., volunteer blood donors in developed countries) would not increase the safety of the blood supply.

In numerous patients the acute phase of HIV infection is commonly followed by a period of clinical latency that may last up to 10 years or more, during which time few cells in the peripheral blood (1 in 1000 to 1 in 10,000) are infected with HIV, and viremia is minimal or absent.[27,28] During this period disease symptoms are usually mild or not evident while immune deterioration progresses, as seen clearly by the gradual decrease of circulating $CD4^+$ T lymphocytes. Viral antigen or cell-free virus can be detected only intermittently in the circulation, and high titers of antibodies to the viral *env*, *gag*, and *pol* proteins are usually present in the patient's serum or plasma, saliva, and cerebrospinal fluid (CSF).

The final phase of infection is characterized by increased virus expression and distribution and by the emergence of multiple disease symptoms indicative of AIDS. By this time the patients have a severely depleted level of $CD4^+$ T-helper cells, leading to collapse of their immune system. As the patient's level of $CD4^+$ T cells drops below 200 cells/μl of blood, the risk of developing life-threatening opportunistic infections and malignancies increases greatly.[29] There is also a decrease in HIV-specific antibodies (usually anti-p24*gag* antibodies), an increase in p24 antigenemia, and the appearance of easily detectable virus in the peripheral blood. Even though virus-neutralizing antibodies can be identified in sera from certain patients at all stages of infection, their detection does not appear to correlate well with the patient's clinical status.

Although HIV has been isolated from saliva, tears, and urine, the presence of infected cells or cell-free virus is rare; and there is no evidence that exposure to these body fluids represents a significant risk of infection. Virus has been isolated more frequently from peripheral blood mononuclear cells of patients. In general, the percentage of patients producing antibody to HIV-1 who also have positive peripheral blood cultures varies as a function of both the severity of the infection and the methods employed to optimize the detection of positive cultures. It has been shown that with optimization of technique at all levels more than 90% to 95% of HIV-1-infected persons have positive co-cultures regardless of CD4 count or clinical staging.[30]

Lymphoid Organs in HIV Disease

During virus infection the virus is filtered from the blood in the lymph nodes, where it is trapped by thread-like processes of the follicular dendritic cells (FDCs) contained within the germinal centers.[31,32] Germinal center B cells and $CD4^+$ T cells in the paracortical areas are subsequently stimulated, resulting in an immune response. When HIV becomes trapped in the FDC network,[33–35] many $CD4^+$ T cells become susceptible to HIV infection as they migrate to the lymph node in response to this antigenic challenge. In the lymph node, cytokines are released and immune activation develops.[36] The combination of immune activation and cytokine production, together with the presence of a large pool of susceptible $CD4^+$ T lymphocytes, may lead to the propagation of HIV infection in the lymphoid tissue. In asymptomatic patients with more than 500 $CD4^+$ T cells per microliter, a low proportion of peripheral blood mononuclear cells (PBMCs) contain integrated HIV provirus sequences and patients are usually p24 antigen-negative. Using quantitative DNA polymerase chain reaction (PCR) analysis of purified $CD4^+$ T cells, it was found that the viral burden in the peripheral blood in patients with early-stage infection is considerably lower (10- to 100-fold) than in lymphoid tissue taken from the same patient.[37] Electron microscopic examination of a lymph node from a patient at this early stage of infection reveals that there are large amounts of antibody-coated virus attached to the FDC processes and that the architecture of the lymph node and of the FDC network is more or less intact at this stage. It appears that the absence of HIV in the peripheral blood and the low levels of virus-infected circulating $CD4^+$ T cells observed during the early infection may be due to the efficient trapping of virus in the FDC network and the movement of $CD4^+$ T cells to the germinal centers of the lymph nodes.

In contrast, late in the course of the disease, when the numbers of circulating $CD4^+$ T cells have declined significantly, plasma viremia again becomes detectable. Electron microscopic examination of lymph nodes from patients with late-stage infection show marked disruption of the architecture of the germinal center and necrosis of the FDC processes.

The reason for the death of the FDC and the dissolution of the FDC network is unclear. The loss of virus-trapping function likely results in a decrease in virus burden in the lymph node as virus spills over into the bloodstream. The decreased ability of the FDC network to trap virus, along with the accelerated virus replication seen during the later stages of HIV disease, may explain the increase in plasma viremia seen at this stage of disease. Furthermore, the loss of the FDC antigen-trapping and antigen-presenting functions probably intensifies the severe immunosuppression that results from the loss of functional $CD4^+$ T cells. It is now clear that HIV is actively replicating in the lymphoid tissues, even when there is little or no virus concentration in the peripheral blood. Therefore the term latency, as it relates to virus production, should be used cautiously in the context of HIV disease.

Conclusion

It has been a relatively short time since the initial identification of HIV. During that time extraordinary progress has been made in our understanding of the viral and cellular mechanisms governing infection and replication. Major gaps in our knowledge of HIV infection, latency, replication, and pathogenesis persist, however, and many essential questions remain to be answered. They are concerned with (1) the nature of the latent state of infection that is believed to exist in vivo; (2) the nature of viral and host factors involved in viral latency; (3) the nature of the molecular events involved in virus-induced cell killing; and (4) the nature of the genetic differences for the HIV variants involved in differential virus replication, tissue tropism, and cytopathogenicity. The answers to these and other critical questions are vital to our understanding of the HIV disease process and to the development of effective new treatments and vaccine strategies against this devastating viral disease.

References

1. Barre-Sinoussi F, Chermann JC, Rey F, et al. Isolation of a T-lymphotropic retrovirus from a patient at risk for acquired immune deficiency syndrome (AIDS). Science 1983;220:868–871
2. Gallo RC, Salahuddin SZ, Popovic M, et al. Frequent detection and isolation of cytopathic retroviruses (HTLV-III) from patients with AIDS and at-risk for AIDS. Science 1984;224:500–502
3. Popovic M, Sarngadharan MG, Read E, Gallo RC. Detection, isolation and continuous production of cytopathic retroviruses (HTLV-III) from patients with AIDS and pre-AIDS. Science 1984;224:497–500
4. Clavel F. HIV-2, the West African AIDS virus. AIDS 1987;1:135–140
5. Clavel F, Guetard D, Brun-Vezinet F, et al. Isolation of a new human retrovirus from West African patients with AIDS. Science 1986;233:343–346

6. Horsburgh CR, Holmberg SD. The global distribution human immunodeficiency virus type 2 (HIV-2) infection. Transfusion 1988;28:192–95
7. Varmus H. Retroviruses. Science 1988;240:1427–1435
8. Myers G, Korber B, Smith RF, Berzofsky JA, Pavlakis GN. Human Retroviruses and AIDS. Los Alamos, NM: Theoretical Biology and Biology and Biophysics, 1993
8a. Gurtler LG, Hauser PH, Eberle J, von Brunn A, Knapp S, Zekeng L, Tsague JM, Kaptue L. A new subtype of human immunodeficiency virus type 1 (MVP-5180) from Cameroon. J Virol 1994;68:1581–1585
8b. Haesevelde MV, Decourt J-L, Leys RJ, Vanderborght B, van der Groen G, van Heuverswijn H, Saman E. Genomic cloning and complete sequence analysis of a highly divergent African human immunodeficiency virus isolate. J Virol 1994;68:1586–1596
9. Alizon M, Wain-Hobson S, Montagnier L, Sonigo P. Genetic variability of the AIDS virus: nucleotide sequence analysis of two isolates from African patients. Cell 1986;46:63–74
10. Clavel F, Guyader M, Guetard D, et al. Molecular cloning and polymorphism of the human immunodeficiency virus type 2. Nature 1986;324:691–695
11. Tsujimoto H, Hasegawa H, Maki N, et al. Sequence of a novel simian immunodeficiency virus from a wild-caught African mandrill. Nature 1989; 341:539–541
12. Peeters M, Honore C, Huet T, et al. Isolation and partial characterization of an HIV-related virus occurring naturally in chimpanzees in Gabon. AIDS 1989;3:625–630
13. Johnson PR, Myers G, Hirsch V-M. Genetic diversity and phylogeny of non-human primate lentiviruses. In Koff W (ed): Annual Review of AIDS Research (Vol. I). New York: Marcel Dekker, 1991:47–62
14. Haseltine WA. Development of antiviral drugs for the treatment of AIDS: strategies and prospects. J Acquir Immune Defic Syndr 1989;2:311–334
15. Bryant ML, Ratner L. Myristoylation-dependent replication and assembly of human immunodeficiency virus 1. Proc Natl Acad Sci USA 1990;87:523–527
16. Debouck C, Gorniak JG, Stickler JE, et al. Human immunodeficiency virus protease expresses in *Escherichia coli* exhibits autoprocessing and specific maturation of the gag precursor. Proc Natl Acad Sci USA 1987;84:8903–8906
17. Fujiwara T, Mizuuchi K. Retroviral DNA integration: structure of an integration intermediate. Cell 1988;54:497–504
18. Peerlin BM, Luciw PA. Molecular biology of HIV. AIDS 1988;2(suppl 1): S529–S40
19. Cullen BR, Greene WC. Regulatory pathways governing HIV-1 replication. Cell 1989;58:423–426
20. Levy JA. Human immunodeficiency viruses and the pathogenesis of AIDS. JAMA 1989;261:2997–3006
21. Takeda A, Tuazon CU, Ennis FA. Antibody-enhanced infection by HIV-1 via Fc receptor-mediated entry. Science 1988;242:580–583
22. Pantaleo G, Graziosi C, Fauci AS. New concepts in the immunopathogenisis of human immunodeficiency virus (HIV) infection. N Engl J Med 1993;328: 327–335
23. Tindall B, Cooper DA. Primary HIV infection: host responses and intervention strategies. AIDS 1991;5:1–14

24. Gaines H, von Sydow MA, von Stedingk LV, et al. Immunological changes in primary HIV-1 infection. AIDS 1990;4:995–999
25. Albert J, Abrahamsson B, Nagy K, et al. Rapid development of isolate-specific neutralizing antibodies after primary HIV-1 infection and consequent emergence of virus variants which resist neutralization by autologous sera. AIDS 1990;4:107–112
26. Horsburgh CR Jr, Ou C-Y, Jason J, et al. Duration of human immunodeficiency virus infection before detection of antibody. Lancet 1989;16:2637–2640
27. Schnittman SM, Psalidopoulos MC, Lane HC, et al. The reservoir for HIV-1 in human peripheral blood is a T cell that maintains expression of CD4. Science 1989;245:305–308
28. Schnittman SM, Greenhouse JJ, Psallidopoulos MC, et al. Increasing viral burden in $CD4^+$ T cells from patients with human immunodeficiency virus (HIV) infection reflects rapidly progressive immunosuppression and clinical disease. Ann Intern Med 1990;113:438–443
29. Fauci AS, Schnittman SM, Poli G, Koenig S, Pantaleo G. NIH conference: immunopathogenic mechanisms in human immunodeficiency virus (HIV) infection. Ann Intern Med 1991;114:678–693
30. Coombs RW, Collier AC, Allain J-P, et al. Plasma viremia in human immunodeficiency virus infection. N Engl J Med 1989;321:1626–1631
31. Tew JG, Kosco MH, Burton GF, Szakal AK. Follicular dendritic cells as accessory cells. Immunol Rev 1990;117:185–211
32. Steinman RM. The dendritic cells system and its role in immunogenicity. Annu Rev Immunol 1991;9:271
33. Wood GS. The immunohistology of lymph nodes in HIV infection: a review. Prog AIDS Pathol 1990;2:25–32
34. Fox CH, Tenner-Racz K, Racz P, et al. Lymphoid germinal centers are reservoirs of human immunodeficiency virus type 1 RNA. J Infect Dis 1991; 164:1051–1057
35. Spiegel H, Herbst H, Niedobitek G, Foss HD, Stein H. Follicular dendritic cells are a major reservoir for human immunodeficiency virus type 1 in lymphoid tissues facilitating infection of $CD4^+$ T-helper cells. Am J Pathol 1992;140:15–22
36. Rouse RV, Ledbetter JA, Weissman IL. Mouse lymph node germinal centers contain selected subset of T cells with the helper phenotype. J Immunol 1982;128:2243–2246
37. Pantaleo G, Graziosi C, Butini L, et al. Lymphoid organs function as major reservoirs for human immunodeficiency virus. Proc Natl Acad Sci USA 1991;88:9838–9842

3
Immunology of HIV Infection

ALISON C. MAWLE and J. STEVEN McDOUGAL

The human immunodeficiency virus (HIV-1) infects, functionally impairs, and depletes a subpopulation of thymus-derived T lymphocytes that express the cell-surface molecule CD4. CD4 T cells perform critical recognition and induction functions in the immune response to foreign stimuli. HIV-1 infection results in gradual CD4 T cell depletion, progressive immune unresponsiveness with effective paralysis in virtually all arms of the immune system, and increasing susceptibility to opportunistic infections and malignancies. The clinical spectrum of HIV infection ranges from asymptomatic infection to severe immune deficiency with the infectious/malignant complications that are characteristic of the acquired immunodeficiency syndrome (AIDS). This spectrum of clinical severity is reflected in a parallel severity of CD4 T cell depletion. Thus HIV-1 immunodeficiency is viewed from the immunologic perspective as a graded severity of CD4 T cell depletion, distributed on a continuum of time, influenced by incompletely understood cofactors, and associated with a spectrum of increasing clinical severity that reflects the severity of CD4 T cell depletion.

The other major clinical sequela of HIV infection, subacute encephalopathy, is not necessarily related to the degree of CD4 T cell depletion. However, a common immunopathologic mechanism is involved: infection of a cell type in the brain (monocyte-derived microglia) that expresses the CD4 molecule.

Individuals infected with HIV mount and sustain a vigorous immune response to HIV. It remains unclear, however, what role the immune response has in controlling HIV infection and, more importantly, which components of the immune response effectively combat infection and progression. The study of the immune response in this setting is difficult because the cells involved in initiating an immune response are the same cells that are infected by HIV. Nevertheless, an understanding and the identification of immune mechanisms that may be effective in controlling or preventing infection are beginning to emerge.

Organization of the Immune System

The function of the immune system is to discriminate self from nonself. As a consequence of nonself recognition, cellular and humoral interactions are set in motion that result in the sequestration and elimination of that which is recognized as foreign or antigenic. The immune response has the features of specificity (ability to discriminate related antigens), memory (a rapid, heightened response on secondary exposure to antigen), recruitment (induction of secondary antigen-nonspecific mechanisms of antigen disposal), and regulation (dampening or amplification).

The immune system is derived from two cell lineages: monocytic and lymphocytic. Monocytes and their tissue counterparts, macrophages, seed structures of the mononuclear phagocytic system (e.g., Kupffer's cells, sinusoidal cells in the spleen, Langerhans cells in the skin, microglial cells in the brain, dendritic cells in lymph nodes and thymus). These cells have no innate antigen specificity but do scavenge antigens (especially particulate antigens) and present them in an immunogenic form to lymphoid cells. By virtue of receptors for immunoglobulin, complement, or lymphokines, monocyte-derived cells respond to mediators of the immune response and are recruited to mount an inflammatory attack on foreign antigens.

TABLE 3.1. Lymphocyte cell-surface markers.

Cell type	Marker	Function
T cell	CD3	T-cell-receptor complex
	CD2	Cell-cell adhesion; sheep red blood cell receptor
	CD4	T-helper cell; interacts with MHC class II
	CD8	T-cytotoxic and T-suppressor cells; interact with MHC class I
	CD4/CD8	Immature T cell; predominately in thymus
T-cell subsets	CD4 CD45RA	Naive T-helper cells
	CD4 CD29	Memory T-helper cells
	CD8 CD11b$^-$	Cytotoxic T cells
	CD8 CD11b$^+$	Suppressor T cells
Activated T cell	CD25	IL2 receptor
	CD38	T-cell activation marker
	MHC class II (Ia)	
B cell	MHC class II (Ia)	Presents antigen to CD4 cells
	sIg	Surface immunoglobulin; binds antigen
	CD20	B-cell accessory molecule
	CD19	B-cell accessory molecule
Natural killer (NK) cell	CD56	Unknown
	CD16	Fc receptor
All lymphocytes	MHC class I	Presents antigen to CD8 cells

Lymphocytes are of two major types (Table 3.1): T cells and B cells. B cells are derived from bone marrow (and fetal liver). They synthesize immunoglobulin and express it on their cell surface as integral membrane immunoglobulin (sIg). Although sIg defines B cells, other surface structures, CD19 and CD20, detected with monoclonal antibodies (mAbs), are often used for enumerating this population. B cells are not actively engaged in the secretion of immunoglobulins. Rather, they are precursor cells that, after appropriate antigen recognition and stimulation, proliferate and differentiate to become memory B cells or plasma cells. The latter are actively engaged in secretion of immunoglobulin and are the source of serum immunoglobulins. B cells comprise about 10% of peripheral blood lymphocytes, but they comprise 50% of cells that populate lymphoid organs (e.g., lymph nodes), where they are organized into follicles.

The T cells are also bone-marrow-derived cells. They require the presence of an intact thymus for development of immune competence. After migration to the thymus, they undergo a series of differentiation and selection steps. Here self-reactive cells are either eliminated or paralyzed (negative selection or tolerance induction) or the cells acquire immune competence and emerge with cell-surface markers characteristic of mature peripheral blood T cells (positive selection). The selection process is highly abortive: Only 10% of precursor cells that populate the thymus emerge as peripheral T cells.

Mature T cells represent nearly 90% of peripheral blood lymphocytes. These cells circulate from blood to tissue to lymph. They are relatively long-lived, and their recirculation pattern is ideally suited for immune surveillance. Mature T cells possess a variety of cell-surface structures that can be detected with monoclonal antibodies. Two T cell markers, CD2 and CD3, are found on most mature T cells and are referred to as pan-T cell markers. CD2 is involved in intercellular adhesion and antigen-independent activation of T cells. CD3 is a polypeptide structure that is physically associated with the T cell receptor (TCR) for antigen. It mediates intracellular signal transduction and activation following antigen recognition by the TCR.

CD2/CD3 T cells can be further subdivided by two other phenotypic markers, CD4 and CD8. During T cell ontogeny in the thymus, double $CD4^+/CD8^+$ cells exist, which lose one or the other marker to become mature T cells. These markers are on mutually exclusive T cell populations in the periphery. (Rare double-positive cells are sometimes found in the periphery.) Historically, CD4 and CD8 T cell populations have been associated with helper/inducer and cytotoxic/suppressor functions, respectively. It remains a predominant association, although it is by no means absolute. A more stringent association of CD4 or CD8 phenotype with function is in the mode by which they recognize antigen rather than in their subsequent effector functions. CD4 T cells recognize antigen that is processed and presented in the context of class II structures of the major histocompatibility complex (MHC), that is, HLA-DR, HLA-DP, or HLA-DQ. CD8 T cells recognize antigen in the context of class I MHC structures, that is, HLA-A, HLA-B, or HLA-C.

When antigen is introduced into a host, a humoral (antibody) response, a cellular immune response, or both may ensue; both involve cell–cell interactions between multiple cell types. The B cell (antibody) response is a case in point. Although the specificity of the surface immunoglobulin on a given B cell is identical to the specificity of the antibody secreted by its progeny, the plasma cell, simple interaction of antigen with B cells is not sufficient to induce antibody formation. For all but a few so-called T-independent antigens, there is a requirement for T cell recognition of processed antigen and T cell stimulation of the B cell response.

Antigen recognition by B cells and T cells is fundamentally different. B cell-surface immunoglobulin or serum antibody binds directly to native antigen. In contrast, T cells generally do not recognize native antigen; rather, they recognize antigen that is processed and presented in association with class I or II MHC molecules by antigen-presenting cells (APCs). Antigen processing is poorly understood. There is a preferential presentation of endogenously produced antigens (i.e., antigens produced by intracellular pathogens such as viruses) with class I MHC. Exogenously introduced antigens tend to be presented in the context of class II MHC. This distinction is by no means rigid. T cells recognize both processed antigen and MHC; that is, their response to antigen is restricted by the MHC (MHC restriction). Antigen-MHC recognition is performed by the TCR, which recognizes antigen bound in a pocket of the MHC molecule. Most nucleated cells express class I molecules and can present antigen to or be recognized by CD8 T cells. Class II MHC has a more limited cellular distribution, being confined to cells of monocyte lineage and B cells. The latter are potent APCs for CD4 T cells (a particularly potent combination is B cells and CD4 T cells specific for different epitopes on the same antigen). Some cell types that normally do not express class II MHC do so when activated (T cells) or stimulated by an inflammatory response (epithelial cells). After antigen recognition by the T cell, signals are transmitted to the interior via CD3, which sets in motion the biochemical events required for the effector phase of the particular T cell's function.

CD4 and CD8 T cells exist in peripheral blood in a ratio of 2:1 (range 1:1 to 3:1). CD4 T cells are a pivotal cell in the orchestration of an effective immune response. They mount the initial specific response to processed antigen (presented in the context of class II MHC) and, in turn, induce other cells to perform their respective immunologic functions. They do it through cell–cell contact or the elaboration of mediators (lymphokines). They induce cytotoxic CD8 T cells, which recognize antigens presented on class I MHC and lyse the presenting cell. Through the elaboration of lymphokines, CD4 T cells can recruit inflammatory cells, particularly macrophages (delayed-type hypersensitivity response). They induce antigen-specific B cells to proliferate and differentiate into plasma cells secreting specific antibody. Finally, they induce CD8 suppressor cells that regulate the amplitude of the immune response at multiple levels.

CD4 and CD8 T cells have been further subdivided using monoclonal antibodies directed at other cell-surface markers: CD45R (2H4) or CD29 (4B4) identify helper cells that induce CD8 cells or B cells. CD11b (Leul5) is used to distinguish CD8 suppressor cells ($CD11b^+$) from cytotoxic cells ($CD11b^-$). At this time, these subpopulation markers do not have particularly relevant clinical or diagnostic usefulness. After antigen-specific or nonspecific (mitogenic) stimulation, T cells of either the CD8 or CD4 T cell subclass express new markers not present on resting T cells. Among them are class II MHC structures (HLA-DR) and a receptor for the T cell growth factor interleukin-2. These markers have been used to assess the degree of T cell stimulation.

A numerically minor proportion of lymphocytes do not fit any of the three phenotypic categories already discussed, namely sIg-positive B cells and the CD4 or CD8 subsets of CD2/CD3 T cells. These cells are commonly (and inappropriately) referred to as null cells. Most null cells have receptors for complement or for immunoglobulin, and most null cells are natural killer (NK) cells. NK cells are large granular lymphocytes that can kill a variety of tumor cell lines in vitro. This activity is nonspecific in the sense that it does not require prior sensitization, nor is it MHC-restricted. These features distinguish NK cells from CD8 cytolytic T cells. NK cell activity is also distinguished from that of another group of "null" cells, K cells, which mediate antibody-dependent cellular cytotoxicity (ADCC). K cells have receptors for immunoglobulin (FcR). They recognize and lyse antibody-coated target cells. NK and K cells defy easy phenotypic classification. Most NK cells express the NKH1 (CD56) marker, are $CD3^-/CD4^-/CD8^-$, and do not use antibody or the conventional TCR for cellular recognition. A few express low levels of CD3, CD8, and a TCR and are distinguished functionally from cytolytic CD8 cells in that they are MHC-unrestricted in their activity.

Cellular Tropism of HIV-1

All mammalian cells that have been tested have an inherent capacity to replicate infectious HIV provided HIV is introduced as cloned and integrated proviral DNA by transfection.[1–3] Therefore the apparent preferential tropism of HIV for certain cell types must relate to events that occur before replication (attachment, penetration, reverse transcription, or integration). Infection by exogenous HIV has been most reliably and reproducibly demonstrated in cells that express CD4. Preferential infectivity for $CD4^+$ cells has been demonstrated using separated normal human cell populations as well as continuous cell lines,[4–7] and virus is most readily isolated or detected in CD4 cells from HIV-infected individuals, although infectivity for some $CD4^-$ cell lines has been reported.[4–7]

The CD4 molecule has binding avidity for the gp120 outer envelope protein of HIV-1, a characteristic that appears to be the major determinant of cellular susceptibility to natural infection by HIV-1. Evidence that CD4 functions as a receptor for HIV-1 comes from several lines of investigation. Monoclonal antibodies to CD4 prevent infection of CD4 cells by HIV-1.[6,8,9] Virus binds specifically to CD4 cells, and this binding is inhibited by anti-CD4 or anti-gp120 antibodies.[10] Bimolecular complexes of CD4 and gp120 are isolated from HIV-1-exposed CD4 T cells.[11] Human cells that do not express CD4 and cannot be infected by HIV-1 are rendered susceptible by transfection and expression of CD4 cDNA.[12] Finally, soluble forms of the CD4 molecule block HIV binding and infectivity.[13]

All human cells expressing the CD4 molecule have been shown to be infectable with the virus in vitro. In addition to T cells and T cell lines, it includes monocytes that may serve as a reservoir of infection in the body,[14,15] monocyte cell lines,[16] Epstein-Barr virus (EBV)-transformed B cell lines,[17] and glial cell lines.[18] Infectivity for some $CD4^-$ cell lines has also been reported, and alternative receptors have been proposed.[19] Mouse cells expressing human CD4 are not infectable.[12] Experiments with mouse–human somatic cell hybrids indicate that a second (other than human CD4) factor is required for infection.[20] The second receptor, or more appropriately a co-receptor, has not been identified.

Mechanisms of CD4 T Cell Depletion

The precise mechanism by which HIV-1 infection causes CD4 cell death remains controversial. Reasonable hypotheses derived from other viral systems include the following. Massive budding of virus from the membrane results in cell lysis. Massive viral replication commandeers cellular transcriptional/translational machinery required for maintenance of cellular integrity. Viral infection somehow initiates normal cellular processes that result in terminal differentiation and senescence, the so-called programmed cell death that is associated with DNA fragmentation (apoptosis). In vitro and clinical studies support the notion that increased viral replication is associated with more rapid CD4 T cell depletion, which in turn may be determined by the activation state of the infected cells.

The apparent link between increased viral replication and more rapid CD4 T cell depletion suggests a direct effect of HIV infection and replication on CD4 T cell survival. Conversely, any explanation of CD4 T

TABLE 3.2. Numerical depletion and functional impairment of CD4 T cells in HIV-1 infection.

Mechanisms
1. Death of individually infected cells.
a. Requires cell activation.
b. Accumulation of CD4: gp120 complexes.
2. Sequestration of uninfected CD4 cells by HIV-infected cells.
a. Syncytia formation.
b. Supertransmission by dendritic/monocytic cells.
3. Immune recognition and destruction of HIV-1 infected cells.
a. Cytolytic T cell recognition (MHC-restricted CTL).
b. Antibody-dependent cellular cytotoxicity (ADCC)
c. C′-mediated cytolysis.
4. Immune recognition of uninfected cells bearing viral products
a. CTL recognition of CD4 cells that have processed soluble gp120.
b. ADCC of CD4 cells with adsorbed gp120.
5. Immunosuppression by viral products.
a. Interruption of CD4 T cell interaction with MHC class II expressing antigen-presenting cells (APC) by gp120.
6. Autoimmunity or aberrant immune reactivity.
a. Cross reacting viral/cellular antigens (gp120/MHC)
b. Non-MHC restricted cytolysis of activated CD4 T cells
c. Anti-CD4 antibodies.
d. Idiotype-anti-idiotype interactions.
e. Superantigen.

cell depletion that at the same time requires or explains increased viral replication seems attractive. Our bias is that clinically important CD4 T cell depletion occurs as a direct effect of CD4 T cell infection by HIV. Other indirect effects of viral infection, viral products, or the immune response to HIV or self antigens (Table 3.2, items 3–6) may contribute, but it remains to be shown that they are clinically important.

In vitro, infection of normal lymphocytes results in relatively rapid loss of CD4 T cells, a process that is hastened if the T cells are activated.[6] Based on genetic manipulation of virus and its introduction into various cell types, the minimum and sufficient requirements for cytopathicity is *env* gene expression by the virus and CD4 expression by the infected cell. Introduction of HIV-1 into $CD4^-$ cells by transfection results in equivalent viral replication but no cell death.[21] Similarly, certain *env* mutants that cannot bind CD4 replicate if introduced into CD4 cells, but cell death does not ensue.[22] Intracellular complexes of envelope and CD4 proteins accumulate in HIV-1-infected cells, interfering with transport of CD4 to the cell surface, and may somehow be toxic to the cell.[7]

Mechanisms of CD4 T cell depletion that do not require productive infection by the doomed CD4 T cell have also been proposed. Infected cells fuse and form syncytia with uninfected CD4 cells, a phenomenon presumed tantamount to cell death. For CD4 T cell depletion to occur, *env* gene expression is required of the infected cell, and CD4 expression is required of its fusion partner.[23,24] Dendritic cells, specialized APCs that reside in lymph nodes, are particularly potent sources of HIV transmission to CD4 T cells.[25] It has been argued that a mechanism by which infected cells sequester uninfected cells may explain how massive CD4 T cell depletion can occur when so few cells with demonstrable HIV-1 are detected in vivo. However, the relative importance of syncytia formation versus direct viral destruction of individually infected cells in explaining CD4 T cell depletion remains to be determined.

The outer envelope protein gp120 is readily shed in soluble form from virions and infected cells, and it retains CD4-binding activity. Several mechanisms by which soluble gp120 impairs CD4 T cell function or mediates CD4 T cell destruction have been proposed. CD4 cells coated with gp120 react with anti-gp120 antibodies and serve as targets for ADCC.[26] Cells armed with anti-gp120 antibodies that mediate ADCC are frequently found in HIV-1-infected people.[27] By virtue of binding and internalization, CD4 T cells can process and present gp120 peptides for recognition by gp120-specific, MHC-restricted cytolytic T cells.[28] Finally, gp120 bound to CD4 may theoretically interfere with normal functions of CD4, such as class II MHC recognition. Immunosuppressive effects of gp120 on CD4 T cell function have been described in vitro.[29] Indeed, separate cross-linking of CD4 by anti-CD4 mAbs or by gp120 prior to class II MHC-restricted antigen recognition by the TCR–CD3 complex can be a potent tolerizing signal and result in the induction of programmed cell death (apoptosis) of antigen-specific T cells in vitro and, in the case of CD4 mAbs, in vivo.[30]

Autoimmune phenomena may occur during the course of HIV infection and have been suggested as possible mechanisms of T cell destruction. Cross reactions between viral and cellular molecules have been described, particularly cross reactions between regions of HIV envelope and MHC molecules.[31–33] The broad cellular reactivity of these cross-reacting antibodies makes it difficult to explain the selective effect of CD4 cells, although autoantibodies to T cells and autoreactive T cells specific for CD4 cells have been described.[34] The possibility that the antibody response to the binding site of gp120 sets up an idiotypic-antiidiotypic cascade that ultimately results in anti-CD4-reactive antibodies has theoretic appeal but no practical demonstration. Antibodies that react with soluble forms of CD4 can be detected in about 10% of HIV-1-infected people. They bear no relation to disease progression; they do not react with the site on CD4 involved in binding gp120, nor do they react with CD4 in its native conformation on the cell surface.[35]

Superantigens are antigens that react with and bridge MHC molecules on APCs and families of CD4 T cells that share structural features of their TCRs. Superantigen activation of T cells may result in activation and expansion of T cells, depletion by induction of apoptosis, or tolerance induction depending on the maturation state and microenvironment of the responding T cell. Although indirect evidence for a superantigen effect in HIV infection, based on depletion of TCR families in HIV-infected individuals, has been described,[36] there is no consistent or confirmatory evidence that HIV contains a superantigen or initiates a superantigen-like effect in vivo.

Major Components of the Immune System in HIV-1 Infection

CD4 T Cells and CD4/CD8 Ratio

A low CD4/CD8 T cell ratio has been a characteristic and consistent observation since the earliest reports of AIDS. Inverted ratios result from a decrease in the absolute number of CD4 T cells and a variable number (increased, normal, or decreased) of CD8 T cells.[37] The infectivity of HIV-1 for CD4 T cells predicts that numerical depletion of CD4 T cells would reflect the primary lesion more accurately than the CD4/CD8 ratio, although both have their place in the evaluation of HIV-1-infected people.

During early infection and with asymptomatic infection, the ratio is more sensitive to reciprocal changes in CD4 and CD8 cell counts that occur, whereas abnormally low CD4 cell counts are unusual. For example, an infectious mononucleosis-like syndrome has been described in association with acute HIV-1 infection and seroconversion. CD4/CD8 ratios are generally low (because CD8 counts are elevated), whereas CD4 cell counts tend to be normal or low normal.[38] In studies comparing uninfected and infected risk group members who are asymptomatic, the CD4/CD8 ratio is a much more sensitive indicator of HIV-1 infection than the absolute CD4 T cell count. Subjects with common but less severe clinical manifestations have abnormalities intermediate between those of AIDS patients and seropositive, asymptomatic risk-group members.

Although the CD4/CD8 ratio is a more sensitive indicator of HIV-1 infection, the absolute CD4 T cell count is a more specific marker for severe immune deficiency with clinical complications.[39,40] Abnormally low CD4 T cell counts are found in most AIDS patients at diagnosis (65–95%), and virtually all patients who survive long enough eventually develop this abnormality. Low CD4 T cell counts are unusual in asymptomatic, infected people (<5%) and infrequent in those with AIDS-related complex (ARC) (<20%). When a low level of CD4 T cells occurs in these settings, it is a poor prognostic sign. Patients with Kaposi's sarcoma as the sole manifestation of disease tend to have less severe abnormalities than those with opportunistic infections. Those who recover from opportunistic infections generally have more CD4 T cells than those who do not; and once the diagnosis of AIDs is made, progressive depletion of CD4 T cells with time is the rule.

Clinically, a reasonable case can be made for the use of CD4 T cell determinations for monitoring HIV-1-infected patients because these cells are the primary target of infection and their depletion reflects the severity of immune suppression better than any other test. CD4 cell counts have been used for stratifying patients for entrance into clinical trials. As a result, CD4 cell counts are recommended for determining whether to institute antiviral therapy or prophylactic therapy for *Pneumocystis carinii* pneumonia.[41] CD4 T cell determinations are now part of the U.S. National Centers for Disease Control and Prevention's surveillance definition of AIDS and classification systems.[42] CD4 T cell determinations will likely continue to be used for assessing the response to therapy. In therapeutic trials, clinical outcome is the generally accepted measurement of therapeutic response; however, measurement of CD4 cell levels is an objective additional measure that has been promoted as a rational surrogate.

The CD4 T cells are functionally and phenotypically heterogeneous. Preferential loss of certain subsets in vivo has been reported.[37,43] Moreover, surviving CD4 T cells, on a per cell basis, may show selective defects with some in vitro assays, such as antigen-induced proliferation or the autologous mixed lymphocyte reaction, whereas responses to pan-T cell mitogens are intact.[44] Antigen-specific proliferation is believed to be an in vitro correlate of delayed-type hypersensitivity, and skin test responses tend to be impaired in HIV-1-infected patients, especially with advanced disease.

The reason for CD4 phenotypic or functional imbalances early in infection is obscure. There is some evidence for preferential infection of the $CD29^+$ CD4 memory subset in vivo and in vitro, which may account for loss of antigen-specific responses.[45] Some of the functional defects can be reproduced in normal cells with HIV-1 gp120.[29] In any event, these phenotypic/functional imbalances do not appear to have strong prognostic significance. More appropriately stated, they do not outperform enumeration of total CD4 T cells for reflecting clinical status.

CD8 T Cells

The CD8 T cells may be normal, elevated, or depressed in number. Elevations are more frequent early during infection and in asymptomatic subjects, whereas with endstage disease CD8 T cells may decline in parallel with CD4 T cells, resulting in panlymphopenia. There is no evidence that normal CD8 cells are susceptible to natural infection with HIV-1. Although HIV-1 introduced into CD8 cells does replicate, the virus is not cytopathic.[21] CD8 T cells depend on CD4 T cells for induction of effector functions and possibly for maintenance of precursor numbers as well. Thus the changes in this population are most likely a secondary consequence of HIV-1 infection.

A poorly responsive immune system may result from an excess of suppression or a lack of induction. Much earlier work was focused on the possibility that an excess of active suppression occurs in AIDS patients, but no substantial evidence to support this hypothesis has been found.

Phenotypic subsets of CD8 T cells can be identified using two-color immunofluorescence. The cytotoxic subset (identified as $CD11b^-$ CD8 cells) tends to be elevated, and certain markers of activation/immaturity (Ia, CD38) that are elevated in AIDS patients occur within the CD8 compartment.[46,47] The relevance and relations of these markers to pathogenesis and disease prognosis are not clearly understood.

Cytotoxic T cell (CTL) responses to a specific virus, such as EBV, cytomegalovirus (CMV), and influenza virus also decline with progression of disease,[48–50] although it is not clear whether it results from a lack of CD4 help for the response or a loss of response by the CD8 cells themselves. In vitro, the virus does not inhibit the CD8 CTL response to EBV in normal individuals, although it is able to block the CD4 CTL response to another herpes virus, herpes simplex virus 1 (HSV-1).[51] This ability seems to result from a direct interaction between gp120 and CD4 on the effector cells, although a cell loss of CD4 CTL precursors could also play a role. Patients with AIDS often suffer from recurrent fulminating herpes lesions and also have an increased incidence of B cell lymphomas. The loss of the CTL response that normally contains these viruses may explain these clinical findings.

Natural Killer Cells

Cells expressing NK-associated phenotypes and NK function tend to be well preserved in asymptomatic HIV-1-infected individuals but decline with advanced disease.[48] Large granular lymphocyte (LGL) or NK function in HIV-1 infection is of special interest for several reasons. Cells that are ordinarily resistant to NK lysis can be rendered susceptible by HIV-1 infection.[52] A subpopulation of LGLs (lacking NK activity) have been shown to produce interferon-alpha (IFN-α), which potentiates NK function and depresses HIV-1 replication in vitro.[53] The extent to which IFN-α and NK function limit infection in vivo and if preservation of these functions would prevent progression remain unknown.

B Cells

Hypergammaglobulinemia, elevated levels of immune complexes, and polyclonal B cell activation are common findings in HIV-1-infected individuals. Although immunoglobulin levels are high, AIDS patients mount relatively poor antigen-specific antibody responses after vaccination. Primary responses are affected to a greater degree than secondary responses.[54] In asymptomatic individuals with HIV-1 infection who have not yet developed immune deficiency associated with severe clinical manifestations, antibody responses to vaccination tend to be normal, and no excess adverse reactions have been reported.[55]

The reason for polyclonal B cell stimulation in HIV-1 infection remains an enigma. Several postulates have been put forth. In vitro, HIV-1 or HIV gp120 protein has been shown to stimulate B cells polyclonally either directly or indirectly through an action on T cells.[56] EBV is a polyclonal B cell stimulator. Infection with EBV is almost universal in AIDS patients, and it is proposed that loss of T cell control results in reactivation of latent EBV infection, leading to polyclonal B cell activation.[50] Other postulates for which there is less experimental support include a T cell dysregulation phenomenon and the hypothesis that HIV-1-infected cells release B cell-stimulating lymphokines.

Monocytes and Macrophages

Human monocytes and their tissue counterparts, macrophages, express the CD4 molecule and can be infected with HIV-1. HIV-1-infected monocytes appear to be the prime target of HIV-1 infection in the central nervous system.[5] Because monocytes replicate virus at low levels and do not die in vitro, it has been proposed that monocyte-derived cells are a more stable reservoir of infection than T cells. However, virus is inevitably detected in T cells from HIV-1-infected individuals by sensitive DNA amplification techniques and is only occasionally detected in monocytes.[57]

Many defects in monocyte function have been reported with HIV-1 infection. For the most part, these defects are found inconsistently and are more prominent with advanced disease. It is not clear whether the defects are a result of infection of monocytes, immunoregulatory T cell imbalance, or supervening infections. For instance, the number of circulating monocytes is generally normal but may decline with advanced disease. Similarly, defects in phagocytosis, intracellular killing, cytotoxicity, and class II MHC expression are most consistently found in patients with advanced disease. Antigen-specific proliferative responses are suppressed, which is an early, relatively consistent feature of HIV infection.[44] In vitro studies of anti-CD3-induced T cell proliferation, which requires presentation of anti-CD3 by monocytes, have demonstrated a defect in presentation in monocytes from infected individuals.[58] Similarly, an HIV-infected monocytic cell line also shows impaired presentation in this system. However, defects in presentation of soluble antigen have yet to be convincingly demonstrated.

Natural History and Codeterminants of Progressive Disease

The natural history of HIV-1 infection is one of progressive CD4 T cell depletion with increasing likelihood of clinical complications (Fig. 3.1). Although perhaps intuitively expected, it has only recently been appreciated that progressive disease is associated with increased virus replication and virus burden in the chronically infected host.[59] Progressive CD4 T cell depletion is a seemingly inexorable process related to duration of infection. However, there is clearly great heterogeneity in the rate at which infected people develop severe immune deficiency. What determines this rate? In vitro studies of virus replication in acutely infected or latently infected cells clearly indicate that the activation state of the host cell is a major determinant of quantitative virus replication and cytopathicity.[6] From these data, one might infer that cofactors that serve to activate T cells would hasten virus replication, T cell depletion, and clinical deterioration. In clinical studies it has been difficult to identify clinical, immunologic, or virologic features that predict progression. Features associated with progression have been described, but it has been difficult to determine whether they are simply a marker or a consequence of severe immune deficiency rather than a marker for a process that initiates progression. Markers reflecting a poor prognosis include low CD4 T cell counts (the best predictive marker), the onset of nonspecific symptoms (fever, weight loss), duration of infection, low antibody titers to HIV-1 (especially to the p24 core antigen), low levels of antibodies that inhibit reverse transcriptase, elevated levels of p24 antigen, plasma viremia, high in vivo titers of HIV-1, infection with or emergence of more

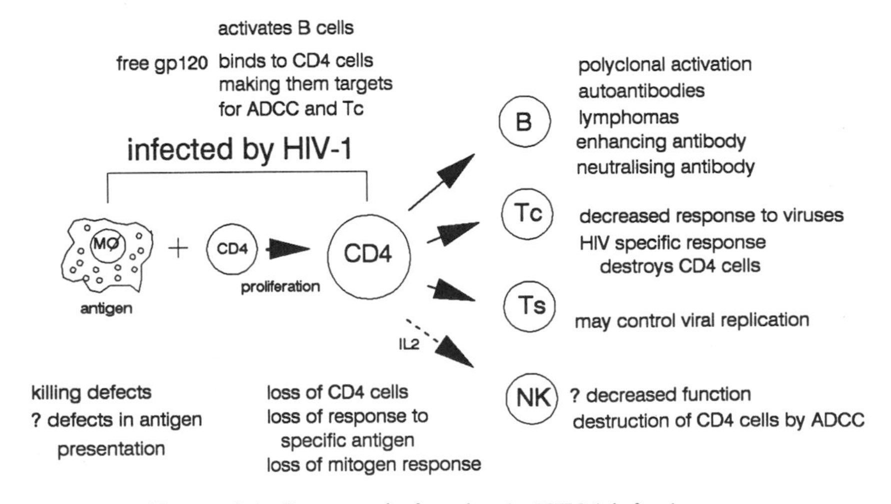

FIGURE 3.1. Immune dysfunction in HIV-1 infection.

virulent strains of HIV-1, elevated β_2-microglobulin levels, and elevated neopterin.[49]

Immune Response to HIV-1 Infection

Individuals infected with HIV-1 mount and sustain a vigorous antibody response to HIV-1. Most generate detectable antibody within 3 months. Reports of prolonged seronegativity have been rare, are difficult to confirm, and are controversial. With advanced immune deficiency, antibody titers to most viral proteins diminish, with the curious exception of antibody titers to the transmembrane protein gp41.[60]

With established infection it is not clear what role antibody plays in limiting infection, nor is it clear how effective antibody would be if it were induced before infection (as by vaccination). Antibody titers decline with progressive disease, suggesting that high antibody levels maintain a stable, low level of infection. However, it is not clear whether declining antibody titers cause progression or are a consequence of it. Human anti-HIV-1 sera inevitably contain antibodies to the envelope protein that inhibit virus binding to CD4 cells and have the capacity to neutralize infectivity.[62] They also mediate ADCC.[63] These same sera may contain antibodies that enhance infection in vitro.[64] Dissection of the antibody response has revealed antigenic determinants that (1) elicit strain-specific and broadly neutralizing antibodies; (2) elicit T cell responses; (3) are normally not immunogenic but potentially elicit a neutralizing response; and (4) have no apparent in vitro activity except possibly to interfere with binding by effective antibodies or perhaps mediate enhanced infection.[62,64–67] Moreover, a dynamic may exist between the infecting strain of HIV and the immune response, reflecting in vivo selection for strains that escape the neutralizing effects of human sera.[68–70] Obviously, the identification of relevant epitopes for induction of functionally important antibody responses is being actively investigated and is of paramount importance for vaccine development. The concept of protection by antibody is certainly feasible because both passive and active immunization have met with limited success in animal models of retrovirus infection and recently in humans.[71,72]

For many viral diseases, antibody induced by vaccination or as a result of primary infection protects against infection on subsequent exposure, but it is the cellular response that is primarily responsible for resolution of infection. HIV-1 is unique in that the very cells that initiate a cellular response to HIV-1 are the cells destroyed by HIV-1. The sustained antibody response to HIV-1, which most certainly requires antigen-specific T cell–B cell collaboration, suggests that HIV-1-specific T cells do exist; moreover, other observations, previously discussed, suggest that there may be phenotypic or functional immune profiles that indicate an attempt by the immune system to eliminate infection. Indeed, cellular responses to HIV-1 have been demonstrated. MHC-restricted, HIV-1-specific cytolytic CD8 T cells have been identified in HIV-1-infected individuals.[73] Noncytolytic CD8 T cells that inhibit viral replication or viral recovery in vitro have also been demonstrated.[74] The relation to disease progression and the role in limiting infection of these cell-mediated responses have not yet been established. Nevertheless, by analogy with other persistent virus infections, it is likely that they are important features of an effective immune response.

Conclusion

The remarkable progress in research on HIV and the immune system has provided a solid foundation for understanding the molecular, cellular, and clinical pathogenesis of HIV-1 infection. HIV-1 has evolved an affinity for the CD4 molecule and a replication apparatus that is cytopathic for host cells: HIV-1 binds, penetrates, replicates in, and destroys CD4 T cells. Numerical/functional depletion of CD4 cells occurs over time, resulting in progressive paralysis of immune responsiveness and rendering the infected person susceptible to opportunistic infections and malignancies. Within this framework, much remains to be learned about factors that govern the extent and control of infection. The hope is that immunologic or other factors that control or prevent infection can be identified and applied successfully in this still incurable disease.

References

1. Fisher AG, Collalti E, Ratner L, Gallo RC, Wong-Stahl F. A molecular clone of HTLV-III with biological activity. Nature 1985;316:262–265
2. Levy JA, Cheng-Meyer C, Dina D, Luciw PA. AIDS retrovirus (ARV-2) clone replicates in transfected human and animal fibroblasts. Science 1986; 232:998–1000
3. Srinivasan A, York D, Jannoun-Nasr R, et al. Generation of hybrid human immunodeficiency virus by homologous recombination. Proc Natl Acad Sci USA 1989;86:6388–6392
4. Klatzman D, Barre-Sinoussi F, Nugeyre MT, et al. Selective tropism of lymphadenopathy-associated virus (LAV) for helper-inducer T lymphocytes. Science 1984;225:59–62
5. Popovic M, Read-Connole E, Gallo R. T4 positive human neoplastic cell lines susceptible to and permissive for HTLV-III. Lancet 1984;2:1472–1473
6. McDougal JS, Mawle AC, Cort SP, et al. Cellular tropism of the human retrovirus HTLV-III/LAV. I. Role of T cell activation and expression of the T4 antigen. J Immunol 1985;135:3151–3162
7. Hoxie JA, Alpers JD, Rackowski JL, et al. Alterations in T4 (CD4) protein and mRNA synthesis in cells infected with HIV. Science 1986;234:1123–1127

8. Klatzmann D, Champagne E, Charmarat S, et al. T-lymphocyte T4 molecule behaves as the receptor for human retrovirus LAV. Nature 1985;312:767–768
9. Dalgleish AG, Beverley PCL, Clapham PR, et al. The CD4 (T4) antigen is an essential component of the receptor for the AIDS retrovirus. Nature 1985;312:763–767
10. McDougal JS, Nicholson JKA, Cosand WL, et al. HIV binding to the CD4 molecule: conformation dependence and binding inhibition studies. In Bolognesi D (ed): Human Retroviruses, Cancer, and AIDS: Approaches to Prevention and Therapy. New York: Alan R. Liss, 1988:269–281
11. McDougal JS, Kennedy MS, Sligh JM, et al. Binding of HLTV-III/LAV to T4^{+} T cells by a complex of the 110K viral protein and the T4 molecule. Science 1986;231:382–385
12. Maddon PJ, Dalgleish AG, McDougal JS, et al. The T4 gene encodes the AIDS virus receptor and is expressed in the immune system and in the brain. Cell 1986;47:333–348
13. Clapham PR, Weber JV, Whitby D, et al. Soluble CD4 blocks the infectivity of diverse strains of HIV and SIV for T cells but not for brain and muscle cells. Nature 1989;337:368–370
14. Nicholson JKA, Cross GD, Callaway CS, McDougal JS. In vitro infection of human monocytes with human T lymphotropic virus type III/lymphadenopathy-associated virus (HTLV-III/LAV). J Immunol 1986;137:323–329
15. Gabudza DH, Ho DD, de la Monte SM, et al. Immunohistochemical identification of HTLV-III antigen in brains of patients with AIDS. Ann Neurol 1986;20:289–295
16. Folks TM, Justement J, Kinter A, Dinarello CA, Fauci AS. Cytokine-induced expression of HIV-1 in a chronically infected promonocyte cell line. Science 1987;238:800–802
17. Montagnier L, Gruest J, Chamaret S, et al. Adaptation of lymphadenopathy associated virus (LAV) to replication in EBV-transformed B lymphoblastoid cell lines. Science 1984;225:63–66
18. Cheng-Mayer C, Rutka JT, Rosenblum ML, et al. Human immunodeficiency virus can productively infect cultured glial cells. Proc Natl Acad Sci USA 1987;84:3526–3530
19. Harouse JM, Bhat S, Spitalnik SL, et al. Inhibition of entry of HIV-1 in neural cell lines by antibodies against galactosyl ceramide. Science 1991;253:320–323
20. Dragic T, Charneau P, Clavel F, Alizon M. Complementation of murine cells for human immunodeficiency virus envelope/CD4-mediated fusion in human/murine heterokaryons. J Virol 1992;66:4794–4802
21. DeRossi A, Franchinin G, Aldorini A, et al. Differential response to the cytopathic effects of human T-cell lymphotropic virus type III (HTLV-III) superinfection in T4^{+} (helper) and T8^{+} (suppressor) T cell clones transformed by HTLV-1. Proc Natl Acad Sci USA 1986;83:4297–4301
22. Arthos J, Deen KC, Chaikin MA, et al. Identification of the residues in human CD4 critical for the binding of HIV. Cell 1989;57:469–481
23. Lifson JD, Reyes GR, McGrath MS, Stein BS, Engleman EG. AIDS retrovirus induced cytopathology: giant cell formation and involvement of CD4 antigen. Science 1986;232:1123–1126

24. Sodroski J, Goh WC, Rosen C, Campbell K, Haseltine WA. Role of the HTLV-III/LAV envelope in syncytium formation and cytopathicity. Nature 1986;322:470–474
25. Cameron PU, Freudenthal PS, Barker JM, et al. Science 1992;257:383–387
26. Lyerly HK, Matthews TJ, Langlois AS, Bolognesi DP, Weinhold KJ. Human T-cell lymphotropic virus III_B glycoprotein (gp120) bound to CD4 determinants on normal lymphocytes and expressed by infected cells serves as target for immune attack. Proc Natl Acad Sci USA 1987;84:4601–4605
27. Weinhold KJ, Lyerly HK, Matthews TJ, et al. Cellular anti-gp120 cytolytic reactivities in HIV-1 seropositive individuals. Lancet 1988;1:902–905
28. Siliciano RF, Lawton T, Knall C, et al. Analysis of host-virus interactions in AIDS with anti-gp120 T cell clones: effect of HIV sequence variation and a mechanism for CD4 cell depletion. Cell 1988;54:561–575
29. Habeshaw JA, Dalgleish AG, Bountiff L, et al. AIDS pathogenesis: HIV envelope and its interactions with cell proteins. Immunol Today 1990;11: 418–425
30. Groux H, Torpier G, Monté D, et al. Activation-induced death by apoptosis in CD4 T cells from human immunodeficiency virus-infected asymptomatic individuals. J Exp Med 1992;175:331–340
31. Atassi H, Atassi MZ. HIV envelope protein is recognized as an alloantigen by human DR-specific alloreactive T cells. Hum Immunol 1992;34:31–38
32. Grassi F, Meneveri R, Gullberg M, et al. Human immunodeficiency virus type 1 gp120 mimics a hidden monomorphic epitope borne by class I major histocompatibility complex heavy chains. J Exp Med 1991;174:53–62
33. Golding H, Robey FA, Gates FT III, et al. Identification of homologous regions in human immunodeficiency virus I gp41 and human MHC class II β 1 domain. Monoclonal antibodies against the gp41-derived peptide and patients' sera react with native HLA classs II antigens, suggesting a role for autoimmunity in the pathogenesis of acquired immune deficiency syndrome. J Exp Med 1988;167:914–923
34. Zarling JM, Ledbetter JA, Sias J, et al. HIV-infected humans, but not chimpanzees, have circulating cytotoxic T lymphocytes that lyse $CD4^+$ cells. J Immunol 1990;144:2992–2998
35. Edelman AS, Zolla-Pazner S. AIDS: a syndrome of immune dysregulation, dysfunction, and deficiency. FASEB J 1989;3:22–30
36. Imberti L, Sottini A, Bettinardi A, Puoti M, Primi D. Selective depletion in HIV infection of T cells that bear specific T cell receptor V beta sequences. Sicence 1991;254:860–862
37. Nicholson JKA, McDougal JS, Spira TJ, et al. Immunoregulatory subsets of the T helper and T suppressor cell population in homosexual men with chronic unexplained lymphadenopathy. J Clin Invest 1984;73:191–201
38. Cooper DA, Gold J, Mallean P, et al. Acute AIDS retrovirus infection: definition of a clinical illness associated with seroconversion. Lancet 1985;1: 537–540
39. Fahey JL, Prince H, Weaver MM, et al. Quantitative changes in the Th or Ts lymphocyte subsets that distinguish AIDs syndromes from other immune subset disorders. Am J Med 1984;76:95–100
40. Fauci AS, Macher H, Longo DL, et al. Acquired immunodeficiency syndrome: epidemiological, clinical, immunologic and therapeutic considerations. Ann Intern Med 1984;100:92–106

41. CDC. Recommendations for prophylaxis against Pneumocystis carinii pneumonia for adults and adolescents infected with human immunodeficiency virus. MMWR 1992;41(RR-4)
42. CDC. 1993 Revised classification system for HIV infection and expanded surveillance case definition for AIDS among adolescents and adults. MMWR 1992;41(RR-17)
43. Wood GS, Burnes BF, Dorfman RD, Warnke RA. In situ quantitation of lymph node helper, suppressor, and cytotoxic T cell subsets in AIDS. Blood 1986;67:596–603
44. Lane HC, Fauci AS. Immunological abnormalities in the acquired immunodeficiency syndrome. Annu Rev Immunol 1985;3:477–500
45. Schnittman SM, Lane HC, Greenhouse J, et al. Preferential infection of $CD4^+$ memory cells by human immunodeficiency virus type 1: evidence for a role in the selective functional defects observed in infected individuals. Proc Natl Acad Sci USA 1990;87:6058–6062
46. Nicholson JKA, Echenberg DF, Jones BM, et al. T cytotoxic/suppressor cell phenotypes in a group of asymptomatic homosexual men with and without exposure to HTLV-III/LAV. Clin Immunol Immunopathol 1986;40:505–514
47. Salazar-Gonzalez JF, Moody DJ, Giorgi JV, et al. Reduced ecto-5′ nucleotidase activity and enhanced OKT10 and HLA-DR expression on CD8 (T suppressor/cytotoxic) lymphocytes in the acquired immune deficiency syndrome: evidence of CD8 immaturity. J Immunol 1985;135:1778–1785
48. Rook AH, Manischewitz JF, Frederick WR, et al. Deficient HLA-restricted cytomegalovirus-specific cytotoxic T cells and natural killer cells in patients with the acquired immune deficiency syndrome. J Infect Dis 1985;152:627–630
49. Shearer GM, Salahuddin SZ, Markham PD, et al. Prospective study of cytotoxic T lymphocyte responses to influenza and antibodies to human lymphotrophic virus-III in homosexual men. J Clin Invest 76:1699–1704
50. Birx DL, Redfield RR, Tosato R. Defective regulation of Epstein-Barr virus infection in patients with acquired immune deficiency syndrome or AIDS-related disorders. N Engl J Med 1986;314:874–879
51. Mawle AC, Thieme ML, Ridgeway MR, McDougal JS, Schmid DS. Inhibition of the in vitro generation of class II-restricted, HSV-1-specific $CD4^+$ CTL bv HIV-1. AIDS Res Hum Retrovir 1990;6:229–241
52. Ruscetti FW, Mikouits JA, Kalyanaraman VS, et al. Analysis of effector mechanisms against HTLV-1 and HTLV-III/LAV infected lymphoid cells. J Immunol 1986;136:3619–3624
53. Ho DD, Hartshorn KL, Rota TR, et al. Recombinant human interferon alpha-A suppresses HTLV-III replication in vitro. Lancet 1985;1:602–604
54. Lane HC, Masur H, Edgar LC, et al. Abnormalities of B cell activation and immunoregulation in patients with the acquired immunodeficiency syndrome. N Engl J Med 1983;309:453–458
55. Huang KL, Ruben FL, Rinaldo CR, et al. Antibody responses after influenza and pneumococcal immunization in HIV-infected homosexual men. JAMA 1987;257:2047–2050
56. Schnittman SM, Lane HC, Higgins SE, Folks T, Fauci AS. Direct polyclonal activation of human B lymphocytes by the acquired immune deficiency virus. Science 1986;233:1084–1086

57. Meltzer MS, Skillman DR, Hoover DL, et al. Macrophages and the human immunodeficiency virus. Immunol Today 1990;11:217–223
58. Prince HE, Moody DJ, Shubin BI, Fahey JF. Defective monocyte function in acquired immune deficiency syndrome (AIDS): evidence from a monocyte-dependent proliferative system. J Clin Invest 1985;5:21–25
59. Nicholson JKA, Spira TJ, Aloisio CH, et al. Serial determinations of HIV-1 titers in HIV-infected homosexual men: association of rising titers with CD4 T cell depletion and progression to AIDS. AIDS Res Hum Retrovir 1989; 5:205–215
60. Hessol NA, Lifson AR, Rutherford GW. Natural history of human immunodeficiency virus and key predictors of disease progression. In: Volberding P, Jacobson MA (eds): AIDS Clinical Reviews. New York: Marcel Dekker, 1989:69–93
61. Sarngadharan MG, Popovic M, Bruch L, Schupbach J, Gallo RC. Antibodies reactive with human T-lymphotropic retroviruses (HTLV-III) in the serum of patients with AIDS. Science 1984;224:506–508
62. Looney DJ, Fisher AG, Putney SD, et al. Type-restricted neutralization of molecular clones of human immunodeficiency virus. Science 1988;241: 357–359
63. Ojo-Amaize EA, Mishanian P, Keith DE, et al. Antibodies to human immunodeficiency virus induce cell-mediated lysis of human immunodeficiency virus-infected cells. J Immunol 1987;139:2458–2463
64. Matsuda S, Gidlund M, Chiodi F, et al. Enhancement of human immunodeficiency virus (HIV) replication in human monocytes by low titers of anti-HIV antibodies in vitro. Scand J Immunol 1989;30:425–434
65. Ho DD, Sarngadharan MG, Hirsch MS, et al. Human immunodeficiency virus neutralizing antibodies recognize several conserved domains on the envelope glycoproteins. J Virol 1987;61:2024–2028
66. Weiss RA, Clapham PR, Weber NJ, et al. Variable and conserved neutralization antigens of human immunodeficiency virus. Nature 1986;324:572–575
67. Krohn K, Robey WG, Putney S, et al. Specific cellular immune response and neutralizing antibodies in goats immunized with native or recombinant envelope proteins derived from human T-lymphotropic virus type III_B and in human immunodeficiency virus-infected men. Proc Natl Acad Sci USA 1987; 84:4994–4998
68. Tremblay M, Wainberg MA. Neutralization of multiple HIV-1 isolates from a single subject by autologous sequential sera. J Infect Dis 1990;162:735–737
69. Albert J, Abrahamsson B, Nagy K, et al. Rapid development of isolate-specific neutralizing antibodies after primary HIV-1 infection and consequent emergence of virus variants which resist neutralization by autologous sera. AIDS 1990;4:107–112
70. Arendrup M, Nielsen C, Hansen JES, et al. Autologous HIV-1 neutralizing antibodies: emergence of neutralization-resistant escape virus and subsequent development of escape virus neutralizing antibodies. J Acquir Immune Defic Syndr 1992;5:303–307
71. Murphy-Corb M, Martin LN, Davison-Fairburn B, et al. A formalin-inactivated whole SIV vaccine confers protection in macaques. Science 1989; 246:1293–1297

72. Karpas A, Hewlett IK, Hill F, et al. Polymerase chain reaction evidence for human immunodeficiency virus I neutralization by passive immunization in patients with AIDS and AIDS-related complex. Proc Natl Acad Sci USA 1990;87:7613–7617
73. Walker BD, Plata F. Cytotoxic T lymphocytes against HIV. AIDS 1990;4: 177–184
74. Walker CM, Moody DJ, Stites DP, Levy JA. $CD8^+$ T lymphocyte control of HIV replication in cultured $CD4^+$ cells varies among infected individuals. Cell Immunol 1989;119:470–475

4
FDA Regulation of HIV-Related Tests and Procedures

JAY S. EPSTEIN

The U.S. Food and Drug Administration (FDA) regulates drugs and medical devices under the Federal Food, Drug and Cosmetic (FD&C) Act (21 U.S.C. §§ 301 *et seq.*). Medical products may also be subject to the licensing provisions of the Public Health Service (PHS) Act (42 U.S.C. §§ 201 *et seq.*) when they meet the definition of a biologic product. Biologic products additionally are regulated either as drugs or medical devices under the FD&C Act and Amendments. To date, the FDA has regulated all human immunodeficiency virus (HIV)-related tests as biologic products, except that the Agency has regulated sample collection systems used in conjunction with HIV-related tests as medical devices. Test procedures provided by clinical laboratories currently are regulated by the Health Care Financing Administration under the Clinical Laboratory Improvement Amendments of 1988 (CLIA '88) (42 U.S.C. §§ 201 note, 263a, 263a note).

For drugs, biologic products, and class III medical devices, the purpose of regulation by the FDA is to determine the safety and efficacy (or effectiveness) of these medical products prior to their commercialization and to protect the public from adulterated and misbranded products. To accomplish its mandate, the Agency reviews information regarding the manufacturing, clinical performance, and labeling of the product. Manufacturers must demonstrate that products are manufactured consistently and in conformance with standards set forth in published Current Good Manufacturing Practice Regulations (cGMPs), that they have been found to be safe and effective for their intended use by current scientific methods including well controlled clinical trials, that all medical claims have been validated, and that the products are properly labeled.

For HIV-related diagnostic tests, safety concerns have included control of biohazards during the manufacturing process, safety during clinical use of kits containing components that have undergone virus inactivation, and the welfare of test subjects with regard to proper interpretation and use of test results. Efficacy concerns have focused on determination of test sensitivity, specificity, and reproducibility in actual intended-use settings. As with other in vitro diagnostic tests, HIV-related tests may be intended for use in epidemiologic surveillance, screening, and medical diagnosis or prognosis. The application of HIV tests as screening tests for HIV antibodies to determine the suitability of blood donors, and hence the safety of blood products, has been closely regulated.

Elements of FDA Regulation

The FDA exercise of statutory authorities over HIV-related tests involves three levels of control: (1) approval of product and of establishment license applications and amendments (now called supplements); (2) surveillance; and (3) enforcement.

Because it is illegal to distribute commercially any unapproved biologic or drug products for clinical use or to promote them for unapproved indications, manufacturers must obtain exemptions to engage in clinical studies. The FDA reviews Investigational New Drug (IND) or Investigational New Device Exemption (IDE) applications to evaluate consistency of product manufacturing, safety and efficacy based on preclinical data, suitable credentials of the investigators, adequacy of the clinical study design, and approval of the study by an Institutional Review Board (IRB).

After clinical investigation, manufacturers of test kits regulated as biologic products must apply for and receive separate licenses for the product and for the establishment where it is manufactured before they engage in commercial distribution. Establishment License Applications are reviewed to evaluate the facility's design including systems validation, assess personnel management and training, and ensure compliance with cGMPs, including equipment validation, quality control procedures, and record keeping. An Environmental Assessment, as specified in 21 CFR 25.1, is also evaluated. Product License Applications are reviewed for validation of manufacturing procedures, including monitoring of the identity, purity, potency, stability, and consistency of components; for validation of product claims through preclinical and clinical studies; and for completeness and accuracy of product labeling. A prelicense inspection is conducted while manufacturing is ongoing as part of the review process. After approval of establishment and product applications, changes in facilities, manufacturing procedures, and product labeling are reviewed through submitted license amendments (supplements). Guidance concerning the content of applications is provided in the FDA's draft of "Points to Consider in the Manufacture and Clinical Evaluation of In-Vitro Tests to Detect Antibodies to the Human Immunodeficiency Virus, Type 1 (1989)," [available from Food and Drug Administration, Center for Biologics Evaluation and Research, Division of Congressional and

Consumer Affairs, HFM-12, Woodmont Office Center, Suite 200N, 1401 Rockville Pike, Rockville, MD 20852-1448 (telefax no. 301-594-1938)].

After a product has been approved, the Agency monitors performance through various surveillance mechanisms, including periodic inspections at least biennially. The Agency may ask that the manufacturer conduct "Phase IV" studies to investigate product performance issues as a condition of approval. The FDA audits mandatory reports of errors, accidents, and adverse reactions (or Medical Device Reports), which may document product failures. Furthermore, the Agency may monitor test kits on a lot-by-lot basis through a release mechanism. Since they were first licensed in 1985, the FDA has performed lot release testing prior to distribution for all lots of HIV tests that are approved for blood donor screening.

The Agency also may exercise control through enforcement actions. It may invoke a variety of administrative and judicial remedies to guard against statutory and regulatory violations, including license suspensions and revocations under the PHS Act, product recalls and seizures, and civil or criminal proceedings under the FD&C Act.

HIV-Related Tests Currently Regulated by the FDA

Since 1985 the FDA has approved a variety of HIV-related tests based on HIV virus antigens or antibodies directed toward these antigens (or both). Some of the currently available tests are approved for use in blood donor screening and medical diagnosis (Table 4.1), and others are approved primarily for use in medical diagnosis (Table 4.2). These tests are characterized as HIV antibody screening tests, supplemental tests used to verify the results of antibody screening tests, and tests for HIV antigen(s). Most approved screening tests for HIV-1 use standard indirect enzyme-linked immunosorbent assay (ELISA) technology with an enzyme-labeled "second antibody," although one test uses an enzyme-labeled antigen in an "antibody sandwich" design. The antigenic substrates have included whole viral lysates from cell culture, purified viral proteins, recombinant DNA-derived viral proteins, and synthetic peptides. Two rapid tests have also been approved: a latex agglutination test and an enzyme immunoassay based on antigen-coated microparticles. Supplemental tests for antibodies to HIV-1 include Western blot and immunofluorescence assays. The approved test for HIV-1 antigen detects viral p24 antigen in a capture ELISA format and includes use of a blocking antibody to identify true-positive results. The FDA has licensed one screening test for HIV-2 antibodies and two screening tests for combined detection of antibodies to HIV-1 plus HIV-2. For each licensed test kit, the criteria used by the FDA to validate the test are provided in a "Summary of Basis for Approval" document [available from the Food and Drug Administration,

TABLE 4.1. HIV test kits currently approved for blood donor screening and medical diagnosis.

HIV screen	Manufacturer	Product trade name	Date licensed
HIV-1 antibodies	Abbott Laboratories, North Chicago, IL	HIVAB HIV-1 EIA	3/1/85
	Cambridge Biotech Corp., Worcester, MA	Recombigen (*env* and *gag*) HIV-1 EIA	5/1/90
		MicroTrak HIV-1 EIA (*env* and *gag*)	5/30/90
	Cellular Products, Buffalo, NY	Retro-Tek HIV-1 ELISA	8/29/86
	Genetic Systems Corp., Redmond, WA	Genetic Systems LAV EIA	2/18/86
	Organon Teknika Corp., Durham, NC	HIV-1 Bio-EnzaBead	4/5/85
		Vironostika HIV-1 MicroElisa System	12/18/87
	Ortho Diagnostic Systems, Raritan, NJ	Ortho HIV-1 ELISA Test System	2/21/86
	United Biomedical, Hauppauge, NY	UBI-Olympus HIV-1 EIA	5/31/89
HIV-2 antibodies	Genetic Systems Corp., Redmond, WA	Genetic Systems HIV-2 EIA	4/25/90
HIV-1 and HIV-2 antibodies	Abbott Laboratories, North Chicago, IL	HIVAB HIV-1/HIV-2 (rDNA) EIA	2/14/92
	Genetic Systems Corp., Redmond, WA	Genetic Systems HIV-1/HIV-2 EIA	9/25/91

TABLE 4.2. HIV test kits currently approved primarily for medical diagnosis.

Purpose of test	Manufacturer	Product trade name	Date licensed
Rapid screen for HIV-1 antibodies	Cambridge Biotech Corp., Worcester, MA	Recombigen HIV-1 LA Test[a]	12/13/88
	Murex Corp., Norcross, GA	SUDS HIV-1 Test[a]	5/22/92
Detection of HIV-1 antigen	Abbott Laboratories, North Chicago, IL	HIVAG-1	8/3/89
Supplemental test for HIV-1 antibodies	Bio-Rad Laboratories, Hercules, CA	Novapath HIV-1 Immunoblot	6/15/90
	Cambridge Biotech Corp., Worcester, MA	HIV-1 Western Blot Kit	1/3/91
	Epitope, Beaverton, OR	EPIblot HIV-1 Western Blot Kit	3/20/91
	Waldheim Pharmazeutika, Vienna, Austria	Fluorognost HIV-1 IFA[a]	2/5/92

[a]The test may be used for blood donor screening only when routine ELISA tests are unavailable or impractical.

Freedom of Information Staff, HFI-35, 5600 Fishers Lane, Room 12A-16, Rockville, MD 20857 (telefax no. 301-443-1726)].

Regulatory Concerns Regarding Screening Tests

The Agency has regarded test sensitivity as the leading issue when giving approval for screening tests for antibodies to HIV-1 or HIV-2. Strategies for evaluating test kit sensitivity have included comparisons with research tests and previously licensed tests. Preclinical studies have used dilutional series of positive human sera as well as seroconversion series to show equivalence to a "state of the art" performance standard such as a Western blot assay. Clinical trials have used "head to head" comparisons with reference tests in high risk populations. Claims for sensitivity have been based on the percent positivity for randomly selected AIDS patients, although reporting of comparative analysis with other test methods has been permitted in test kit package inserts. To ensure that commercial kits continue to meet a uniform minimum standard for sensitivity, the FDA developed and has maintained a lot release panel of sera that is used to test every lot of antibody test kits prior to distribution.

A continuing controversy surrounds the inconsistency of test kits for detection of IgM antibodies to HIV. Most test kit manufacturers have avoided the use of IgM-specific conjugates in the indirect ELISA design because of the familiar problem of false-positive reactions with nonspecifically "sticky" IgM antibodies. Such a problem has been observed for several blood donor screening ELISA tests following immunizations for influenza.[1] Despite early skepticism, there is little doubt that increased sensitivity for IgM can enhance antibody detection during the first few weeks after HIV infection. The "antibody sandwich" ELISA was developed in part to improve sensitivity for IgM.

The validation of tests using recombinant and synthetic antigens of HIV as substrates has presented special challenges. A leading concern is the theoretic possibility that rapid genetic evolution of HIV worldwide could result in the emergence of infectious virus variants eliciting an antibody response that could escape detection by a narrow range of epitopes represented in the antigenic substrate. To address this possibility, the FDA has required that test kits based on neoantigens be validated in premarket studies against approximately 1000 known positive sera, including samples from all of the known major regions of the epidemic. Additionally, some manufacturers have been asked to conduct phase IV surveillance studies to look for false-negative results due to virus variation. To determine the appropriate level of scientific concern, on December 18, 1992 the Agency discussed with its Blood Products Advisory Committee the issue of genetic variation of HIV in the immunodominant region of the viral envelope. The gene sequence in this region is the basis for use of synthetic peptides as HIV antigenic substrates.[2] After presentations of worldwide data, the committee recommended that that FDA approve such products when they meet current standards, but that they continue to monitor virus variation.

Another current issue concerns the validation of tests using samples other than conventional serum or plasma. Testing of serum eluted from spots of capillary blood dried onto filter paper was developed to improve neonatal screening for HIV exposure—as a surveillance tool for monitoring the spread of the epidemic in childbearing women and as a way to identify HIV-exposed children. Three ELISA screening tests have been approved by the FDA for such a procedure. The possibility of testing for HIV antibodies present in samples of urine and oral fluid (gingival transudate) has been pursued. Despite the reduced levels of antibodies in these samples, detection methods have been developed that approach the clinical sensitivity of detection in blood. The advantages of noninvasive sample collection must be weighed carefully against the possibility of reduced test sensitivity for determining the public health implications of these test systems.

The accuracy of rapid tests that require subjective interpretation by the operator is closely linked with training. Poorly trained operators can easily misinterpret test results. The Agency has attempted to address this problem, in cooperation with manufacturers, through labeling to ensure adequacy of instructions for use and by asking manufacturers to provide training in the performance and interpretation of the test. Proficiency monitoring of test performance also is a responsibility of clinical laboratories under CLIA '88 implementing regulations.

The approved test for HIV antigen is labeled "for intended use in medical diagnosis (including prognosis) and for monitoring HIV expression in tissue cultures." The efficacy of screening blood donors for HIV antigen was investigated in a large multicenter trial conducted in 1989.[3] Based on the results and a discussion at a meeting of the FDA's Blood Products Advisory Committee on March 23, 1989, tests for HIV antigen are not recommended for blood donor screening owing to the lack of a demonstrated public health benefit. This position may have to be reexamined if antigen tests with greater sensitivity are developed. Research tests for HIV antigen that utilize procedures to dissociate immune complexes have increased the sensitivity for detection of HIV antigen in the presence of antibodies. In theory, these tests may offer an advantage for monitoring antiviral therapy and as diagnostics in the perinatal setting, but they have not been shown to improve routine screening.

Regulatory Concerns Regarding Supplemental Tests

Since 1985 it has been a policy of the U.S. Public Health Service to recommend that confirmatory testing be performed prior to notification of test subjects concerning their screening test results for HIV. This strategy permits confident identification of true-positive test results and minimizes the impact of reporting false-positive results, which occur inevitably with large-scale screening programs. By far the most widely used procedure for this purpose is the Western blot assay, which is based on electrophoretically separating proteins of the whole virus. Immunofluorescence tests based on HIV-infected and uninfected control cells are less widely used, probably because of the special requirements for equipment and training. The FDA has termed these "additional, more specific tests" to stress their positive predictive value while recognizing the possibility that they, like screening tests, are not definitive when negative. Licensed Western blot test kits for HIV-1 antibodies have been available since 1987, and a licensed HIV-1 immunofluorescence test kit became available in 1992. Validation of these tests has emphasized the need for sensitivity comparable to the approved screening tests and for accuracy of a positive test interpretation. For Western blots, reproducibility of the banding pattern also has been examined closely. These tests are subject to lot release control using the same FDA panel that is applied to screening tests.

The greatest concern regarding Western blot tests has been the high prevalence of nonspecific banding patterns, resulting in indeterminate test results. Unfortunately, this phenomenon is intrinsic to the technology due to a variety of causes that include antibodies in the patient sample that bind to nonviral proteins in the viral antigen preparation and antibodies of unknown specificity that cross-react with viral proteins. Attempts to optimize test performance based on interpretive criteria have necessitated a choice between maximizing sensitivity or specificity. The criteria recommended for the first FDA-approved test were conservative in that they required a major band from each gene group of HIV structural proteins: *gag* (p24), *pol* (p31), and *env* (either gp41 or gp120/160). These criteria maximized test specificity but at the expense of sensitivity in some clinical settings, including early seroconversion and acquired immunodeficiency syndrome (AIDS). Currently used criteria are less stringent and have resulted in tests that are more sensitive but have a higher false-positive rate. The FDA has accepted product amendments for revised Western blot interpretive criteria that are consistent with the prevailing scientific view and the recommendations of the Public Health Service. The agency has attempted to reduce the occurrence of falsely reactive bands by close attention to quality control issues in manufacturing and through lot release criteria. There are future prospects for technology improvements that may further reduce the existing problem of nonspecific bands.

Food and Drug Administration approval of an HIV-1 immunofluorescence test has provided clinical laboratories with an alternative to the Western blot test. Care must be taken to ensure the proficiency of the operator, who must be highly skilled in recognizing both the pattern and the intensity of fluorescence to achieve a correct test interpretation. To address this need, the manufacturer provides materials and educational training for operator qualification and proficiency monitoring as part of the approved product.

The FDA has become increasingly concerned about the commercialization of unlicensed supplemental test kits. The Agency did not take enforcement action against the use of unlicensed tests for a brief period after 1985 owing to the lack of availability of licensed tests and the pressing public health need for a means to validate HIV screening test results. As licensed supplemental tests now have become widely available, the Agency intends to initiate enforcement actions against unlicensed products. The Agency also intends to initiate enforcement action to effect compliance in such areas as the use of alternative samples (e.g., blood spot eluates, oral fluid, and urine) in conjunction with screening and supplemental tests.

Sample Collection Systems for HIV-Related Testing

Many sample collection systems are regulated as class I medical devices, which do not require premarket approval. However, sample collection systems intended for use in HIV-related testing have been regarded as class III medical devices that require premarket approval. Basically, these systems fall into two categories: products intended only for professional use in medical care settings and products intended for lay use. FDA concerns common to both types of product include (1) the adequacy of the instructions for sample collection, storage, and shipping; (2) validation of the adequacy of the sample and its compatability with a licensed screening and supplemental test; and (3) performance of the test by a properly certified clinical laboratory, including proficiency monitoring. For tests intended only for professional use, FDA reviews whether labeling and marketing limit distribution to a health care setting and if a mechanism for reporting results through a health care provider is present.

The FDA reviews collection systems intended for lay use for validation of control of potential biohazards in the home use setting, comprehensibility of the instructions by lay persons, a means of assessing the adequacy of sample collection, and adequacy of the pretest and posttest education and counseling. Additionally, the FDA seeks evidence that sample collection and testing outside a health care setting is in the interest of the public's health. This question has generated much controversy each time it has been brought before an FDA Advisory Committee for public discussion. The FDA's current position is that demonstration of safety for lay use should include a determination that public health benefits outweigh any risks.

FDA Perspective on Emerging Technologies

Despite the excellent performance characteristics of existing tests for HIV, medical and scientific needs continue to accelerate the pace of new technology development in HIV-related diagnostics, especially to improve early detection and to provide results with higher specificity. These needs include improvement in perinatal diagnosis, more sensitive blood donor screening tests, screening and supplemental tests for HIV-2, more cost-efficient test systems, and tests for lay use. Novel tests for HIV disease staging, prognosis, and therapy monitoring may also emerge. For example, research tests for the level of viral expression, the degree of immune compromise, and resistance to antiviral drugs have been developed. Some of these tests may be based on analytes other than HIV components or anti-viral antibodies.

Tests based on amplification of viral gene segments using the polymerase chain reaction or other methods have already been introduced in some clinical laboratories and are under development as commercial products. In clinical studies, gene-based tests have shown promise for development in a variety of uses such as diagnosis in high risk settings (perinatal exposure, acute viral syndrome of HIV) and potentially as supplemental tests following routine screening for antibodies. The FDA's concerns when reviewing applications for these tests focus on consistency in manufacturing and on the manufacturer's validation of well defined clinical claims. The potential use of these tests as the primary test for mass screening presently is limited by practical constraints (e.g., low through-put and the labor-intensive nature of the test) and concerns over the potential for false-positive results due to sample or laboratory contamination. In addition, there is a well recognized need for reagent standards and quality assurance systems to maintain the proficiency of laboratories performing testing by these methods.

Advances in supplemental testing may take a variety of forms. Scientists at the Centers for Disease Control and Prevention (CDC) have developed a modified Western blot procedure that may improve the consistency of that assay.[4] In addition, studies have been done to develop the potential for accurate Western blot testing of blood spot eluates. Other developments may include supplemental test systems based on the use of synthetic peptide antigens and new immunofluorescence assays. FDA evaluates such developments in the context of requests for investigational exemption and new product applications or amendments (supplements); and it works cooperatively with manufacturers, especially in areas that address public health needs.

Considerations of economy and user convenience have led to the development of combined tests for HIV-1 and HIV-2. This trend, which is expected to continue, is facilitated by the ease with which neoantigens can be developed for addition to the existing antigenic substrate. For instance, a manufacturer of an HIV-1 test may seek to develop a combination test by simple addition of an HIV-2 peptide or recombinant protein. In these cases, FDA routinely seeks validation that the modified kit has not lost sensitivity for HIV-1 and that it is as sensitive for HIV-2 as a test based on whole viral proteins of HIV-2. Reduced specificity has been a problem for the development of combination tests.

Conclusion

Regulation of HIV-related diagnostic tests by the FDA is intended to review consistency of manufacturing, validation of claims for intended use, and adequacy of labeling. Through the approval process, surveillance, and enforcement mechanisms, the FDA establishes and maintains manufacturing and product performance standards. This activity complements efforts of other government agencies to ensure proficiency standards in clinical laboratories. Particular discretion is exercised over approval of tests labeled for blood donor screening due to the direct effect of the accuracy of testing on the safety of blood products, which are themselves regulated as biologic products. Manufacturers of HIV-related diagnostic test kits must be aware of their obligations to comply with FDA regulations. Similarly, users of these kits should be aware that unapproved tests may violate the law. The process of FDA review and approval of new test technologies operates best with the cooperation of industry and the full participation of the medical community in developing and maintaining appropriate product standards.

References

1. MacKenzie WR, Davis JP, Peterson DE, et al. Multiple false-positive serologic tests for HIV, HTLV-I and hepatitis C following influenza vaccination, 1991. JAMA 1992;268:1015–1017
2. Wang JJ, Stell S, Wisniewolski R, Wang CY. Detection of antibodies to human T-lymphotropic virus type III by using a synthetic peptide of 21 amino acid residues corresponding to a highly antigenic segment of gp41 envelope protein. Proc Natl Acad Sci USA 1986;83:6159–6163
3. Alter H, Epstein JS, Swenson S, et al. Determination of the prevalence of HIV-1 p24 antigen in U.S. blood donors and an assessment of the potential efficacy of this market for blood screening. N Engl J Med 1990;323:1312–1317
4. Parekh BS, Pau CP, Granade TC, et al. Oligomeric nature of transmembrane glycoproteins of HIV-2: procedures for their efficient dissociation and preparation of western blots for diagnosis. AIDS 1991;5:1009–1013

5
Detection of HIV Infection Using Serologic Techniques

J. RICHARD GEORGE and GERALD SCHOCHETMAN

Human immunodeficiency virus (HIV) infection, regardless of clinical stage, produces many biologic indicators of virus infection, replication, or both (Fig. 5.1). Such indicators include viremia, antibodies against viral proteins, circulating viral proteins, and nonspecific markers of infection such as neopterin, β_2-microglobulins, and changes in the absolute number of and the ratio of CD4 and CD8 cells. Most of the markers can be detected and in many cases semiquantified by serologic tests (Table 5.1). These tests have been used to protect the blood supply, diagnose infections, monitor the progression of disease, monitor the efficacy of drug therapy, and diagnose infections in infants born to HIV-infected mothers. The availability of highly specific, inexpensive tests for HIV antibody, such as the enzyme immunoassay (EIA), have permitted public health agencies to conduct-large scale seroprevalence surveys to define the epidemic.

Of the tests listed in Table 5.1, EIAs for HIV antibody are most often used to diagnose HIV infections. To date, the U.S. Food and Drug Administration (FDA) has licensed several EIAs for both diagnostic and screening purposes. In the United States most licensed EIA tests detect antibodies only to HIV-1. However, in April 1990 the FDA licensed an EIA test for the detection of HIV-2 antibodies in plasma or serum. In September 1991 the FDA licensed the Genetic Systems Corporation combination HIV-1/HIV-2 EIA[1], and in February 1992 it licensed the HIVAB HIV-1/HIV-2 (rDNA) EIA (Abbott Laboratories, Abbott Park, IL). These combined tests permit simultaneous testing for HIV-1 and HIV-2 in a single test. Licensure of the combination assays was followed by a recommendation by the FDA that all blood banks screen donated

---

Use of trade names is for identification only and does not constitute endorsement by the Public Health Service or the U.S. Department of Health and Human Services.

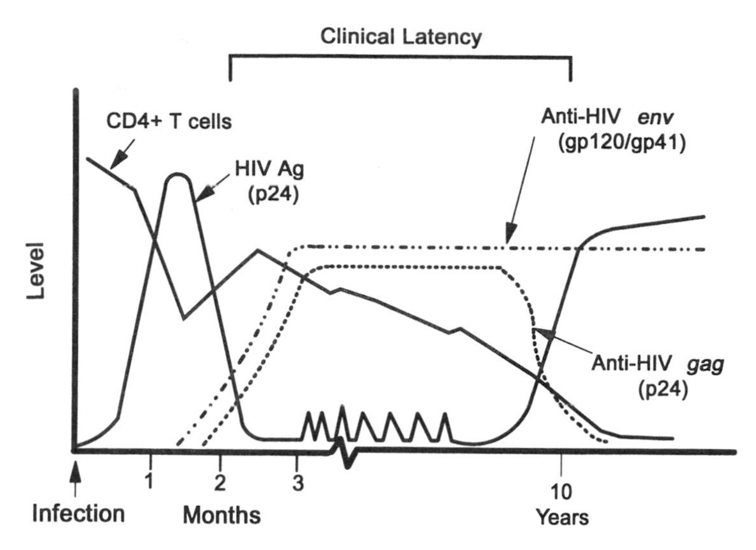

FIGURE 5.1. Natural history of HIV-1 infections.

blood for both viruses. Testing of donated whole blood, blood components, and source plasma for HIV-2 antibodies began June 1, 1992.

Outside the United States most countries routinely test for both HIV-1 and HIV-2 antibodies. The World Health Organization (WHO) recommended that only combination assays be used. As the combined assays

TABLE 5.1. Tests for the detection of HIV infections.

Antibody tests
Enzyme immunoassays (EIAs)
Whole virus lysate antigens
Recombinant protein antigens
Chemically synthesized antigens
Other types of screening assay
Particle agglutination assays
Hemagglutination assays
Latex agglutination assays
Solid-phase immunoassays
Supplemental tests
Immunoblotting (Western blot)
Immunofluorescence assay (IFA)
Radioimmunoprecipitation assay (RIPA)
Virus neutralization
Tests for virus or viral antigens
Enzyme immunoassays (antigen capture)
Radioimmunoassay
Antigen capture
Competitive inhibition
Virus culture

TABLE 5.2. African countries with a high prevalence of HIV-2 infection.

West African nations
Benin
Burkina Faso
Cape Verde[a]
Cote d'Ivoire[a]
Gambia[a]
Ghana
Guinea
Guinea-Bissau[a]
Liberia
Mali[a]
Mauritania[a]
Niger
Nigeria[a]
Sao Tome
Senegal
Sierra Leone[a]
Togo
Other African countries
Angola[a]
Mozambigue[a]

[a] Prevalence reported to exceed 1% in the general population.

have become more readily available, many private and public health laboratories in the United States have adopted them for routine testing. However, because of the low prevalence of HIV-2 in the United States, the U.S. Public Health Service still does not recommend routine testing for HIV-2 in settings other than blood banks unless demographic or behavioral information suggests that HIV-2 infection might be present. Those at risk for HIV-2 infection include individuals from a country in which HIV-2 is endemic or who are sexual partners of such persons. As of July 1992, HIV-2 was endemic in parts of West Africa (Table 5.2), and an increasing prevalence of HIV-2 had been reported in Angola, France, Mozambique, and Portugal. Additionally, HIV-2 testing should be conducted where there is clinical evidence or suspicion of HIV disease in the absence of a positive test for antibodies to HIV-1.

Use of combined assays for HIV-1 and HIV-2 has resulted in major modifications of the testing algorithm originally recommended by the Centers for Disease Control and Prevention (CDC)[2–5] for use with HIV-1-specific assays. It is likely that as the prevalence of HIV-2 infections becomes more common in the United States and other parts of the world monospecific HIV-1 assays will be discontinued. The development and acceptance of combined supplemental assays[6] will further change testing algorithms and simplify testing.

Screening Tests for HIV Antibody Detection

Enzyme Immunoassay

In the United States and worldwide, EIAs for HIV antibody employ a variety of antigens representing various structural proteins of HIV-1 and HIV-2. These antigens include whole virus lysate, recombinant proteins (rDNA), and chemically synthesized peptides. Some assays employ a combination of antigen types to achieve the desired sensitivity. These antigens are used in a variety of assay formats, including indirect assays, competitive assays, antigen sandwiches, and antibody capture assays. All of these formats and antigens have been used to produce tests with excellent sensitivity and specificity.

The indirect EIA (Fig. 5.2) is the most common assay configuration for antibody detection. The whole virus lysate, recombinant protein, or synthetic peptide is bound to a solid support, and the patient's serum or plasma is incubated with the fixed antigen. HIV-specific antibody bound to the HIV antigen is then detected by first adding anti-human immunoglobulin (Ig) conjugated to an enzyme, usually alkaline phosphatase or horseradish peroxidase. The second step of the detection reaction consists in reacting the bound enzyme with an appropriate substrate, producing a color change in the substrate that can be quantified by spectroscopy. The intensity of the color is proportional to the amount of bound antibody.

Competitive assays (Fig. 5.3) use the same types of antigen as the indirect methods. These antigens are also bound to a solid support, usually a plastic microplate well. Patient's serum or plasma is added to the well simultaneously with a standard enzyme-labeled HIV antibody. For competitive assays to be accurate and precise, the same amount of

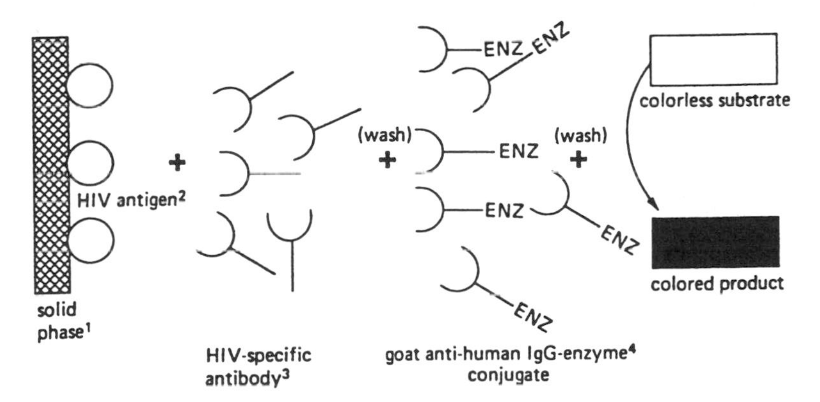

FIGURE 5.2. Configuration of enzyme-linked immunosorbent assay (ELISA) or indirect enzyme immunoassay (EIA). [1] Microplate well or other solid phase surface; [2] whole virus lysate, recombinant, or synthetic peptide HIV antigen; [3] patient's serum or plasma; [4] horseradish peroxidase or alkaline phosphatase enzyme.

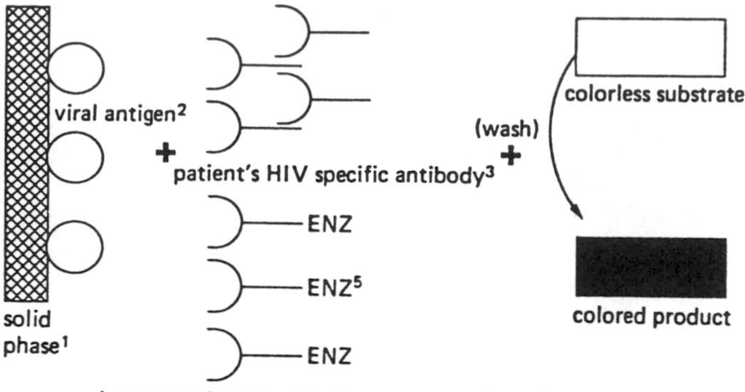

FIGURE 5.3. Configuration of the competitive enzyme immunoassay for antibodies against HIV. [1]Microplate well or other solid phase surface; [2]whole virus lysate, recombinant, or synthetic peptide HIV antigen; [3]patient's serum or plasma; [4]human monoclonal or polyclonal antibody; [5]horseradish peroxidase or alkaline phosphatase enzyme.

antigen and conjugated antibody must be present in each well. Usually the amount of antibody–conjugate added is sufficient to occupy approximately 50% of the available antigen-binding sites. During incubation, the patient's antibodies compete with the enzyme-labeled antibodies for the limited number of antigen-binding sites on the wells of the microplate. If there are no HIV antibodies present in the patient's sample, all antigen-binding sites are occupied by the enzyme-labeled antibody. After incubation with an appropriate substrate, samples negative for HIV antibody give the maximum possible absorbance reading. As the concentration of antibody present in the patient's serum increases, the enzyme-labeled antibody is displaced from the antigen-binding sites, and color development is less than that observed with antibody-negative specimens. In contrast to the indirect assay, the intensity of the color reaction observed in the competitive assay is inversely related to the amount of HIV antibody present in the test sample.

Many of the new combination HIV-1/HIV-2 assays that have been introduced are being referred to by their manufacturers as "third generation assays," a designation inferring that these assays represent different generations of increasing sensitivity. One such assay, HIVAB HIV-1/HIV-2 (rDNA) EIA (Abbott Laboratories), an antigen sandwich assay (Fig. 5.4), has been found to be more sensitive, though less specific, than the previous assays from that manufacturer. The HIVAB uses recombinant HIV-1 and HIV-2 *env* antigens and *gag* from HIV-1 bound to the solid phase. HIV-specific antibody of all classes can bind to the antigen. Bound HIV antibody is detected by first adding the same HIV antigens as those used to capture the HIV antibodies conjugated to the enzyme

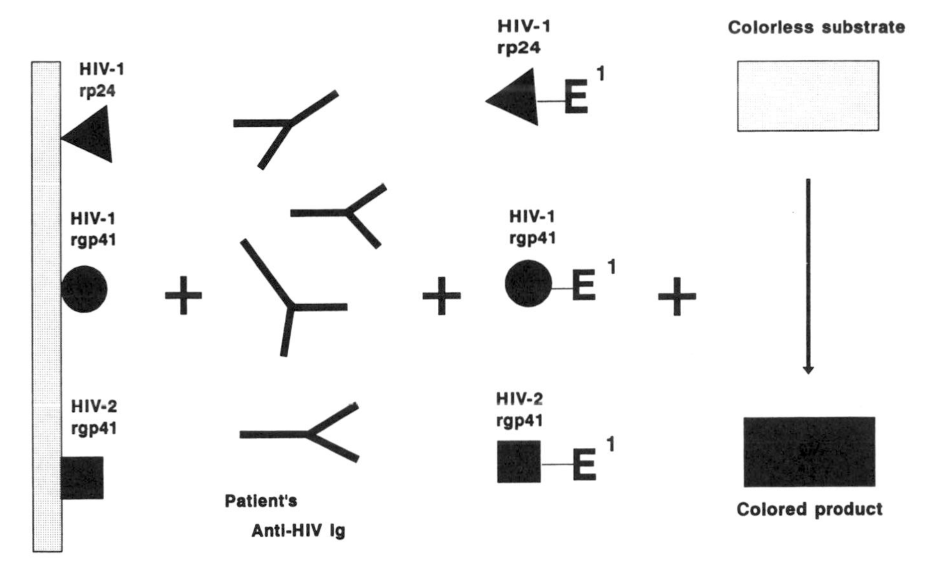

FIGURE 5.4. Abbott HIVAB HIV-1/HIV-2 (rDNA) enzyme immunoassay is an example of an antigen sandwich assay with "third generation" sensitivity. [1] Horseradish peroxidase.

horseradish peroxidase. Bound conjugate is reacted with an appropriate substrate (OPD–hydrogen peroxide), and a color reaction is produced and measured spectrophotometrically. As with the indirect method, absorbance is directly related to the concentration of antibody in the patient's serum. This assay has demonstrated improved sensitivity for the detection of early infection due to its ability to detect IgM and, to a lesser extent, IgA with greater sensitivity than either indirect or competitive EIAs.

An antibody-capture assay, HIV 1+2 GACELISA (Murex Corporation, Dartford, UK) is a versatile test that has been shown to be highly sensitive. Because of its excellent sensitivity for detecting low concentrations of HIV antibody, it has become the preferred test for testing such specimens as urine or oral fluid. The improved sensitivity of GACELISA is accomplished by using a high affinity anti-human IgG serum bound to the microplate well to capture a representative sample of the IgG of all specificities present in the patient's serum (Fig. 5.5). The antibody specific for HIV is then detected by adding recombinant HIV antigens (core + *env*) conjugated to alkaline phosphatase. After removing unbound materials by washing, bound conjugate is reacted with an amplifying substrate (NADP), which is dephosphorylated to NAD. The assay can detect HIV antibody contained in samples where the total IgG is less than 0.1 μg/ml.[7]

The sensitivity and specificity of commercial EIA tests are between 98.1% and 100.0%[8] when compared to the Western blot assay. It has

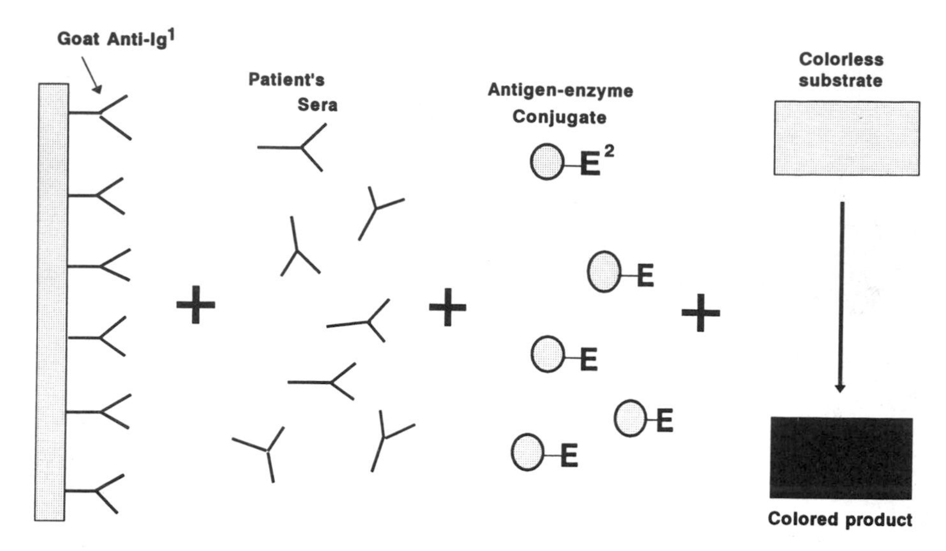

FIGURE 5.5. Configuration of the antibody-capture enzyme immunoassay. This assay has been shown to be useful for testing samples such as oral fluid and urine that have low concentrations of antibody. [1]Can be anti-heavy chain for a particular immunoglobulin class; [2]alkaline phosphatase or horseradish peroxidase.

been estimated that when properly performed, the average sensitivity and specificity of EIA tests is 99.8%.[4] In routine blood bank practice, 0.21% to 0.66% of samples are initially reactive and 0.03% to 0.33% are repeatedly reactive.[9] The percentage of repeatedly reactive specimens found to be positive by supplemental tests varies according to the true antibody prevalence of the tested population. As a U.S. national average, the prevalence of Western blot-positive samples among blood donors has been estimated to be 0.017%.[8]

Although commercial EIA tests are sensitive, false-negative results may occur. Seroconversion may be seen as early as 2 weeks after infection and almost always by 2 to 3 months after exposure to the virus.[9–14] In some unusual cases, seroconversion does not occur for 5 months [15] or longer if the individual is immunocompromised.[16] Most EIAs currently licensed, with the notable exception of the HIVAB HIV-1/HIV-2, do not detect other classes of immunoglobulin with the same sensitivity as IgG and may be negative in early infections before the switch from IgM to IgG antibody production.[17,18] Class-specific assays are discussed later in this chapter as they relate to the diagnosis of infections in neonates.

Recent publications[18a,18b] have suggested another possible cause for false-negative screening test results. These reports have identified persons from Cameroon infected with a highly divergent strain of HIV-1, tentatively classified as subtype O (see Chapter 2 for additional information on global variants of HIV). Persons infected with the subtype O strain of HIV-1 have been identified in West Africans residing in France. To date, similar infections have not yet been reported in the United States. This viral strain is the most divergent of the HIV-1s that has yet been identified with overall genetic distances of 65% to HIV-1 and 56% to HIV-2. The *env* gene of this isolate exhibited genetic distances to HIV-1 and HIV-2 of 53% and 49%, respectively. It will be important to evaluate various HIV immunoassays for their sensitivity to antibodies produced in response to infections with this virus. Current information suggests that antibody assays which utilize whole virus lysate antigens will be sensitive for subtype O infections. However, it is possible that some assays based on synthetic peptides and recombinant antigens will miss some subtype O infections. When sera from subtype O HIV-1 infected persons become more readily available, HIV antibody tests licensed by the FDA will need to be evaluated and modified where necessary to ensure that they can readily detect persons with subtype O antibodies. Meanwhile, it is unlikely that infected persons from regions of Africa where this virus is present will represent a serious public health problem for the United States. Few persons infected with this virus are likely to volunteer to donate blood and those who do would be deferred since they would have recently come from a malaria endemic area. However, we must remain vigilant for such highly divergent viruses.

False-positive EIA reactions also occur and can cause considerable anxiety if these results are reported to the patient independently of a supplemental test. In general, the probability of an EIA positive result being falsely positive is inversely related to the intensity of the reaction. In a low prevalence population (blood donors), 86.7% of the strongly positive EIA results were also Western blot- or culture-positive. On the other hand, sera that were weakly or moderately positive by EIA were positive by Western blot or culture only 1.9% of the time.[19] False-positive EIA results have been attributed to many conditions and practices. Most false-positive results arise from technical errors associated with the performance of the test: Samples may be placed in the wrong wells; wells containing negative specimens may be contaminated by positive samples from adjacent wells; plate washers may malfunction; or other similar errors may ocur. In addition, heat-treated, lipemic, and hemolyzed sera have been implicated in false positivity. HLA antibody for HLA DQ3 and HLA DR4 antigens can give false-positive results in EIA tests that use viral antigens produced in H-9 cell lines.[20,21] False-positive results have been reported to occur in 19% of hemophilia patients,[21] 13% of alcoholic patients with hepatitis,[22] 4% of hemodialysis patients,[23] and 24% of patients with a positive reagin test.[24] For these reasons and because of the considerable psychological and social implications of being HIV antibody-positive, it is inappropriate to counsel people or to make medical decisions about them based on EIA testing alone. Testing by the sensitive EIA is done to identify those persons who need additional, more specific supplemental testing. Counseling and medical decisions are made based on the results of the supplemental assay, not on those of the screening test alone.

Agglutination Tests

Other than EIA, the particle agglutination test SERODIA-HIV (Fujirebio, Japan) is probably the most widely used screening test for HIV antibody. The test is available as either a combined HIV-1/HIV-2 assay or as a monospecific HIV-1 test. SERODIA-HIV is not licensed by the FDA but is widely available in most other countries, particularly those in Asia, Australia, Latin America, and Africa. The complexity of this assay is comparable to that of the EIA except that it does not require any instrumentation to set up or read, although automated equipment is available for those laboratories that desire it. The assay, illustrated in Figure 5.6, is well suited for large-scale seroprevalence surveys or for low-volume remote laboratories. The cost per assay is comparable to that of EIA.

Diluted serum is mixed in the well of a microplate with a suspension of gelatin particles coated with purified HIV antigen. During the 2-hour room temperature incubation, the particles settle into distinctive patterns determined by the presence or absence of agglutination of the particles by HIV-specific antibodies in the patient's sera. In the absence of anti-HIV antibodies, the particles settle rapidly to the bottom of the wells and form a sharp button with well defined outer boundaries. In the presence

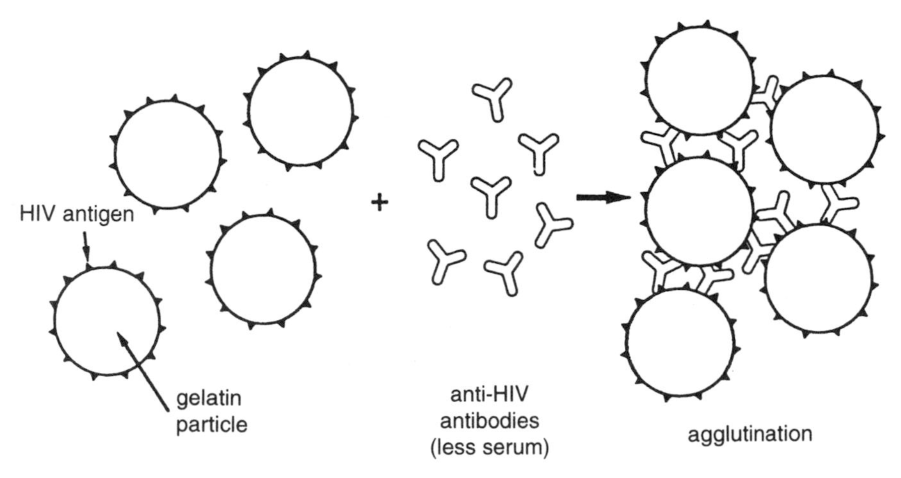

FIGURE 5.6. All agglutination tests have essentially the same mechanism. Particles such as gelatin (shown) or latex are coated with viral antigen. In the presence of HIV antibody, particles are cross-linked, and agglutination occurs.

of antibody, the particles agglutinate and settle more slowly, smoothly covering the entire bottom of the well. An agglutination control is run for each serum in a separate well containing serum and gelatin particles that do not contain HIV antigen. Such controls should always give a negative pattern. If both the coated and uncoated particles agglutinate, the serum should be absorbed with the uncoated particles and the test repeated. In the combined assays, separate HIV-1 and HIV-2 particles are provided and separate tests performed.

SERODIA-HIV has been found to be slightly less sensitive[25,26] and less specific than EIA, but the differences are not large; and with some specimens (seroconverters) the assay may be more sensitive owing to its ability to detect IgM antibodies prior to the production of detectable levels of IgG. False-positive reactions by SERODIA-HIV seem to be due to conditions different from those that cause false-positive EIA tests. Samples that were falsely positive by EIA—including heat-treated specimens, specimens containing HLA antibodies, specimens containing autoantibodies, and specimens from persons with a history of multiple blood transfusions—were found to be negative by SERODIA-HIV.[27]

A second agglutination test, the Recombigen HIV Latex Agglutination (LA) Test, has been licensed by FDA for use as a diagnostic test. This test is monospecific for HIV-1. Recombigen LA has not gained wide acceptance because of its cost, difficulty in reading, and reported lack of sensitivity (95.2%) and specificity (96.1%) compared to the EIA.[25] The assay itself is technically simple and rapid (<10 minutes), but inexperienced personnel have found it to be difficult to read even when using the high-intensity light and magnification recommended by the manufacturer. When performed by well trained, experience technologists, however, the results have been found to be reliable.[28]

Simpli-RED HIV-1 AB is an agglutination assay with an unusual design that is used to test for HIV-1 antibody in whole blood. It uses a chemical conjugate of a monoclonal antibody that binds to the surface of the human red blood cell (but itself does not cause agglutination) and a synthetic gp41 peptide. The antibody–peptide conjugate coats the red blood cells but does not cause agglutination in antibody negative samples. HIV-1 antibody present in a blood sample then causes cross-linking between cells that results in visible agglutination. This assay can also be used to test serum and plasma, but red blood cells must be added to the reaction. This assay has found some acceptance in Asia and Australia and seems to have acceptable sensitivity and specificity.

Another agglutination test from Abbott Laboratories, RETROCELL HIV-1, is used in some Latin American countries. In fact, this test has been found to have sensitivity and specificity comparable to those of the EIA when used as a screening assay.[29] RETROCELL HIV-1 is similar to the SERODIA-HIV in that they both require no special equipment to perform. With the RETROCELL test, HIV-1 antigen is covalently bound to a glutaraldehyde-preserved human type O erythrocyte. Uncoated erythrocytes are provided as controls for nonspecific agglutination. Reading is the same as for SERODIA-HIV.

The Quick PHT-HIV agglutination test (Salck Industries, São Paulo, Brazil) is intended for the same uses as SERODIA-HIV and RETROCELL HIV-1. However, the Salck test has been found to be insensitive and less specific[30] than SERODIA-HIV and has not been used much outside Brazil.

Other Rapid Tests

Most rapid tests, other than agglutination tests, are either solid-phase enzyme immunoassays or they visualize HIV antibody bound to HIV antigen immobilized on a filtration device by staining with a colloidal metal (e.g., gold) conjugated to protein A. Such assays are rapid (10–15 minutes), are simple to perform, require no special equipment, and are ideal for remote laboratories where only a few samples are tested at one time. They also are usually expensive ($3–10), not suited for high volume situations, and require additional, more-specific testing of positive samples. Many of these assays have been found to be as sensitive and specific as the EIA.[31] Other versions of these assays have been developed and offered for sale at low prices in developing countries without being validated through clinical trials or reviewed by any recognized institution such as the FDA, the WHO, or the Pan American Health Organization (PAHO).

The solid-phase enzyme immunoassay is the format favored by most manufacturers. HIV-1 and HIV-2 antigens are bound to the membrane or deposited onto the membrane of the filtration device. When antibodies of HIV-1 or HIV-2 are present in the test serum, they react with the HIV proteins and also are bound to the membrane. Nonreactive antibodies are washed through the membrane with the wash solution. During the second reaction step, an anti-human IgG and IgM enzyme conjugate is added to the cartridge. If antibody to the HIV antigens was bound in the previous reaction, the conjugate binds to the antibody–antigen complex. Unbound conjugate is removed by the wash step. An appropriate substrate is added to the cartridge, and a color reaction develops on the membrane if antibody was present in the test sample. Most devices contain a reactive control spot that also develops color if the reagents have the appropriate potency and were added in the correct sequence during the test procedure. Some devices also contain nonspecific binding control spots to detect nonspecific protein–protein binding. This control should always be colorless. An evaluation of one such test, Genie HIV-1 and HIV-2, found it to be 99.3% sensitive and 99.5% specific.[32] One test of this type, SUDS (Murex Corporation, Norcross, GA), has been licensed as a diagnostic test by the FDA.

The second assay type is performed by immobilizing the HIV antigen on a membrane in the test cartridge. The patient's serum is filtered through the device, and if HIV antibody is present it binds to the antigen immobilized on the membrane. Nonreactive antibodies are filtered through with the wash buffer. Human immunoglobulins bound on the membrane are visualized by staining with a colloidal gold–protein A conjugate. Protein A specifically binds to IgG subclasses I, II, and IV. One assay of this type, HIVCHEK 1+2 (Ortho Diagnostics), has been evaluated and found to be 99.3% sensitive and 100% specific.[31] Tests using other metal colloids and binding systems are being developed and may provide inexpensive, sensitive assays for HIV antibody for use where the more technically difficult EIAs are not viable choices for either surveillance or diagnostic testing for HIV. Tests of this type are also more rugged than those using enzyme amplification systems and can be exposed to ambient temperatures for extended periods.

Supplemental Tests

Western Blot

The Western blot assay[33] is a relatively specific and informative supplemental test to be performed on specimens that have previously tested positive by a screening test. The Western blot test uses HIV proteins separated by electrophoresis and immobilized on nitrocellulose paper to determine the specificity of antibodies in the sample that were responsible for the reactivity observed in the screening test. Once the strips have been prepared, the assay procedure is similar to that previously described for the solid-phase enzyme immunoassay. An appropriate dilution of serum or plasma is incubated with the strip. Antibodies to HIV, if present, bind to the viral antigens bound to the surface of the nitrocellulose. After washing away unbound materials, the strips are incubated with antibody (usually prepared in goats) against human IgG that has been conjugated with an enzyme (horseradish peroxidase or alkaline phosphatase). If anti-HIV antibody was present in the patient's sera, the conjugate binds to the antibody–antigen complex. Unbound conjugate is removed by washing. An appropriate substrate is added, and the presence of antibody is visualized by conversion of the colorless substrate to a colored product that stains the strip at the positions where antibody was bound. The identities of these proteins are established by comparison to the positions of bands produced by a reference antiserum. The diagnostic Western blot bands for HIV-1 and HIV-2 are listed in Table 5.3. Production of Western blot strips is not a precise process, and differences in protein concentrations, identities, and positions are observed between manufacturers and even between lots from the same manufacturer.

Interpretation of Western blot tests rely on the structural proteins coded for by the *env*, *pol*, and *gag* genes (Table 5.3). Most reports indicate that antibodies to *gag* proteins appear first in the serum following seroconversion. Antibodies to *gag* proteins may persist, decrease, or become undetectable with the onset of symptoms.[10,34–38] Antibodies to the transmembrane glycoprotein can be detected in nearly all HIV-infected persons regardless of clinical stage.[37–40] Current Western blot interpretive criteria (Table 5.4) place heavy emphasis on the presence of *env* glycoproteins and the *gag* protein p24 to establish a positive pattern.

It is interesting to note that considerable confusion remains over the accurate identification of *env* glycoproteins present on commercial HIV-1 and HIV-2 Western blots. A number of commercial Western blots continue to refer to 120 and 160 kDa bands as representing the external envelope glycoprotein and the precursor of *env* glycoprotein, respectively. In fact, these bands usually represent trimeric and tetrameric forms of the transmembrane protein (TMP) gp41.[41,42] The external envelope glycoprotein and the precursor of *env* protein may in fact also be present at those positions but is usually not distinguishable from the more intense oligomeric forms of gp41.

Similar observations have been made for the *env* glycoproteins of HIV-2.[43] In fact, the identification of glycoproteins can be even more complicated for HIV-2 than for HIV-1 because multiple forms of the TMP are

TABLE 5.3. Identification of major structural gene products of HIV-1 and HIV-2 by molecular weight in Western blot; recommended nomenclature.

Gene	HIV-1		HIV-2	
	Gene product	Nomenclature	Gene product	Nomenclature
env	gp160	Precursor[a]	gp125[b]	External envelope
	gp120	External envelope	gp36/41[c]	Transmembrane
	gp41	Transmembrane	gp70[d]	Dimeric form of transmembrane
pol	p66	Reverse transcriptase	p68	Reverse transcriptase
	p51	Reverse transcriptase	p56	Reverse transcriptase
	p32	Endonuclease	p55	Reverse transcriptase
			p34	Endonuclease
gag	p55	Precursor		
	p40	Precursor		
	p24	Core	p26	Core
	p17	Matrix	p16	Matrix

[a] Usually not present. Often confused with oligomeric form of gp41.
[b] Identified as gp105 by some manufacturers. Usually present but often obscured by oligomeric forms of transmembrane glycoprotein.
[c] Different virus strains may produce one or both forms of the transmembrane glycoprotein.
[d] Sometimes seen on some commercial reagents.

TABLE 5.4. Criteria for positive interpretation of Western blot tests.

Organization	Criteria for HIV-1[a]	Criteria for HIV-2[a]
CDC/ASTPHLD	Any two of p24 gp41 gp120/160	None
FDA-licensed Biotech/ DuPont Test	p24, p31, and gp41; or gp120/160	None
American Red Cross	One band from each gene product group (total of 3): *gag* and *pol* and *env*	None
Consortium for Retrovirus Serology Standardization	Two or more bands: p24 or p31 plus gp41 or gp120/160	None
WHO	Two *env* bands with or without *gag* or *pol*	Two *env* bands with or without *gag* or *pol*

[a] All of the criteria, except those established by WHO, score blots with *no* bands as negative. WHO scores blots as negative if bands are present that do not correspond to the structural proteins listed in Table 5.2. All other patterns not scored as positive are indeterminate.

produced by some virus strains used to produce Western blot tests. For example, strain NIH_Z produces a full-length 41 kDa glycoprotein as well as a truncated 36 kDa form. Multiple forms of the monomers and oligomers of these proteins may be present on the Western blot strips, and the correct identification of individual bands becomes almost impossible. This problem can be made less confusing by controlling reduction of the viral protein by 3-mercaptoethanol and sodium dodecyl sulfate (SDS). Mild reduction leaves most TMP in the oligomeric form, whereas harsher reduction drives the TMP into the monomeric form. The HIV-2 Western blot test currently available in the United States from Cambridge Biotech Corporation is optimized to present most TMP in the trimeric (gp105) form.

Since the first recommendation[4] by the U.S. Public Health Service (PHS) that EIA-positive specimens should be further tested by Western blot, there has been extensive debate over the appropriate interpretive criteria for this assay. The selection of such criteria is complicated by the special needs of different testing situations and the perceived need to apply a single interpretive criterion in every case. When testing populations at risk for HIV infection [i.e., intravenous drug users (IDUs) or male homosexuals], it is important that the positive criterion be sensitive to early infection, where only a few bands may be present. If the prevalence of HIV infection is high, the positive predictive value of minimal banding patterns is also high. In a population of blood donors having a low prevalence, it may be desirable to use more stringent criteria. In this population, a Western blot pattern with only a few bands present has a much lower positive predictive value.

The selection of interpretive criteria is more complicated in populations where both HIV-1 and HIV-2 are present and the Western blot test is used to type the infecting virus. Interpretive criteria were sought that not only were sensitive and specific for infection but also were serotype-specific. Considerable cross-reactivity exists between HIV-1 and HIV-2 antibodies and all viral proteins found on the Western blot assays.[44] This point has been found to be especially true in West African countries centering around Cote d'Ivoire (Ivory Coast). Cross-reactivity was predicted between antibodies against HIV-1 and HIV-2 *gag* and *pol* proteins based on the sequence homology. It was, however, somewhat surprising that extensive cross-reactivity occurred between the *env* transmembrane glycoprotein. In a study of 1362 consecutive tuberculosis (TB) patients and 2127 consecutive blood donors in Abidjan, Cote d'Ivoire, it was found that 73% and 83%, respectively, of all EIA-positives specimens were dually reactive by Western blot.[44]

Various interpretive criteria for HIV-1 and HIV-2 Western blot tests are presented in Table 5.4. In the United States, the CDC/Association of State and Territorial Public Health Laboratory Directors (CDC/ASTPHLD) criteria[45] are the most frequently applied standards. Some examples of both common and unusual HIV-1 Western blot banding patterns are illustrated in Figure 5.7. A specimen is interpreted as positive

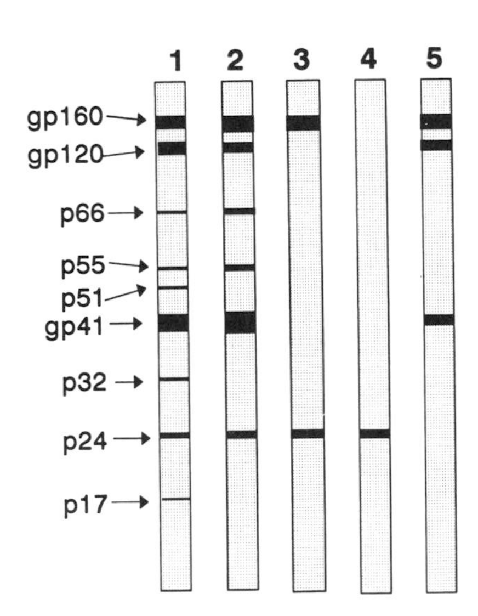

FIGURE 5.7. HIV-1 Western blot patterns. (1) Positive control. (2) Typical positive pattern. (3) Minimal positive pattern that occurs with early seroconversion. (4) Indeterminate pattern. (5) Rare *env* (only positive pattern that has been associated with a false-positive interpretation in a few blood donors).

by this standard when two of three bands (p24, gp41, gp120/160) are present. Bands gp120 and gp160 are considered as a single band because they sometimes run together when strongly reactive sera is being tested. Bands are identified by comparison to a positive control serum run on a strip from the same kit that is used for the unknown specimen. A negative interpretation requires the absence of all bands. Any other pattern not meeting the positive criteria is considered to be indeterminate. As presented in Table 5.4, several other investigative groups have presented interpretive criteria, some of which are no longer recommended. For example, the FDA/Biotech/Du Pont criteria were originally proposed in the product insert for use with that particular kit. In 1992 the recommended criteria in the kit brochure was changed to the CDC/ASTPHLD criteria. In most cases these criteria are more restrictive than the CDC/ASTPHLD criteria. In the case of the restrictive FDA/Ortho/Du Pont recommendation that required the presence of *gag*, *pol*, and *env* products to be present for a positive interpretation, patients with acquired immunodeficiency syndrome (AIDS) who had lost some of these bands would have been classified as indeterminate.[35,46]

The HIV-2 Western blot kits from different sources are different in terms of the number and position of *env* glycoprotein bands for the reasons previously discussed. Therefore a single interpretive criterion that can be applied to kits from all manufacturers seems impractical. The HIV-2 interpretive criterion recommended by the WHO[47] was based on data collected from assays performed using tests from Diagnostics Pasteur. The WHO criterion classifies a person as positive based on the presence of two glycoprotein bands (gp36, gp105, or gp140). Applying these criteria to other tests could result in misclassification of infected persons, as some tests, such as the Cambridge Biotech HIV-2 Western blot, contain only a single glycoprotein band (gp105, the trimeric form of gp36) in quantities sufficient to be used for diagnosis.

Indeterminate Western blot patterns occur frequently in persons who are not infected by HIV. Indeterminate patterns—*gag* bands p24 and p17 being most common—appear in 10% to 15% of persons found by Western blot not to be infected.[8] Fortunately, most of these patterns do not cause the EIA test to be positive and do not cause problems when the appropriate testing algorithm is used. However, in blood donors approximately 5% of EIA repeatedly positive samples are indeterminate by Western blot. People with these patterns are deferred from further blood donations and suffer considerable anxiety over the uncertainty of this test result. Cumulative experience from nearly 8 years of testing suggests that persons with indeterminate Western blot patterns that remain stable for months or years probably are not infected.[48–50] In addition, two other studies[51,52] have followed populations of blood donors with persistently indeterminate results (6 months) and found them to be free from HIV infection. It was shown that none of these donors had detectable HIV proviral DNA when analyzed by the polymerase chain reaction (PCR), nor were they found by virus culture to be infected. From these studies it was concluded that persons without risk factors for HIV infection but with persistent, stable indeterminate Western blot patterns should be monitored for at least 6 months. If no additional bands develop during that time, the patients should be advised that they are not infected by HIV. Individuals with HIV-1 indeterminate patterns who are at risk for HIV-2 should be tested for antibodies against that virus.

Immunofluorescence Assay

Many laboratories find the Western blot assay to be expensive and laborious, especially in high volume and high prevalence situations. The cost of the Western blot is particularly critical in developing countries, where funds for HIV testing are severely limited. The immunofluorescence assay (IFA) offers an alternative method that has been demonstrated to possess sensitivity and specificity equal to that of the Western blot.[53–57] The cost of this test is higly variable. The cost of the Fluorognost HIV-1 IFA (Waldeheim Pharmazeutika, Vienna, Austria), licensed by the FDA, is comparable to the cost of the Western blot assay. Laboratories capable of performing virus culture can produce IFA reagents at a relatively low cost, although use of "home-made," nonstandardized tests is not encouraged.

Typically, the IFA test for HIV antibodies is based on specific antibody binding to HIV antigens expressed on the surface of immortalized human T cells fixed to a glass slide. T cells not expressing HIV antigens serve as controls for each test. Slides are processed to inactivate any virus associated with infected cells, including complete drying, refrigerated acetone treatment to remove the lipid content of the virus and dehydration through evaporation of the acetone solvent. This inactivation process fixes both HIV-infected and uninfected cells to their respective cell wells on the glass IFA slide. Both HIV-1 and HIV-2 IFA tests can be prepared in this manner.

To perform the assay, a serum or plasma is diluted and placed in the infected and uninfected cell well of the IFA slide and incubated. If HIV antibodies are present in the sample, they bind to the HIV antigens present on the infected cell surface. Unbound materials are removed by vigorous washing. Antibody to human immunoglobulin conjugated to fluorescein isothiocyanate (FITC) is added to the infected and uninfected cell wells, and the solutions are again incubated. After a second vigorous wash, a glass coverslip is mounted to the IFA slide and the slide is viewed under a microscope with ultraviolet light. If antibodies are present in the serum or plasma specimen, a characteristic pattern of fluorescence becomes visible. The interpretation of this fluorescence is evaluated by comparing and differentiating the pattern and intensity of fluorescence in the uninfected and infected cell well for each unknown sample.

Like the Western blot test, all specimens cannot be classified as positive or negative using the IFA test. A specimen is interpreted as indeterminate when there is fluorescent staining in both the HIV-infected and the uninfected cell wells or when it is not possible to clearly differentiate the intensity of fluorescent staining and the pattern of fluorescence between HIV-infected and uninfected cells. Results are also considered indeterminate when duplicates are discordant. In the case of indeterminate samples, the test should be repeated on the original sample. If the indeterminate result persists, it may be necessary to obtain a fresh sample. In some cases the indeterminate IFA result persists and the sample must be tested by another supplemental test, such as the Western blot test. It is important to realize that the indeterminate IFA interpretation does not imply that HIV antibodies are, or are not, present in the test specimen. It simply means that the HIV antibody status of that particular serum or plasma cannot be resolved through use of IFA on that test run. In some cases the nonspecific staining can be prevented by absorbing the serum or plasma with uninfected cells of the same type as that used to produce the IFA slides.[57]

Radioimmunoprecipitin Assay

The radioimmunoprecipitin assay (RIPA) is another test that is sometimes performed as a supplemental test for HIV antibody. Actively growing cells infected with HIV-1 or HIV-2 are exposed to a growth medium containing a radioactive amino group that is incorporated into the viral proteins.[38,57–60] Cell lysates prepared from these cells are first cleared of proteins that bind non-virus-specific antibodies by absorption with an HIV-negative human control serum bound to protein A–sepharose beads. The lysates are then mixed with the patient's serum and incubated until equilibrium is reached. The immune complexes are absorbed onto protein A–sepharose beads. Radioactive antibody–antigen complexes are eluted, and proteins are separated based on their electrophoretic mobility on SDS-polyacrylamide gel electrophoresis (SDS-PAGE) to determine the specific viral patterns. The banding patterns are similar to those seen with the Western blot assay.

This technique is labor-intensive and expensive, and it is used primarily as a research tool. It may be useful for testing Western blot indeterminate samples because it is more sensitive than Western blot for antibodies against the *env* glycoproteins. The increase in sensitivity is due to higher concentrations of *env* proteins in cell lysate versus the lysate of pelleted virus used for Western blot antigen. In addition, cell lysates used for RIPA are richer in precursor glycoprotein (gp160) than are the lysates prepared from pelleted virus.

Alternative Samples for HIV Antibody Tests

Blood Dried on Filter Paper

For nearly 30 years, state laboratories in the United States and many foreign public health laboratories have screened newborns for congenital metabolic disorders by testing blood specimens dried on a special collection paper [Schleicher and Schuell (S&S) No. 903, Keene, NH]. In 1988 it was reported that these same samples could be used to test for HIV antibodies.[61] Because HIV IgG antibodies are passively transferred from the mother and reflect the mother's HIV status rather than that of the infant, these specimens can be used to measure the prevalence of HIV infections in women bearing liveborn infants.[62] More than 10 million infants (mothers) have been tested by this method in the CDC Survey of Child Bearing Women. A procedural manual has been published[63] describing the laboratory methods used in this study. The methods were validated by studies comparing results from matched serum and blood spots collected from the same person.[64] Subsequent studies by Genetic Systems Corporation (Seattle, WA) and Organon Teknika (Durham, NC) have further demonstrated that testing blood dried on S&S paper collected by fingerstick gives the same sensitivity and specificity as testing serum and plasma. Based on data from these studies, the FDA has licensed the EIAs from these two companies for use with dried blood spot samples. No supplemental test has been approved by the FDA for use with dried blood spot samples.

Testing a dried blood spot requires that the antibodies be eluted from the sample in the diluent supplied by the manufacturer of the EIA kit to be used for testing. Eluates have been proved to be stable at 4°C for at least 1 week. A 0.25 inch spot punched from the blood spot contains approximately 5.0 μl of serum. The eluate is usually more concentrated than that used in the EIA and requires further dilution before testing. Afterward, the appropriate dilution is made, and the eluates are tested using the same protocol as that used for serum and plasma.

As recommended for serum and plasma specimens, repeatedly reactive blood spot eluates should be confirmed by supplemental tests such as Western blot. In the CDC Survey of Childbearing Women, a miniaturized Western blot is used that permits testing of 70 μl of 1:100 dilution of eluate. If sufficient sample is available, four 0.25 inch punches containing 20 μl of serum can be added to 2 ml of diluent (1:100 dilution) and tested by the full-size Western blot strips. However, this modification is not approved by the FDA and should not be used for diagnostic purposes.

Collection of blood samples on absorbent paper has unique advantages for settings where the collection of blood and the subsequent shipping, processing, and storage of serum and plasma would be difficult or impossible. Blood spots can be prepared from heel, finger, or ear punctures, spotted onto the S&S 903 paper, dried, and stored. When protected from moisture, samples can be stored for at least 45 days at ambient temperature,[65] even in tropical areas, without loss of test sensitivity. For long-term storage, the samples should be placed in gas-impermeable bags with small packets of desiccant and stored at −20°C.

The disadvantages of collecting and testing dried blood spots for HIV antibody are few but are important to note. Samples must be collected in a precise way to ensure a uniform, reliable sample. Circles on the collection paper must be completely saturated so a punch taken anywhere within the circle contains the expected quantity of blood. Once collected, the blood must be allowed to dry completely before being placed in plastic bags for shipping or storage. Residual moisture permits bacterial and fungal growth, which can cause assay inaccuracy. Complete guidelines for collecting blood on special collection paper such as S&S 903 have been published by the U.S. National Committee for Clinical Laboratory Standards.[66] Testing of samples should include filter paper controls for each test run. Laboratories testing dried blood samples should participate in a proficiency testing program that distributes samples containing blood dried on the same collection material.

Oral Fluid

Oral fluid (saliva) has been proposed as a noninvasive alternative to blood for testing for antibodies to a variety of viral agents, including hepatitis A virus, hepatitis B virus, rubella virus,[67–69] and hepatitis C virus.[70] Furthermore, oral fluid has been used to detect the presence of a variety of substances including hormones, therapeutic drugs, cocaine, caffeine, and tumor markers. Currently there is considerable interest in the use of oral fluid as specimens for HIV antibody testing, and many organizations are seeking advice related to the accuracy of testing with these fluids.

The first step in this evaluation of testing oral fluid for HIV antibody is to understand the types and sources of immunoglobulins present in oral fluid. The term *saliva* has been used loosely to describe fluid taken from the oral cavity. In fact, saliva is a specific fluid taken from the submandibular, parotid, sublingual, and labial salivary glands. Pure saliva contains a small number of immune and epithelial cells, small amounts of immune globulin (primarily secretory IgA), along with a variety of digestive enzymes, including proteases. Proteases from the salivary glands, along with those from bacteria that are a part of the oral flora, may degrade salivary proteins including immunoglobulins. *Oral fluid* is a term proposed to describe the fluid that is usually collected for the purpose of antibody testing. Oral fluid contains salivary gland secretions, products of the oral mucosa, and gingival crevicular fluid (CF). Most of the immunoglobulins, other than secretory IgA, are introduced into the oral fluid as a component of the CF. Immunoglobulins contained in the CF are derived from plasma components that are passively transported from the capillary bed underneath the buccal and gingival mucosa to the oral cavity. IgG is the immunoglobulin class present at highest concentrations in oral fluid. However, the IgG concentration can vary from less than 0.1 mg/L to more than 50 mg/ml.[71] These figures represent an IgG concentration that can range from 10^2- to 10^5-fold less than that present in serum or plasma. IgG levels at the lower end of this range of concentrations may be well below the detection level of some assays optimized for use with serum and plasma. However, reports using a particular highly sensitive antibody-capture assay (GACELISA) suggest that it may not be a problem.[72–74]

Testing for HIV antibodies present in oral fluid began during the mid-1980s.[67,75–78] Since that time more than 15 reports have appeared in the literature describing the sensitivity and specificity of various EIAs, rapid assays, agglutination tests, and Western blot assays for the detection of antibodies in oral fluid. Oral fluids were collected using various collection devices, such as Orasure (Epitope, Beaverton, OR), Omni-Sal (Saliva Diagnostics Systems, Vancouver, WA), or Salivette (Sarstedt, Leicester, UK); in some cases the fluid was collected by simply dribbling into a collection cup. All methods seem to be equally satisfactory for producing samples for HIV antibody testing. However, the sensitivities reported for various test methods ranged from 32% to 100%. The 32% sensitivity[79] seems totally discordant with the findings of other investigators, which clustered between 88% and 100% sensitivity. Later studies using assays that have been designed for oral fluid testing[72–74] or that were optimized to detect the lower antibody concentrations in oral fluid have produced improved sensitivities.[75–77] Successful modifications of serum EIAs include increasing the sample volume, decreasing the test dilution, increasing incubation times, increasing conjugate concentration, or lowering the cutoff values. Specificity has been high: 98% to 100% in most of the studies.

The Western blot is the supplemental test most frequently used with oral fluid. In the early studies, antibody profiles were weaker than those seen when testing serum. Reactivities to *env* antigen were most frequently seen, whereas reactivity to *gag* was often absent in comparison to the corresponding blots performed on serum or plasma. Sensitivities for the Western blot test from these earlier studies ranged from 95% to 100%.[76,78,79] However, a more recent study using a Western blot specifically developed for oral fluid produced banding patterns identical to those present in the corresponding serum for the same patient.[71] Use of these blots thus permits use of the same interpretive criteria used for serum and plasma.

Studies evaluating HIV antibody testing using oral fluid have been encouraging, especially when assays were used that were developed for use with these samples. The use of oral fluid for HIV testing offers several potential advantages over the use of blood. Collection of oral fluid is safer, as the risk from needlesticks, disposal of blood, needles, and cuts from broken glass is eliminated. In addition, the reported load of infectious virus is low compared to that in blood.[76] Samples collected by the various collection devices (e.g., Orasure) are made safer by including detergents in the collection fluid that inactivates the virus by disruption within a few minutes after collection. Another reported advantage is that individuals are reported to be more accepting of HIV testing when oral fluid is collected than when a sample is obtained by venipuncture or fingerstick. Collecting oral fluid is simpler in children, individuals with collapsed veins, or those who are resistant to blood collection for cultural or religious reasons. Oral fluid can be held and transported at room temperature for 2 weeks without loss of activity when using the commercial collection devices (owing to the inclusion of microbial growth and protease inhibitors).

Testing of oral fluid also has several disadvantages, paramount of which is the lack of an FDA-licensed test for use with oral fluid. In addition, there is a concern that some oral fluid samples may not contain sufficient concentrations of IgG for sensitive testing. When testing oral fluid, quality control and proficiency control materials in sufficient volumes are more difficult to obtain than when using blood. Finally, the use of oral fluid may raise concerns about exposure to certain infectious agents such as *Mycobacterium tuberculosis*.

Urine

Like oral fluid, urine has been proposed as an noninvasively obtained alternative to blood for HIV antibody testing. The advantages noted for oral fluids apply also to urine—as do the disadvantages, with the exception of the problem of collecting sufficient volumes of control material.

There are fewer reports comparing testing of urine to that of serum and plasma for HIV antibody. Early studies using tests developed for use with serum and plasma reported sensitivities ranging from 75% for Western blot to 100% for Abbott EIA (03A11) and specificities ranging from 100% and 92%.[80] More recently, similar studies using assays such as GACELISA and Calypte (Calypte Biomedical, Berkeley, CA) have reported sensitivities equivalent to those found when testing serum or plasma: 98.8% and 99.6%, respectively.[81] In a separate report, the urine of some people tested positive for HIV antibodies when the serum was negative.[82] Several theories have been offered to explain this unexpected finding. One such theory is that some people produce a compartmentalized immune response confined to tissue that has been exposed to the virus, in this case the urinary tract. Such theories are highly speculative but have added to the interest in testing urine for HIV antibody.

Testing Algorithms for HIV Antibody Testing

Most laboratories in the United States still use monospecific HIV-1 assays because of the low prevalence of HIV-2. As of December 1993, only 49 cases[83] of HIV-2 had been confirmed in the United States, all of which had occurred in West Africans, sexual partners of West Africans, or those who had lived in West Africa.

Testing for HIV-1 has changed little since the original PHS recommendations published in 1985.[2–5] The most current recommendation[84] from the ASTPHLD is presented in Figure 5.8. Sera found to be repeatedly reactive by EIA or some other screening test is tested by an additional, more specific, FDA-licensed assay, such as Western blot assay or IFA. If a specimen is found to be "indeterminate" by IFA, it should be tested by Western blot. "Indeterminate' Western blot specimens are more complicated to resolve. Under the FDA license granted to the Florognost IFA test, Western blot "indeterminates" can be tested by IFA, and positive or negative results can be reported to the patients. An alternative protocol is to follow individuals with "indeterminate" Western blot results and no identifiable risk for at least 6 months. If the Western

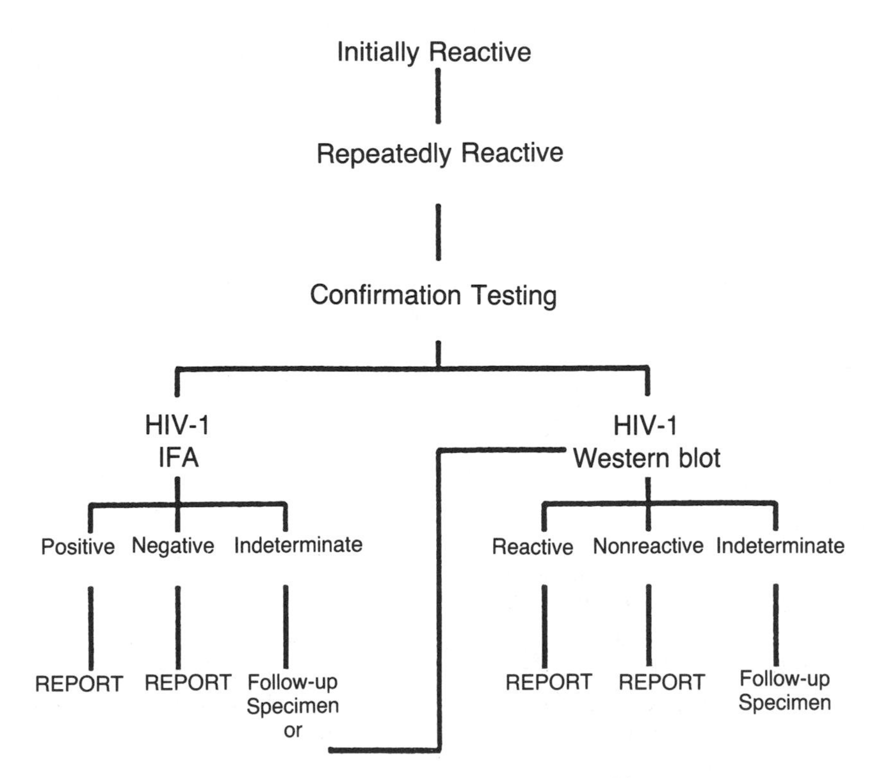

FIGURE 5.8. HIV-1 testing algorithm.

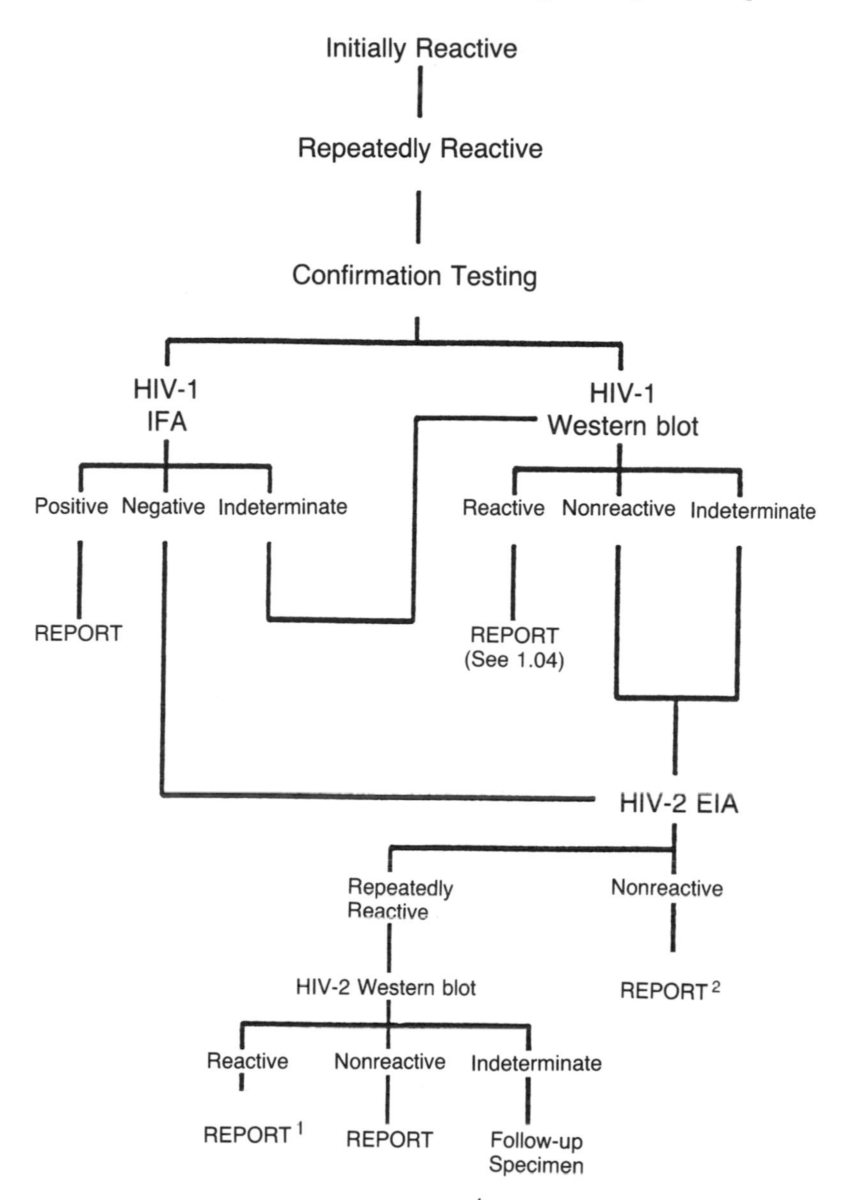

FIGURE 5.9. HIV-1/HIV-2 testing algorithm. [1] Each state must determine appropriate wording regarding the reporting of reactive HIV-2 Western blot results. [2] Report must consider the HIV-1 result; that is, if HIV-1 Western blot is indeterminate, follow-up is recommended.

blot patterns disappear or remain stable (no additional bands appear), the subject can be advised that he or she is not infected by HIV.

Testing for both HIV-1 and HIV-2 requires a more complicated algorithm[84] (Fig. 5.9). Screening can be done using either separate monospecific HIV-1 and HIV-2 assays or a combined HIV-1/HIV-2 assay. Obviously, the second choice is less expensive and more efficient. In the United States, where the prevalence of HIV-2 is low, repeatedly reactive combined EIAs are then tested using HIV-1 supplemental tests such as Western blot or IFA. The rules for using these assays are the same as described for the HIV-1 testing algorithm. If a specimen tests positive by one of the HIV-1 supplemental tests, in the absence of any risk for HIV-2 infection, testing stops and the individual is told that he or she is HIV-positive. Specimens that are repeatedly reactive by the HIV-1/HIV-2 combined assay that test negative or indeterminate by the HIV-1 supplemental tests, or any sera from a person at risk for HIV-2, are tested by HIV-2 EIA. Specimens that test positive by HIV-2 EIA are tested by an HIV-2 supplemental test. At the present time, there are no FDA-licensed HIV-2 supplemental tests. Both HIV-2 Western blot and HIV-2 IFA tests have been submitted for licensure and are currently available for research use only.

The algorithms in Figs 5.7 and 5.8 were developed for use in the United States and other developed countries where the prevalences of HIV-1 and HIV-2 are low. The strategy of both these algorithms is to identify HIV infections sensitively and specifically in these low prevalence populations as economically as possible. It is understood that many HIV-2 infections cannot be differentiated by this algorithm. Initially reactive screening tests are repeated in duplicate rather than proceeding directly to the supplemental test because most initial positive EIA results are false positives in low prevalence populations. When testing high prevalence populations, where most positives are true positives, it might be cost-effective to proceed directly to the supplemental test. Simple calculations of positive and negative predictive values (described later) can guide the laboratorian in deciding the best algorithm for every situation. Remember, the decision of whether an individual is infected or uninfected is based on the results of the supplemental test (confimatory test). The screening assays are used to sensitively select those specimens to be tested by the more specific and expensive confirmatory procedures.

Because of the expense of current confirmatory tests, most developing countries, where the prevalence of HIV infection is high, find the algorithms described here to be too expensive. In some African and Latin American countries, Western blot tests can cost as much as U.S. $60 to $100. Therefore the WHO has recommended alternative strategies[85] for HIV antibody detection that use a single EIA, rapid or simple assays, or combinations of screening tests in place of the EIA plus Western blot strategy. Three testing strategies aimed at maximum accuracy but minimum cost are recommended. The most appropriate strategy in a particular setting depends on the objective of the test and the prevalence of HIV in the population to be tested (Table 5.5).

Strategy I, recommended for screening donated blood, tests the blood

TABLE 5.5. WHO recommendations for HIV testing strategies.

Objective of testing and prevalence of infection	Testing strategy[a]
Transfusion/donation safety	
Any prevalence	I
Surveillance	
>10%	I
≤10%	II
Diagnosis: clinical signs/symptoms of HIV infection/AIDS	
Any prevalence	II
Diagnosis: symptomless infection	
>10%	II
≤10%	III

[a] See text for explanation of strategies.

with a single EIA, a rapid or simple test. Blood found to be reactive is discarded. Donors found to be positive by a single test should not be notified of their HIV status. This strategy is also recommended for surveillance when HIV prevalance is >10%.

Strategy II is recommended for diagnosis when there are symptoms of HIV infection. By this strategy, anyone found positive by strategy I would be retested by a second EIA, rapid or simple test, based on a different antigen preparation, a different test principle, or both. Samples found to be reactive by both tests are taken to be HIV antibody-positive, and the patients are so advised. Individuals negative by the second test after being positive by the first are told that they are HIV antibody-negative.

Strategy III is recommended for diagnosis in individuals having no symptoms of HIV infection. They are first tested as described for strategies I and II. If they are positive by the first and second assays, they are tested by a third assay using a different antigen preparation or test method, or both. If the serum is reactive in all three tests it is taken to be HIV antibody-positive. Samples that are negative in either of the first two tests is taken to be antibody-negative. However, samples that are positive in the first two tests but negative in the third is judged to be equivocal for HIV antibody. Individuals producing equivocal results should be rebled and tested within a minimum of 2 weeks.

Early Detection of HIV Infection in Neonates by Serologic Methods

Detection by serologic methods of infection in infants born to HIV-seropositive mothers is difficult because of interference from the anti-HIV IgG antibodies passively transferred from the mother. Passively acquired antibodies may persist for up to 18 months in the infant,[86,87] although studies have shown that in most infants maternal antibodies are undetectable by 9 months and are absent in essentially 100% of uninfected infants by 15 months.[88] Consequently, until 15 months only those assays that are able to distinguish between antibodies of maternal and infant origin are of diagnostic value. Assays that are specific for IgM or IgA seem to be best suited for that purpose, as it is thought that these antibodies do not cross the placenta.

Assays for IgM have not proved to be as useful for the detection of HIV infections in neonates as has been demonstrated in other diseases, such as measles, cytomegalovirus (CMV) or herpes infections, and toxoplasmosis. Several investigators have reported that IgA antibodies are diagnostic for HIV infection in some infants at birth.[89,90] The assay most frequently used is a modification of Western blot that first absorbs all IgG antibody using a recombinant form of streptococcal protein G (GammaBind, Genex Corp, Gaithersberg, MD). In some cases as many as four absorptions were required to remove the IgG completely. Once the IgG antibodies have been removed, IgA antibodies produced by the infant can be specifically detected using a class-specific anti-human IgA–enzyme conjugate. A second type of assay, an IgA-capture assay, did not require that IgG be removed before testing and was found to have equivalent sensitivity and specificity.[89]

Both assays are useful for early detection of infection in neonates, but they do not detect all HIV-infected infants. Only about 15% of HIV-infected infants are positive at birth by these methods.[89] Both assays detect 70% to 80% of infected infants after 4 months. Only one child gave a false-positive result in the study by Parekh et al.,[89] although other studies[90] have reported as many as 15% of tests in these infants may be falsely positive at birth and then become negative by 1 month. It has been suggested that there may be some "leakage" of IgA from the mother to the infant by some mechanism yet to be discovered. Currently, IgA assays offer a serologic alternative to PCR, but the sensitivity and specificity is less than that seen with direct detection of proviral DNA.

Sensitivity, Specificity, and Predictive Values of Serologic Tests

How reliable is the positive or negative test result produced by the sensitive and specific HIV assays previous described? Although it seems that this question would be simple to answer, in fact the answer is complex and requires that we understand several other concepts. First, we must introduce the concept of predictive value of both positive and negative test results. Predictive value was defined by Galen and Gambino[91] as "how accurately a test predicts the presence or absence of

TABLE 5.6. Predictive value of EIA tests for HIV antibody applied to a population containing infected and uninfected subjects.

Disease state	Test results		Total
	Subjects with positive test result	Subjects with negative test result	
With disease	TP	FN	TP + FN
Without disease	FP	TN	FP + TN
Total	TP + FP	FN + TN	TP + FP + FN + TN

TP = true positives, number of sick subjects correctly classified by the test; FP = false positives, number of healthy subjects misclassified by the test; TN = true negatives, number of healthy subjects correctly classified by the test; FN = false negatives, number of sick subjects misclassified by the test.

$$\text{Sensitivity (positivity in disease)} = \frac{TP}{TP + FN} \times 100$$

$$\text{Specificity (negativity in health)} = \frac{TN}{TN + FP} \times 100$$

$$\text{Predictive value of a positive result} = \frac{TP}{TP + FP} \times 100$$

$$\text{Predictive value of a negative result} = \frac{TN}{TN + FN} \times 100$$

disease." The answer to this question is based on the analysis of three variables: (1) sensitivity, the incidence of true-positive results in patients with disease; (2) specificity, the incidence of true-negative results in patients without disease; and (3) prevalence, the number of cases of disease in the population being tested (cases per 100,000).

The predictive value of a positive test (PPV) is defined as the percentage of positive results that are true positives when the test is applied to a population containing both healthy and diseased subjects. The predictive value of a negative test (NPV) is the percentage of negative results that are true negatives. As previously suggested, these figures depend on the number of false positives and false negatives produced by the test being used in relation to the number of true positives and true negatives and the prevalence of infection in the population being tested. These relations are illustrated in Table 5.6. The ideal test would establish the presence or absence of infection in every individual tested, and there would never be false-positive or false-negative results. Unfortunately, there are no perfect tests. As large numbers of blood donors were screened, it became clear that the distribution of test values determined in infected and uninfected specimens overlapped to a small degree.

Licensed tests for HIV antibody are highly sensitive (>99.8%) and specific (>99.8%). In the United States most populations being screened have a low prevalence (>0.1%). In fact, the prevalence in blood donors is

TABLE 5.7. Predictive value of EIA tests for HIV antibody applied to a population with a prevalence of 0.02% infection.[a]

Disease state	Test results		Total
	Subjects with positive test result	Subjects with negative test result	
With disease	199	1	200
Without disease	2000	997,800	999,800
Total	2199	997,801	1,000,000

$$\text{Predictive value of a positive result} = \frac{199}{199 + 2000} \times 100 = 9.0\%$$

$$\text{Predictive value of a negative result} = \frac{997,800}{1 + 997,801} \times 100 = 100\%$$

[a] Assay sensitivity = 99.8%, assay specificity = 99.8%.

less than 0.02%. Among a population of blood donors, assays that are 99.8% sensitive and 99.8% specific have a PPV of only 9.0% (Table 5.7), which means that for every true positive sera, nine HIV-negative individuals will also test positive at the initial screening. As the prevalence increases, the percentage of false-positive test results decreases (Fig. 5.10). At a prevalence of 1.0%, the same test has a PPV of 83.4% (Table 5.8). Even at this level, if people were counseled on the basis of their screening results, 16.6% would be incorrectly informed of their HIV status.

The problem of low PPV was addressed by serial testing initially reactive subjects by additional EIAs and supplemental tests such as the Western blot, so the PPV of the individual tests becomes less important. The critical issue is the probability of a false-positive test (P_{FP}), the

TABLE 5.8. Predictive value of EIA tests for HIV antibody as applied to a population with a prevalence of 1.0% infection.[a]

Disease state	Test results		Total
	Subjects with positive test result	Subjects with negative test result	
With disease	9,980	20	10,000
Without disease	1,980	998,020	990,000
Total	11,960	988,040	1,000,000

$$\text{Predictive value of a positive result} = \frac{9980}{9980 + 1980} \times 100 = 83.4\%$$

$$\text{Predictive value of a negative result} = \frac{998,020}{998,020 + 20} \times 100 = 100\%$$

[a] Assay sensitivity = 99.8%, assay specificity = 99.8%.

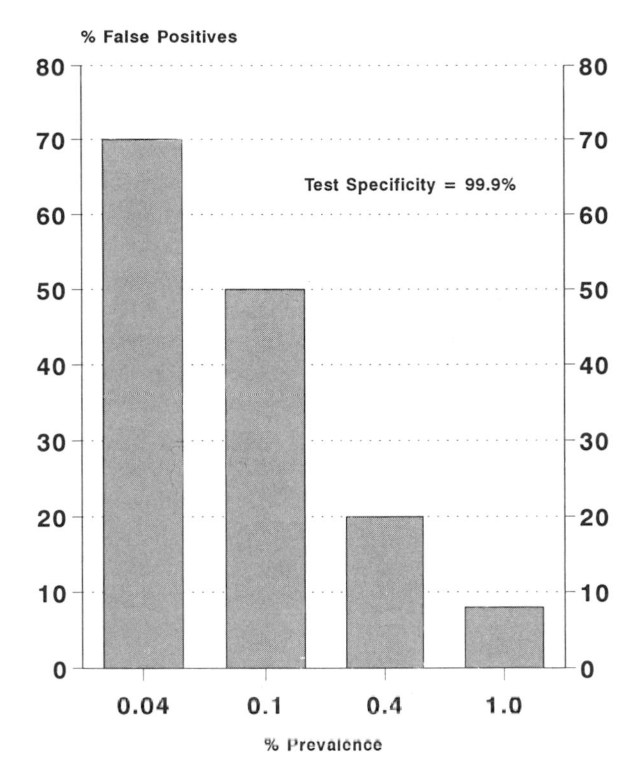

FIGURE 5.10. Relation of prevalence and false positivity. Percentage of false-positive results decreases as the prevalence of infection in the population increases.

probability of a false-negative test (P_{FN}), and the PPV of the recommended HIV testing algorithm. For example, let S^+ stand for a positive subject and S^- for a negative subject; let E^+ be a positive EIA test result and E^- a negative test result; and let W^+ and W^- represent the same for Western blot results. The prevalence is then the probability that the subject is positive: $P(S^+) = 0.001$, or 0.1%. The sensitivity of the EIA is $P(E^+/S^+) = 0.998$; and the specificity is $P(E^-/S^-) = 0.998$. For the Western blot, $P(W^+/S^+) = 0.996$ and $P(W^-/S^-) = 0.996$. The formula for the P_{FP} for one test is as follows:

$$P_{FP} = \frac{P(T/S^-)[1 - P(S^+)]}{P(T/S^-) + P(S^+)[P(T/S^+) - P(T/S^-)]}$$

We need the $P(T/S^+)$ and $P(T/S^-)$ for our entire four-test algorithm to substitute in the formula.

The $P(T/S^+)$ is the probability that the algorithm gives a positive result given that the subject is a true positive. The possible outcomes of the test that lead to a positive conclusion are ++++ or +−++ or ++−+.

Assuming that the tests are independent, given that the subject is positive:

$$\begin{aligned} P(T/S^+) &= P_{T+} \\ &= P(++++ \text{ or } +-++ \text{ or } ++-+) \\ &= 0.998 \times 0.998 \times 0.998 \times 0.996 + 2 \times 0.998 \\ &\quad \times 0.998 \times (1 - 0.998) \times 0.996 \\ &= 0.9940040239 \end{aligned}$$

If the subject is negative:

$$\begin{aligned} P(T/S^-) &= (1 - 0.998)^3 \times (1 - 0.996) + 2 \times (1 - 0.998)^2 \\ &\quad \times 0.998 \times (1 - 0.996) \\ &= 0.000000031968 \end{aligned}$$

Substituting in the equation for calculating P_{FP}:

$$P_{FP} = 0.00003212761425452$$

or 32 false-positive results per 1 million tests.

$$PPV = 0.999967679$$

or essentially 100%. The probability of a false-negative test is

$$P_{FP} = 0.000007987$$

or 8 false-negative results per 1 million tests. The NPV is essentially 100%.

HIV Antigen Testing

The first use of HIV antigen testing was for monitoring virus growth in cell culture, thereby providing direct confirmation of reverse transcriptase assay results.[92] Commercial EIA tests for HIV core antigen (p24) became available in 1986 (Abbott Laboratories and E.I. du Pont de Nemours) for detecting p24 antigen in serum, plasma, and cerebrospinal fluid. In 1988 Abbott Laboratories became the first to receive FDA licensure for diagnostic testing, followed a few years later by Coulter. Several other companies offer p24 antigen tests for research only.

Assay Methods

The most common configuration for p24 antigen tests is illustrated in Figure 5.11. The patient's test sample is incubated with the anti-HIV-1 p24 capture antibody bound to a solid support, such as polystyrene beads or a microplate well. After washing, the antigen–antibody complex bound to the solid phase is incubated with a second anti-HIV antibody prepared in rabbits. After removal of unbound reagent by washing, enzyme–goat

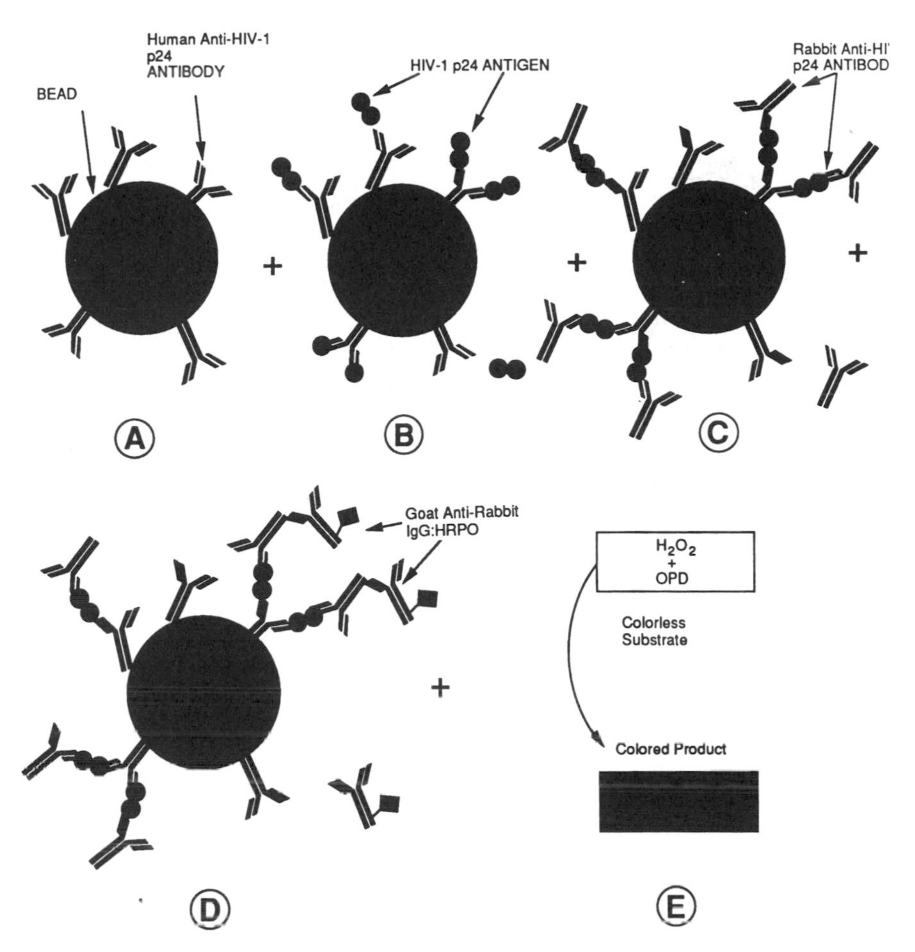

FIGURE 5.11. Configuration of the HIV-1 p24 antigen test.

anti-rabbit IgG is added. After the final wash, the initial binding of p24 antigen is detected and quantified by adding the appropriate substrate. The change in color (absorbance) is measure spectrophotometrically. Another version of the p24 antigen tests eliminates one step by conjugating the second antibody with the enzyme. In either example, assays are made quantitative by comparing the absorbance of the patient's sample to a standard curve prepared from known amounts of p24 antigen. The detectable limits of these assays has been reported to be between 10 and 30 pg/ml.[93–95]

All reactive specimens should be confirmed by a specific neutralization assay to verify the presence of HIV antigen. The neutralization test is performed by first incubating the serum with anti-HIV p24 antigen and then performing the assay as described. If the specimen contains p24 antigen, as indicated by the initial test, it is removed by the preincubation with anti-HIV p24 antibody, and the subsequent test is negative or the absorbance less intense. Reactive specimens are confirmed as p24 antigen-positive when the absorbance is reduced by at least 50% when compared to the same specimen preincubated with normal serum and tested in the same assay run.

Applications for p24 Antigen Testing

The HIV-1 p24 antigen is an important prognostic marker of HIV infection. It is also used to evaluate antivirus drugs and vaccines. In addition, considerable effort has been expended to demonstrate that antigen testing should be added to the battery of tests used to screen blood used for transfusion. It is proposed that including p24 antigen testing can shorten the "window" period (Fig. 5.1) between infection and seroconversion. The usefulness of the p24 antigen test for this purpose depends on the incidence rate among donor populations. In Europe and the United States[96,97] it was found that antigen detection did not detect additional HIV infections that would be missed by antibody testing; however, similar studies done in areas having higher incidence rates has detected HIV infections prior to seroconversion.[98] Antigen testing has also been shown to be useful for early detection of HIV infection in neonates.

The usefulness of antigen assays has been limited by the fact that all current commercial antigen detection kits measure free p24 antigen in serum and plasma. Unfortunately, most p24 antigen present in serum after seroconversion exists as an antigen–antibody complex and is not detected with the commercial kits. The usefulness of antigen assays would be significantly extended if any anti-p24–p24 antigen immune complexes could be dissociated and the previously complexed antigen detected. Methods such as pretreatment of serum samples with HCl[99,100] or a combination of heat and acid treatment[101,102] have been used to dissociate these complexes and enhance the detection of p24 antigen. These modifications have not been generally applied because of variations in results obtained using these methods. The variation is due to the use of reagents that are not buffered; therefore small changes in the volume of HCl or NaOH can lead to significantly lowered levels of antigen detected. Presumably, the loss of sensitivity is due to denaturation of the antigen.

More recently, Coulter Immunology (Hialeah, FL) has introduced a modification of their HIV p24 antigen test that furnishes an acid dissociation reagent (1.5 M glycine buffer, pH 1.8–2.2). After dissociation by incubation with this acid reagent, the mixture is neutralized by adding 1.5 M Tris (pH 8.6–9.0). Use of this buffered system has provided a consistent, sensitive method for detecting p24 antigen following the dissociation of immune complexes. In adults representing all stages of disease, detection of antigen was substantially improved in this system when compared to the results of testing untreated samples.[103] Acid treatment also dramatically improved the value of antigen testing for the early detection of infection in neonates by improving the sensitivity in infants older than 1 month from 18.0% to 89.5%.[104] Clearly, after licensure by the FDA, antigen testing following dissociation of the immune complexes will become the standard procedure.

Conclusion

Sensitive, specific, inexpensive tests for HIV antibodies have permitted the implementation of effective screening programs to protect the blood supply, to conduct seroprevalence studies to measure the spread of disease, and to accurately diagnose HIV infection in asymptomatic individuals. Although these programs have reduced the rate of spread in most of the industrialized world, current tests remain either too technically difficult or too expensive for developing countries. As the AIDS epidemic continues to spread, there will be increased demand for testing. Much of the increase is likely to come from Asia, where it is estimated that the number of cases will be greater than that found in Africa by the beginning of the next century. A less expensive, simpler algorithm is needed to permit all blood to be tested. Tests for early detection of infection in neonates must be further improved, especially if effective intervention strategies are developed.

Direct tests that measure viral antigen or, more important, the viral genome have improved dramatically. The tests have progressed from experimental procedures that require several days to perform and cost several hundred dollars to methods that can be completed in a single day at a fraction of the cost. These tests continue to improve and may one day challenge serologic tests as the front-line diagnostic test for HIV. In the meantime, antibody testing will retain the dominant position in the diagnosis and control of HIV. Considerable progress can be expected in the development of simple, rapid, inexpensive tests for HIV antibody, ensuring that all blood will be tested and diagnosis will be available to all concerned persons worldwide.

References

1. Food and Drug Administration. Revised recommendations for the prevention of human immunodeficiency virus (HIV) transmission by blood and blood products [memorandum]. Bethesda: Center for Biologics Evaluation and Research, FDA, April 23, 1992
2. Centers for Disease Control. Provisional Public Health Service inter-agency recommendations for screening blood and plasma for antibody to the virus causing acquired immunodeficiency syndrome. MMWR 1985;34:1–5

3. Centers for Disease Control. Public Health Service guidelines for counseling and antibody testing to prevent HIV infections and AIDS. MMWR 1987; 36:509–515
4. Centers for Disease Control. Update: serologic testing for antibody to human immunodeficiency virus. MMWR 1988;39:833–840
5. Martin PW, Burger DR, Caouette S, Goldstein AS, Peetoom T. Importance of confirmatory tests after strongly positive HTLV-III screening tests. N Engl J Med 1986;317:1577–1578
6. Pau C-P, Holloman DL, Lee-Thomas S, Schochetman G, George JR. A combination immunoblot for the simultaneous detection and differentiation of HIV-1 and HIV-2. Clin Diagn Virol 1993;1:201–205
7. Parry J. Personal communication. London: Public Health Laboratory Service
8. Cooper DA, Gold J, Maclean P, et al. Acute AIDS retrovirus infection: definition of a clinical illness associated with seroconversion. Lancet 1985;1: 537–540
9. Schochetman G, Epstein JS, Zuck TF. Serodiagnosis of infection with the AIDS virus and other human retroviruses. Annu Rev Microbiol 1989;43: 629–659
10. Estaban JI, Shih JW, Tai CC, et al. Importance of Western blot analysis in predicting infectivity of anti-HTLV-III/LAV positive blood. Lancet 1985;2: 1083–1086
11. Gaines H, Von Sydow M, Sonnerberg A, et al. Antibody response in primary human immunodeficiency virus infection. Lancet 1987;2:1249–1253
12. Ho DD, Sargadharan MG, Resnick L, et al. Primary human T-lymphotropic virus type III infection. Ann Intern Med 1985;103:880–883
13. Kumar P, Pearson JE, Martin DH, et al. Transmission of human immunodeficiency virus by transplantation of a renal allograph, with development of the acquired immunodeficiency syndrome. Ann Intern Med 1987;106: 244–245
14. Marlink RG, Allan JS, McLane MF, et al. Low sensitivity of ELISA testing in early HIV infection. N Engl J Med 1986;315:1549
15. Ulstrup JC, Skaug K, Figenschau KJ, et al. Sensitivity of Western blotting (compared to ELISA and immunofluorescence) during seroconversion after HTLV-III infection. Lancet 1986;1:1151–1152
16. Anderson KC, Gorgone BC, Marlink RC, et al. Transfusion-acquired human immunodeficiency virus infection among immunocompromised persons. Ann Intern Med 1986;105:519–527
17. Bedarida G, Cambie G, D'Agostino F, Ronsivalle JB. HIV IgM antibodies in groups who are seronegative on ELISA testing. Lancet 1986;2:570–571
18. Parry JV, Mortimer PP. Place of IgM antibody testing in HIV serology. Lancet 1986;2:1979–1980
18a. Gurtler LG, Hauser PH, Eberle J, von Brunn A, Knapp S, Zekeng L, Tsague JM, Kaptue L. A new subtype of human immunodeficiency virus type 1 (MVP-5180) from Cameroon. J Virol 1994;68:1581–1585
18b. Haesevelde MV, Decourt J-L, Leys RJ, Vanderborght B, van der Groen G, van Heuverswijn H, Saman E. Genomic cloning and complete sequence analysis of a highly divergent African human immunodeficiency virus isolate. J Virol 1994;68:1586–1596

19. Ward JW, Grindon AJ, Feorino PM, et al. Laboratory and epidemiologic evaluation of an enzyme immunoassay for antibodies to HTLV-III. JAMA 1986;256:357–361
20. Blanton M, Balakrishnan K, Dumaswala U, Zelenski K, Greenwalt TJ. HLA antibodies in blood donors with reactive screening tests for antibody to the human immunodeficiency virus. Transfusion 1987;27:118–119
21. Kuhnl P, Seidl S, Holzberger G. HLA DR4 antibodies cause positive HTLV-III antibody ELISA results. Lancet 1985;1:1222–1223
22. Mendenhall CL, Roselle GA, Grossman CJ, et al. False-positive tests for HTLV-III antibodies in alcoholic patients with hepatitis. N Engl J Med 1986;314:921–922
23. Peterman TA, Lang GR, Mikos NJ, et al. HTLV-II/LAV infection in hemodialysis patients. JAMA 1986;255:2324–2326
24. Fleming DW, Cochi SL, Steece RS, Hull HF. Acquired immunodeficiency syndrome in low incidence areas. JAMA 1987;258:785–787
25. World Health Organization. Operational characteristics of commercially avaialble assays to determine antibodies to HIV-1. Global Programme for AIDS/Biomedical Research Unit 1989;89:4
26. Croft JN, Maskill WJ, Healy DS, Gust ID. Particle agglutination test for anti-HIV. Lancet 1987;2:797–798
27. Yoshida T, Matsui T, Kobayashi S, Yamamoto N. Evaluation of passive particle agglutination test for antibody to human immunodeficiency virus. J Clin Microbiol 1987;25:1433–1437
28. Riggens CH, Thorn RM. Visually read HIV immunoassays. Lancet 1989;1: 671–672
29. George JR, Stetler H, Granade T, et al. Evaluation of alternative testing methods for use in developing countries. Manuscript in preparation, 1994
30. Spielberg F, Ryder RW, Harris J, et al. Field testing and comparative evaluation of rapid, visually read screening assays for antibody to human immunodeficiency virus. Lancet 1989;1:580–584
31. World Health Organization. Operational characteristics of commercially available assays to determine antibodies to HIV-1, and/or HIV 2 in human sera. Report 2. Global Program for AIDS/Biomedical Research Unit 1990; 90:1
32. DeCock KM, Maran M, Kouadio JC, et al. Rapid tests for distinguishing HIV-1 and HIV-2. Lancet 1990;336:757
33. Tsang VCW, Peralta JM, Simons AR. Enzyme-linked immunoelectrotransfer blot techniques (EITB) for studying the specificities of antigens and antibodies separated by gel electrophoresis. Methods Enzymol 1983;92:377–391
34. Goudsmit J, Lang JMA, Paul DA, Dawson GJ. Antigenemia and antibody titer to core and envelope antigen in AIDS, ARC, and subclinical HIV infection. J Infect Dis 1987;155:558–560
35. Lange J, Coutinho RA, Krone WJA, et al. Distinct IgG recognition patterns during progression of subclinical and clinical infection with lymphadenopathy associated virus/human T lymphotropic virus. BMJ 1986;292:228–230
36. Lange J, Paul DA, Huisman HG, et al. Persistent HIV antigenemia and decline of HIV core antibodies associated with transition to AIDS. BMJ 1986;293:1459–1462

37. McDougal JS, Kennedy MS, Nicholson JKA, et al. Antibody response to human immunodeficiency virus in homosexual men: relation of antibody specificity, titer, and isotype to clinical status, severity of immunodeficiency and disease progression. J Clin Invest 1987;80:316–324
38. Barin F, McLane MF, Allan JS, Lee TH. Virus envelope protein of HTLV-III represents major target antigens for antibodies in AIDS patients. Science 1985;118:1094–1096
39. Essex M, Allan J, Kanki P, et al. Antigens of human T-lymphotropic virus type III/lymphadenopathy-associated virus. Ann Intern Med 1985;103:700–703
40. Kitchen LW, Barin F, Sullivan JL, et al. Aetiology of AIDS-antibodies to human T-cell leukemia virus type III in hemophiliacs. Nature 1984;312:367–369
41. Pinter A, Honnen WJ, Tilly SA, et al. Oligomeric structure of gp41, the transmembrane protein of human immunodeficiency virus type 1. J Virol 1989;63:2674–2679
42. Zoller-Pazner S, Gorney MK, Honnen WJ, Pinter A. Reinterpretation of human immunodeficiency virus Western blot patterns. N Engl J Med 1989; 320:1280–1281
43. Parakh BS, Pau C-P, Granade TC, et al. Oligomeric nature of transmembrane glycoproteins of HIV-2: procedures for their efficient dissociation and preparation of Western blots for diagnosis. AIDS 1991;5:1009–1013
44. DeCock KM, Porter A, Kouadio J, et al. Rapid and specific diagnosis of HIV-1 and HIV-2 infections and evaluation of testing strategies. AIDS 1990;4:875–878
45. Centers for Disease Control. Interpretation and use of the Western blot assay for diagnosis of human immunodeficiency virus type 1 infections. MMWR 1989;38(S-7):1–7
46. Weber JN, Clapham P-R, Neiss RA, et al. Human immunodeficiency virus infection in 2 cohorts of homosexual men-neutralizing sera and association of anti-gag antibody with prognosis. Lancet 1987;1:119–122
47. World Health Organization. Proposed criteria for interpreting results from the Western blot assasy for HIV-1, HIV-2, and HTLV-1/HTLV-II. Wkly Epidemiol Rec 1990;37:381–383
48. Klienman S, Fitzpatrick L, Secord K. Follow-up testing and notification of anti-HIV Western blot atypical (indeterminate) donors. Transfusion 1988; 28:280–282
49. Dock NL, Lamberson HV, O'Brien TA, et al. Evaluation of atypical human immunodeficiency virus immunoblot reactivity in blood donors. Transfusion 1988;28:412–418
50. Josephson SC, Swack NS, Ramirez MT, Hausler WJ. Investigation of atypical Western blot (immunoblot) reactivity involving core proteins of human immunodeficiency virus type 1. J Clin Microbiol 1989;37:922–927
51. Jackson JB, MacDonald KL, Caldwell J, et al. Absence of HIV infection in blood donors with indeterminate Western blot tests for antibody to HIV-1. N Engl J Med 1990;322:217–222
52. Dock NL, Klienman SH, Rayfield MA, et al. Status of human immunodeficiency virus infection in individuals with persistently indeterminate Western

blot patterns: prospective studies in a low prevalence population. Arch Intern Med 1991;151:525–530

53. Blumberg RS, Sandstrom EG, Paradis TJ, et al. Detection of human T-cell lymphotropic virus type III-related antigens and anti-human T-cell lymphotropic virus type III antibodies by anticomplementary immunofluorescence. J Clin Microbiol 1986;23:1072–1077
54. Gallo D, Diggs JL, Shell GR, Dailey PJ, Hoffman MN. Comparison of detection of antibody to the acquired immune deficiency syndrome virus by enzyme immunoassay, immunofluorescence and Western blot methods. J Clin Microbiol 1986;1049–1051
55. Kaminaky LS, McHugh T, Stites D, et al. High prevalence of antibodies to acquired immune deficiency syndrome (AIDS)-associated retrovirus (ARV) in AIDS and related conditions but not in order of disease state. Proc Natl Acad Sci USA 82:5535–5539
56. Gallo D, Hoffman MN, Yeh ET, George JR, Hanson CV. Comparison of indirect immunofluorescence and membrane fluorescence for the differentiation of antibodies to human immunodeficiency virus types 1 and 2. J Clin Microbiol 1992;30:2275–2278
57. Allan JS, Ciligan JE, Lee TH, Sodroski JG. Immunogenic nature of a pol gene product of HTLV-III/LAV. Blood 1987;69:331–333
58. Barre-Sinoussi F, Mathur-Wagh U, Rey F, et al. Isolation of lymphadenopathy-associated virus (LAV) and detection of antibodies from US patients with AIDS. JAMA 1985;253:1737–1739
59. Kanki PJ, Barin F, M'Boup S, et al. New human T lymphotropic retrovirus to simian T-lymphotropic virus type III (STLV-III)$_{AGM}$. Science 1986;232:238–243
60. Schupbach J, Popovic M, Gilden RV, et al. Serologic analysis of a subgroup of human T-lymphotropic retroviruses (HTLV-III) associated with AIDS. Science 1984;224:503–505
61. Hoff R, Berardi V, Weiblen BJ, et al. Seroprevalence of human immunodeficiency virus among child-bearing women. N Engl J Med 1988;318:525–530
62. Pappaioanou M, George JR, Hannon WH, et al. HIV seroprevalence surveys of childbearing women-objectives, methods, and uses of the data. Public Health Rep 1990;105:147–152
63. George JR, Hannon WH, Jones W, et al. Serologic Assays for Human Immunodeficiency Virus Antibody in Dried-Blood Specimens Collected on Filter Paper from Neonates. Bethesda: US Department of Health and Human Services, Public Health Service, Centers for Disease Control and the National Institute of Child Health and Human Development, NIH 1989
64. Pappaioanou M, Kashamuka M, Behets F, et al. Accurate detection of maternal antibodies to HIV in newborn whole blood dried on filter paper. AIDS 1993;7:483–488
65. Behets F, Kashamuka M, Pappaioanou M, et al. The stability of human immunodeficiency virus type 1 (HIV-1) antibody in whole blood dried on filter paper and stored under varying tropical conditions in Kinshasha, Zaire. J Clin Microbiol 1992;50:1179–1182

66. National Committee for Clinical Laboratory Standards. Blood Collection on Filter Paper for Neonatal Screening Programs; Approved Standard. NCCLS publ. LA4-A2. Villanova, PA: NCCLS, 1992:1–25
67. Parry JV, Perry KR, Mortimer PP. Sensitive assays for viral antibodies in saliva: an alternative to tests run on serum. Lancet 1987;2:72–75
68. Parry JV. Detection of viral antibodies in saliva specimens as an alternative to serum. J Clin Chem Clin Biochem 1989;27:245–246
69. Parry JV. A specimen of convenience. MLW 1991
70. Thieme T, Yoshira P, Piacentini S, Beller M. Clinical evaluation of oral fluid samples for diagnosis of viral hepatitis. J Clin Microbiol 1992;30:1076–1079
71. Mortimer PP, Parry JV. Non-invasive virological diagnosis: are saliva and urine specimens adequate substitues for blood? Rev Med Virol 1991;1:73–78
72. Klokke AH, Ocheng D, Kalluvya SE, et al. Field evaluation of immunoblobulin G antibody capture tests for HIV-1 and HIV-2 antibodies in African serum, saliva, and urine. AIDS 1991;5:1391–1392
73. Gershey-Damet GM, Koffi K, Abouya L, et al. Salivary and urinary diagnosis of human HIV-1 and 2 infections in Cote d'Ivoire using two assays. Trans R Soc Trop Med Hyg 1992;86:670–671
74. Covell R, Follett E, Coote I, et al. HIV testing among drug users in Glasgow. J Infect 1993;26:27–31
75. Groopman JE, Salahuddin SZ, Sarhgadharan MG, et al. HTLV-III in saliva of people with AIDS-related complex and healthy homosexual men at risk for AIDS. Science 1984;226:310–313
76. Archibald DW, Zon L, Groopman JE, McLane MF, Essex M. Antibodies to human T lymphotropic virus type III in saliva of acquire immunodeficiency syndrome (AIDS) patients and persons at risk for AIDS. Blood 1986;67: 831–834
77. Archibald D, Essex M, McLane MF, et al. Antibodies to HTLV-I in saliva of seropositive individuals from Japan. Viral Immunol 1987;1:241–246
78. Archibald DW, Barr CE, Torasian JP, McLane MF, Essex M. Secretory IgA in the parotid saliva of patients with AIDS and AIDS-related complex. J Infect Dis 1987;155:793–796
79. Van den Akker R, van den Hoek JA, van den Akker WM, et al. Detection of HIV antibodies as a tool for epidemiologic studies. AIDS 1992;6:953–957
80. Desai S, Bates H, Michalski FJ. Detection of antibody to HIV-1 in urine. Lancet 1992;337:183–187
81. Sterne JAC, Turner AC, Connel JA, et al. Human immunodeficiency virus: GACPAT and GACELISA as diagnostic tests for antibodies in urine. Trans R Soc Trop Med Hyg 1993;87:181–183
82. Urnovitz HB, Clerici M, Shearer GM, et al. HIV-1 antibody serum negativity with urine positivity. Lancet 1993;342:1458–1459
83. Metler RP, Schable CA, Weisfuse IB, et al. Surveillance for HIV-2 in the United States. Presented at The First National Retrovirus Conference on Human Retroviruses and Related Infections, December 12–16, 1993, Washington, DC, abstract 171
84. Eighth Annual Conference on Human Retrovius Testing. Association of State and Territorial Public Health Laboratory Directors, Atlanta, March 5–7, 1993

85. Tamashiro H, Maskill W, Emmanuael J, Sato P, Heymann D. Reducing the cost of HIV antibody testing. Lancet 1993;342:87–91
86. European Collaborative Study. Mother-to-child transmission of HIV infection. Lancet 1988;1:1039–1042
87. Mok JQ, Giaquinto C, DeRossi A, et al. Infants born to mothers seropositive for human immunodeficiency virus. Lancet 1987;1:1164–1167
88. Adjorlolo G, DeCock KM, Ekpini E, et al. Prospective comparison of mother-to-child transmission of HIV-1 and HIV-2 in Abidjan, Cote d'Ivoire. 1993, submitted for publication
89. Parekh B, Shaffer N, Coughlin R, et al. Human immunodeficiency virus type 1-specific IgA capture enzyme immunoassay for early Diagnosis of human immunodeficiency virus type 1 infection in infants. Pediatr Infect Dis J 1993;12:908–913
90. Quinn TC, Kline RL, Halsey N, et al. Early diagnosis of perinatal HIV infection by detection of viral-specific IgA antibodies. JAMA 1991;266: 3439–3442
91. Galen RS, Gambino SR. Beyond Normality: The Predictive Value and Efficiency of Medical Diagnosis. New York: Wiley, 1975
92. Feorino P, Forrester B, Schable C, Warfield D, Schochetman G. Comparison of antigen assay and reverse transcriptase assay for detecting human immunodeficiency virus in culture. J Clin Microbiol 1987;252:344–2346
93. Goudsmit J, DeWolfe F, Paul DA, et al. Expression of human immunodeficiency virus antigen (HIV-Ag) in serum and cerebrospinal fluid during acute and chronic infection. Lancet 1986;2:177–180
94. Goudsmit J, Lange JMA, Paul DA, Dawson GJ. Antigenemia and antibody titers to core and envelope antigens in AIDS, AIDS-related complex, and subclinical human immunodeficiency virus infection. J Infect Dis 1987;155: 558–560
95. Barr PJ, Stimer KS, Sabin EA, et al. Antigenicity and immunogenicity of domains of the human immunodeficiency virus (HIV) envelope polypeptide expressed in the yeast Saccharomyces cerevisiae. Vaccine 1987;5:90–101
96. Alter H, Epstein JS, Swensen SG, et al. Collaborative study to evaluate HIV Ag (HIV-Ag) screening for blood donors [abstract 5202]. Transfusion 1989;29(suppl 7s):56
97. Backer U, Weinauer F, Gathof AG, Eberle J. HIV antigen screening in blood donors. In: Abstracts, IVth International Conference on AIDS, 1988; 2:364
98. Kessler H, Blaaw B, Spear J, et al. Diagnosis of human immunodeficiency virus infection in seronegative homosexuals with acute viral syndrome. JAMA 1987;258:1196–1199
99. Nishanian P, Huskins KR, Stehn S, Detels R, Fahey JL. A simple method for improved assay demonstrates that p24 antigen is present as immune complexes in most sera from HIV infected individuals. J Infect Dis 1990; 162:21–28
100. Von Sydow M, Gaines H, Sonnerborg A, et al. Antigen detection in primary HIV infection. BMJ 1988;296:238–240
101. Kageyama S, Yamada O, Mohammad SS, et al. An improved method for the detection of HIV antigen in blood of carriers. J Virol Methods 1988;22: 125–131

102. Kesten L, Hoffd G, Gigas PL, Deley R, van der Groen G. HIV antigen detection in circulating immune complexes. J Virol Methods 1991;31:67–76
103. Vasudevachari MB, Salzman N, Woll D, et al. Clinical utility of an enhanced human immunodeficiency virus type 1 p24 antigen capture assay. J Clin Immunol 1993;13:185–188
104. Quinn TC, Kline R, Moss MW, Livingston RA, Hutton N. Acid dissociation of immune complexes improves diagnostic utility of p24 antigen detection in perinatally acquired human immunodeficiency virus infection. J Infect Dis 1993;167:1193–1196

6
Quality Control for HIV Testing

BRUCE J. MCCREEDY and J. RICHARD GEORGE

Quality assurance is the dynamic and ongoing process of monitoring the diagnostic laboratory's testing system for reproducibility that permits corrective action when established criteria are not met. Those techniques include statistical quality control procedures and procedures for method selection, method evaluation, preventive maintenance, in-service training, and laboratory management. Quality control is the study of those errors that are the responsibility of the laboratory and of the procedures used to recognize and minimize them. This study includes all errors arising within the laboratory between the receipt of the specimen and the dispatch of the test results report. On some occasions the responsibility of the laboratory may extend to the collection of the appropriate specimen, the method and time of collection, and the type of collection tube used.

Quality control requires the involvement of each technician and supervisor working in the laboratory. Quality control relies on the technicians to identify and solve problems, so supervisors and employees must work closely together. Every employee should have an understanding of the quality control system, and everyone should be working to achieve zero errors. *There is no minimum average acceptable quality*. Do it right the first time. Do not make mistakes. *Quality is free if produced the first time*.[1] This chapter describes the issues of quality control as it relates to serologic and polymerase chain reaction testing for the human immunodeficiency virus (HIV) of HIV infection.

Quality Control for Serologic Testing

The development and subsequent licensure by the U.S. Food and Drug Administration (FDA) of enzyme immunoassays (EIAs) for HIV antibody provided a low cost test that was well suited for diagnosis, mass screening of populations, and testing blood donations for HIV infection. Each year the number of tests performed and the number of laboratories performing tests increase in the United States. Yet programs for quality control of laboratory testing have not kept pace. Standard panels of serum for quality control and evaluation of EIA and Western blot tests are not generally available. Quality control programs for serologic tests are crude compared to the sophisticated computer-driven programs used in clinical chemistry laboratories. In fact, some serology laboratories do no more than run the controls supplied by the kit manufacturer. However, they have adapted the quality control procedures similar in design to those used for clinical chemistry to monitor testing for HIV infections. This section describes these adaptations and recommends a simple, inexpensive system for quality control of HIV serology tests.

Preparation of Quality Control Materials

One of the objectives of quality control is to ensure that the results obtained with an EIA or Western blot test performed today are qualitatively and quantitatively the same as those obtained last week or last year. Furthermore, that result must accurately reflect the infection status of the patient. Such consistency must pertain even if the laboratory has changed testing methods during that time. An essential component required for that assurance is a pool of quality control materials on which a continuum of results is available. For HIV serology, the control material required would be pools of positive and negative sera for use in the testing algorithms. These pools should be large enough that they last 6 to 12 months before new serum pools must be prepared.

Most laboratories find it difficult to obtain a single serum in sufficient quantity to last the required time. Therefore the easier solution is to prepare pools of serum from unused samples sent in for analysis. In some ways these pools may be superior to that from a single donor, as they represent individuals in different stages of infection (asymptomatic HIV infection to full-blown acquired immunodeficiency syndrome, or AIDS), and these donors may represent different levels of EIA and Western blot reactivity. Preparation of serum pools has been well described[2] and begins by selecting only clear, straw-colored serum specimens. Plasma specimens should not be mixed with serum as they clot even after recalcification. Turbid, chylous, hemolyzed, or icteric serum should not be used.

First, a pool of sera negative for HIV-1 antibody by EIA and Western blot is prepared. This pool is used as a diluent for the HIV-1 antibody-positive pool as well as for the negative control for the EIA and Western blot assay. All specimens are pooled; 0.01% sodium ethylmercurithiosalicylic acid is added; and the pool is mixed on a stirrer at 4°C overnight to ensure homogeneity. The serum pool is then filtered through a 10 μm pore size glass prefilter and then a 0.22 μm pore size sterile membrane filter. The final pool is tested by Western blot assay and must not produce any bands on HIV-1 and HIV-2 strips.

The HIV-1 antibody-positive pool is prepared by pooling sera found to be positive by EIA and Western blot. The pool of sera is filtered as described above. Dilutions of the positive pool from 1:2 to 1:2048 are prepared in the antibody-negative diluent. Each dilution of the pool should be tested in quadruplicate by EIA and plotted as shown in Fig. 6.1. The dilutions found to be on the "straight line" part of the curve (1:8–1:128) are then tested by Western blot. This analysis is used to select the dilution factor for the control pool. An appropriate positive control should have an absorbance close to the cutoff point but sufficiently above it so it does not often test negative due to the normal variability of the test. The positive control pool (1) should have a sample/cutoff ratio of approximately 3 [this recommendation might have to be modified for assays where the cutoff values tend to be high (>0.5)]; and (2) should

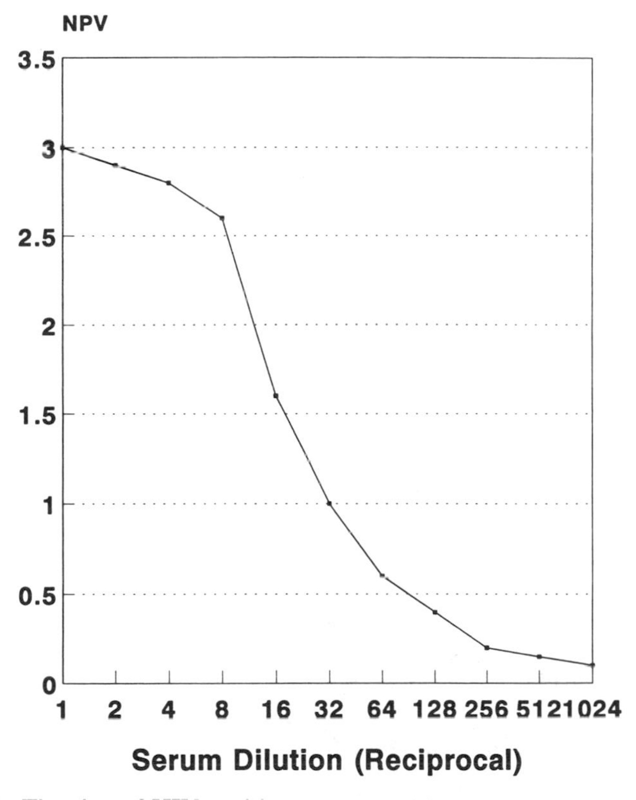

FIGURE 6.1. Titration of HIV-positive serum pool by enzyme immunoassay.

give clear, distinct bands for the HIV-1 structural proteins (gp160/120, p65, p55, p51, gp41, p32, p24, p17) on the Western blot assay.

Once prepared, the positive and negative control sera should be dispensed into quantities sufficient for 1 week of testing when controls are run in each assay microplate. Aliquots should be stored at −20°C or colder. Once thawed, the controls are used for 1 week and then discarded. In-house controls should be placed at several positions throughout the plate to monitor such variables as position effect, variability among wells, and timing differences that arise from when the first sample is added to the last. Calculations for control charts should be performed using absorbance values normalized by subtracting the cutoff value from the absorbance of the control. The net positive value (NPV) can then be monitored by plotting the results on quality control charts as described in the next section.

Monitoring EIA Performance with Quality Control Charts

Statistical Quality Control

Each assay run must contain the following controls: an in-house positive control, an in-house negative control, and the controls supplied by the manufacturer with each kit. It is important to understand the intended use of these various control materials. The manufacturer's controls are supplied to ensure that the kit is performing as described in the kit brochure. Values obtained for the positive and the negative control sera are used to calculate the positive/negative cutoff values and the sensitivity (positive control) and specificity (negative control) of each assay run. The in-house positive and negative controls are used to ensure that the assay performs at the same level of sensitivity and specificity run to run, lot to lot, and year to year. The latter is accomplished by preparing large quantities of control materials and standardizing new lots by overlapping runs before the previous lot is exhausted.

The best way to use quality control charts for serologic tests is still being debated. Some authors favor a statistical approach similar to that used for such clinical chemistry analytes as digoxin, thyroxine, or other hormones. These charts, called Shewhart or Levy-Jennings charts,[1,3] are prepared by analyzing duplicate samples of a control pool for at least 20 independent runs of the assay. Once this number is reached, the mean and standard deviation of the mean values are calculated. These values are used to produce a chart such as is presented in Fig. 6.2. This chart was drawn using the mean values of duplicate measurements of an HIV-1 antibody-positive control obtained from successive runs with an FDA-licensed EIA.

Quality control charts prepared in this way are based on the assumption that data are normally distributed, which may not be the case. With a

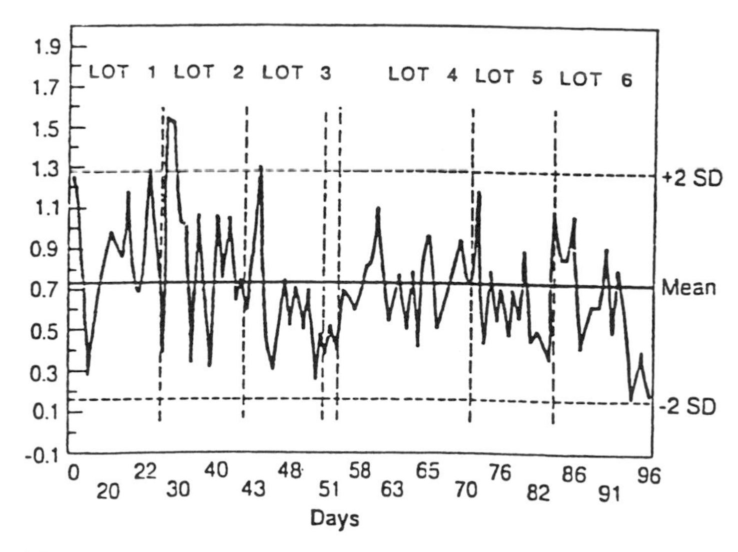

FIGURE 6.2. Quality control chart for an enzyme immunoassay for HIV-1 antibody. NPVs ($n = 39$) from the analysis of the positive control in the same laboratory using different lots of the kit.

Gaussian, or normal, distribution, a fixed portion of the data falls within the interval given by the mean ± any multiple of the standard deviation, termed "s." Fig. 6.3 shows a Gaussian curve and the percentage of observations found within certain distances from the mean. For example, 68.3% of the observations are within ±1.00s of the mean, 95.5% within ±2.00s of the mean, and 99.7% within ±3.00s of the mean. It is common to talk about a 95% control and to estimate it by the mean ±2.00s, even though the correct multiplier would be 1.96 for a 95% limit. The same is true of the 99% control limit, where ±3.00s is often used when ±2.58s is the correct value. Charts prepared as described offer visual evidence of whether new measurements made on standard materials are from the distribution that we have observed previously in the laboratory over many runs. Variability in assay values is expected and inevitable. The analyst uses quality control charts to monitor and record the variability in the analytic process. Quality control charts present a history of the process in a diagrammatic form that is easy to follow.

Interpretation of Control Charts

The common statistical theory is that if a stable pattern continues to exist, in the long run approximately 99 of 100 future-run means will be between the upper and lower 99% control limits (mean ±2.58 standard deviations)

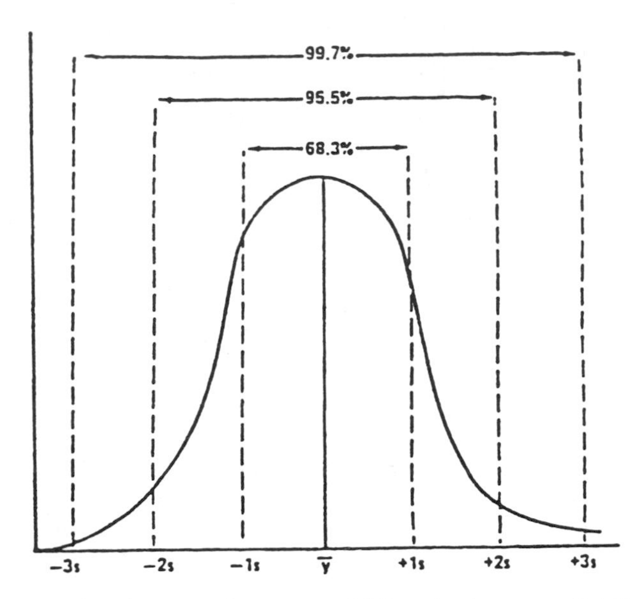

FIGURE 6.3. Fractional area under a normal (Gaussian) sample distribution.

and 19 of 20 will be between the upper and lower 95% control limits (mean ±1.96 standard deviations). This frequency (long run) is of interest, but the operater will be faced with deciding after each run whether the determinations are consistent with a stable pattern. If the mean should fall outside the 99% control level, what course of action should be taken? One must select from two alternatives: (1) a stable pattern exists, and a rare event has just been observed (it would occur 1 of 100 times in the long run); or (2) an unstable pattern has developed, and the occurrence is not just a rare event. Because the odds are against the rare event it would be sensible to check for condition 2. Suppose a value fell outside the 95% control limits. Probability would say that this situation would occur 1 of 20 times when the system is stable. Two successive values, either above or below the 95% control limit, would indicate that the system is out of control, statistically speaking. Upon observing these signals, all samples from these runs should be repeated.

Although the statistical approach seems to satisfy our urge to be quantitative, it can be misleading and difficult to apply to the task of HIV antibody testing. If confidence intervals were calculated for the linear portion of Fig. 6.1, it would be obvious that the relation between antibody titer and absorbance is not linear, and the variability of the assay measured at various points along this curve is different. The variability at the high absorbance value is much greater than that observed at values close to the cutoff. Therefore control materials giving high absorbance values may not reflect the stability of the assay at the cutoff. Furthermore, EIA tests for HIV antibody are not intended to be quantitative. In fact, they are designed to tell you whether a sample is positive or negative for HIV antibody.

Quality control charts can be used in a less rigid manner, without the use of control limits, to monitor the performance of HIV antibody EIAs. Quality control charts that plot the assay values of several controls on the same chart are informative. In Fig. 6.4, the NPV for the in-house low positive control and negative control is plotted simultaneously with the cutoff value of the assay. On the right edge of the chart overlapping runs of new control pools are run with the current controls, indicating that new control materials are being introduced. This chart allows a visual presentation that permits trends in data to be detected early. Some of the sources of systematic error that can be detected are as follows.

1. Tendencies for the lines to move toward lower values could indicate the deterioration of some assay component, such as conjugate or the antigen-coated microplate.
2. Upward trends could be indicative of an unstable chromagen or instrument (e.g., plate reader, incubator).

Even though EIAs exhibit considerable variability (15–20% among-runs), within a given assay all absorbance values should vary in the

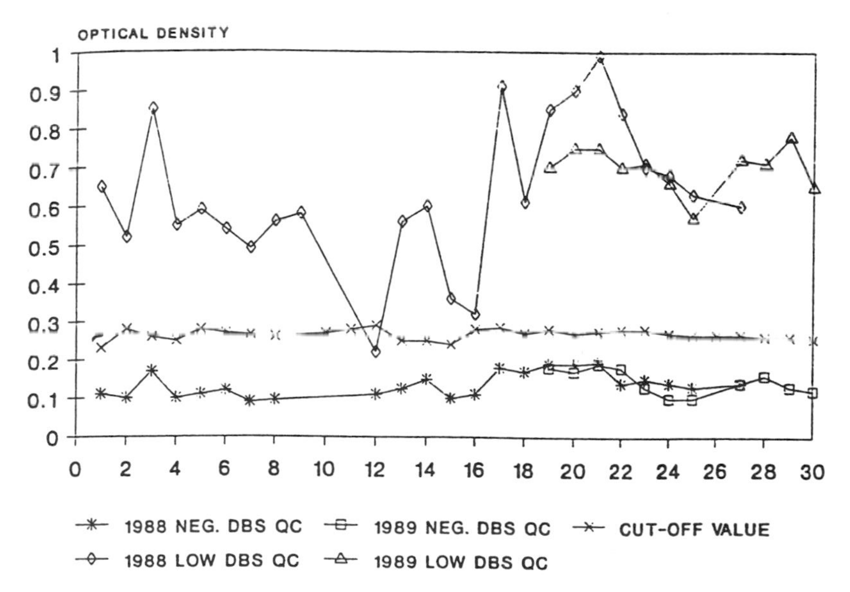

FIGURE 6.4. NPV of the in-house low positive control and kit negative control are plotted together with the positive/negative cutoff values. Overlapping lines show how new lots are calibrated by overlapping runs using old and new lots of control sera.

same direction. If absorbances are lower than normal, all control values including the cutoff should be proportionally low to maintain the sensitivity of the assay. High background for negative samples should correspond to higher absorbances for the negative kit control. When control values vary independently, a random source of error should be suspected (e.g., pipetting, improper mixing of reagents, improper plate washing). Such random errors usually can be eliminated by examining one's technique. Careful pipetting of samples and assay components can dramatically improve the reproducibility of the assay values and tighten the scatter of control points.

Charts that monitor the variability of the kit controls and the in-house controls permit changes in the assay introduced by the manufacturer to be identified. For example, if only kit controls are run and monitored, a shift downward in kit sensitivity could be masked by increasing the potency of kit controls. This problem would be immediately detected by a chart that monitors the in-house and kit positive control (Fig. 6.5). Such a situation would produce a downward shift in values for the in-house control while the kit control values remain at their usual levels.

H.G. Wells, in the essay "Mind at the End of Its Tether" said, "We live in reference to past events." It is unlikely that Wells was thinking of quality control when he wrote these words, but he did describe the principle of quality control charts. These charts allow us to monitor the performance of an assay over time using the same pool of control

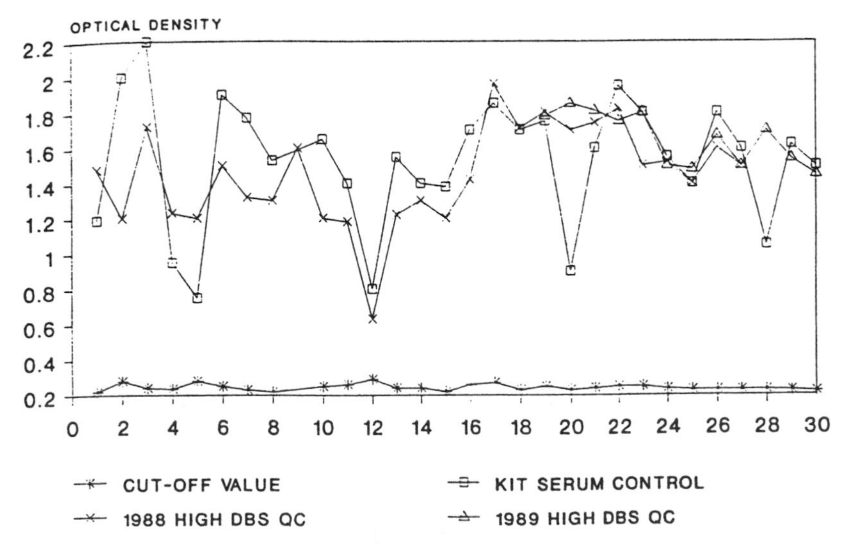

FIGURE 6.5. NPV of the kit positive control sera are plotted with the external positive control.

materials. When the values produced by an assay deviate from the norm established by our past experience, one should look for the source of the variability. Furthermore, within a given population one would soon learn how many positive results to expect for a given run or how many low positive results that do not repeat as positives normally occur within a given plate. Sudden or even gradual shifts in the expected can reflect changes in the population or, more usually, these can be changes in the test system.

Quality Control of the Western Blot Assay

Quality control of the Western blot test for HIV antibody is more difficult than that for the EIA. The major reason for the difficulty is that the quality control recommendations must be a compromise between what is scientifically necessary and what is economically feasible. Commercial Western blots cost between $18 and $45 per test for the FDA-licensed reagents. For each assay run, it is necessary to run a strongly positive, a weakly positive, and a negative control supplied by the manufacturer. For a run of 27 strips (the number of strips in the Biotech/Du Pont/Ortho kit) approximately 10% of the reagents are used for quality control. It is difficult to recommend that additional quality control samples be run because it is doubtful that they would significantly improve the quality of the test results. Occasional use of in-house positive controls may be helpful to demonstrate that the reactivity against various proteins is consistent in intensity and position among various lots of reagents.

It is important, however, that the quality control samples from the manufacturer be carefully examined and the results appropriately analyzed. The strongly positive control is used primarily to demonstrate that the antigen strips contain all of the proteins required to test for HIV infection. It also provides a reference for the position of these bands on the strips. Runs for which critical bands are missing must be rejected. The weakly reactive control is a reference for the intensity. In all licensed kits the bands are assigned a reactivity score based on a comparison with a band on the weakly reactive control. These reference bands should be carefully examined to ensure that their intensity has not significantly increased or decreased for a particular run. Reference bands that are weaker than normal increase the sensitivity of the tests and could cause bands to be scored as reactive that normally would be scored as equivocal. Such occurrence could increase the number of indeterminate results. Reference bands that are stonger than normal decrease the sensitivity of a test. Sera that normally would be positive may become indeterminate. Negative control strips should show *no* bands. The presence of bands on the negative control strip could be due to splash-over from adjacent wells or, more seriously, indicate deterioration of the kit.

Proficiency Testing

Every laboratory performing HIV testing should be enrolled in one or more performance evaluation programs. Such programs are available from the College of American Pathologists, the American Association of Bioanalysts, and the Centers for Disease Control and Prevention. These programs permit laboratories to be tested to see how well they are performing in relation to other laboratories that use the same and different tests. Unfortunately, proficiency testing panels are not particularly challenging and usually detect only gross deficiencies in laboratory performance. Certainly such programs are no substitute for rigorous quality control programs within each laboratory.

Other Quality Control Requirements

Specimen Collection

All assays require that the proper specimen be collected and handled in the prescribed manner. The manufacturers' instructions as contained in the kit brochure should be carefully studied and all recommendations followed. Some important points in specimen collection are as follows.

1. Should plasma or serum be collected? If plasma is to be tested, the anticoagulant could be important for some tests.
2. Do hemolyzed, lipemic, or icteric samples interfere with the tests to be performed?
3. Does the test require fresh samples? Certain assays (Recombigen Latex Agglutination) require samples that have never been frozen. Other tests work perfectly well with frozen samples.

Equipment

Equipment should be subjected to daily monitoring and routine maintenance.

1. The temperature of refrigerators, freezers, waterbaths, and incubators must be checked daily.

2. Microplate readers (spectrophotometers) should be periodically cleaned and calibrated. Linearity of microplate readers should be checked. Microplate readers are probably the most neglected instruments in the laboratory, mainly because the manufacturers do not emphasize the necessity of preventive maintenance and calibration to their customers. However, when asked, most manufacturers can provide protocols and standards for these calibrations.

3. Microplate washers must be subjected to a rigorous daily regimen of cleaning and, when necessary, adjustment. Washers must be flushed and disinfected, and waste materials carefully disinfected and discarded. Washers are especially subject to corrosion due to the high salt content of

the buffers they dispense. The situation is further aggravated by poor design that allows spilled wash solution to reach the internal parts of the machine.

4. Pipettes should be periodically cleaned and calibrated. Even the most careful technologist eventually draws liquid into the pipette. Certain companies provide maintenance and calibration service at the user's laboratory for approximatly $30 per pipette. Each technologist should be carefully instructed in proper pipetting techniques and be required to demonstrate proficiency before being allowed to begin actual testing.

Unreliable HIV antibody test results reported to the physician can be detrimental to the physical and mental well-being of the patient. To recognize inaccurate assay results, therefore, a well conceived and well executed program of quality control is essential. In addition, the laboratorian must also understand the complex physical and immunologic principles involved in the assay. Through this understanding, signals of inaccuracy generated by quality control are recognized, the source of inaccuracy identified, and the appropriate corrective action taken.

Quality Control of PCR Amplification Procedures

Since its introduction in 1985[4,5] the polymerase chain reaction (PCR) has become a mainstay of technology within the field of molecular biology (see Chap. 8 for a detailed description of PCR). The PCR procedure is simple in principle: repetitive cycles of in vitro oligonucleotide primer-directed DNA synthesis. The power of PCR comes from the repeated cycles of target replication, the end result of which is an exponential amplification of a selected genetic target yielding millions of exact copies of a desired nucleic acid sequence. The combined use of PCR amplification and DNA probe detection strategies produces diagnostic assays with exquisite analytic sensitivity. Improvements in PCR technology, such as the use of a thermostable DNA polymerase,[6] introduction of large-sample-capacity thermocycling machines capable of rapidly and accurately attaining target temperatures, and the incorporation of innovative new primer designs with optimized target-specific cycling protocols[7] have facilitated the use of PCR technology in clinical laboratories. However, the ability to detect low levels of organisms through selective amplification of conserved genetic sequences can also lead to problems controlling reaction specificity due to the accumulation of large quantities of specific amplification products. Carryover contamination of PCR reactions with previously amplified reaction products (amplicons) can produce false-positive results for clinically negative specimens. Basic guidelines for reducing the incidence of carryover contamination helped to minimize the potential for false-positive PCR results,[8] but many laboratories still

TABLE 6.1. Considerations for quality assurance of PCR testing in clinical laboratories.

Facility and equipment
Separate pre- and post-PCR work stations
Dedicated equipment for each work station
Positive displacement pipettors or aerosol-resistant pipette tips
Routine preventive maintenance and QC of equipment
Reagents
Synthesize or purchase: multiple small volume aliquots
Consistency of lots (QC)
Personnel
Adequate training for PCR procedures
Control movement between pre- and post PCR areas
Document proficiency and performance of QC procedures
PCR protocols (standard operating procedures)
Optimize reactions, document sensitivity and specificity
Use only low copy number positive controls
Incorporate enzymatic or chemical PCR product inactivation methods
Multiple negative controls and reagent controls
Result interpretation
Analysis of assay positive and negative controls
Documentation of specimen competency (absence of inhibitors)
Reporting algorithm followed

PCR = polymerase chain reaction; QC = quality control.

experienced problems with reaction specificity, making the routine use of PCR testing too risky for most clinical laboratories.

Improvements in PCR technology and associated equipment have significantly changed the way PCR procedures are performed today. Reproducible, user-friendly PCR assays incorporating amplification product inactivation technology for carryover control have been described for use in clinical laboratories. This section is intended as a review of the principles and procedures that should be considered for adequate quality control of "in-house" PCR protocols intended for diagnostic applications. Some of the methods and procedures described in this chapter may not be applicable for laboratories performing PCR testing using commercially available PCR test kits. An outline of the factors to consider in the overall design of a quality assurance program for PCR laboratories is given in Table 6.1.

Facility Design

Basically, the design of a PCR laboratory should focus on segregating the areas of the laboratory where amplified products may be handled from areas where reagents are prepared or specimens are processed. The ideal

PCR laboratory design would include physically separate areas for reagent production, sample preparation, PCR reaction setup (template addition), and post-PCR product analysis. However, few laboratories have either the existing space or the funds to construct such a dedicated facility. With the exception of a dedicated room for preparation of reagents to be used in PCR procedures, establishing dedicated areas within the existing facility for the pre- and post-PCR processes greatly reduces the potential for transfer of amplified products outside of designated product-containing areas (for review see McCreedy and Callaway[9]). Figure 6.6 illustrates a practical arrangement that can be set up in most laboratories. It is desirable, though not essential, that the pre-PCR area not be "downstream" of the post-PCR area with respect to the direction of forced air from the facility's heating and cooling systems. Experiments should always be planned in a unidirectional work flow concept, proceeding from pre-PCR to post-PCR processes in order to minimize the possibility of transferring amplified products from post-PCR areas to areas where specimens are prepared or PCR reactions are set up.

Equipment

All equipment used for routine PCR testing should be dedicated to the area in which the equipment is needed for use. Scheduled maintenance, periodic checks for performance, and calibration (when applicable) should be performed as part of routine quality control. Speeds for centrifuges (actual revolutions per minute) should be verified against the displayed value or speed setting of the instrument. Temperatures of heating blocks

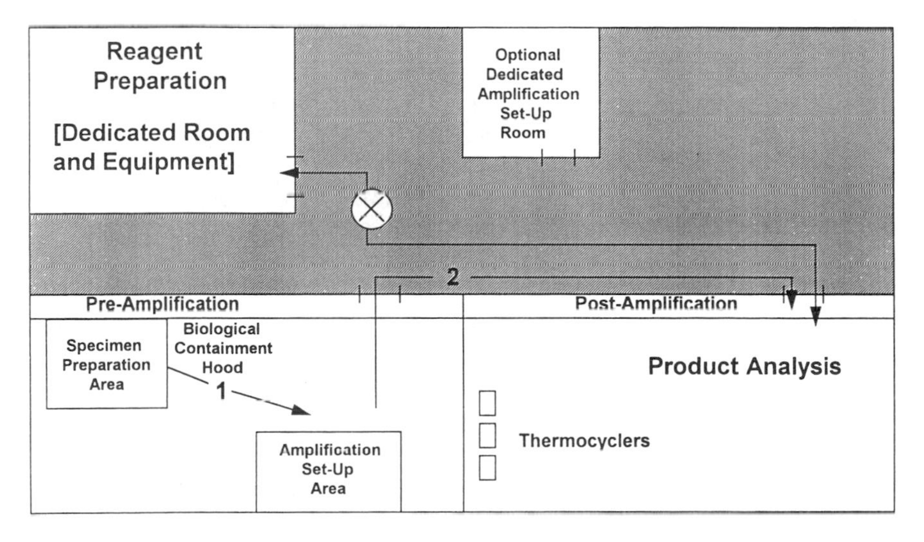

FIGURE 6.6. PCR laboratory setup.

and waterbaths should be checked daily using a thermometer standardized against a National Bureau of Standards thermometer. When heating blocks are used instead of waterbaths, it is a good idea to fill the wells of the heat block with silica sand, which distributes heat evenly when specimen tubes of various sizes are placed into the wells containing the sand. It is imperative that the performance of thermocyclers by verified. Fluctuation in temperature set points, ramp times, and cycling times can significantly alter the reproducibility and accuracy of PCR reaction protocols. In addition to the internal diagnostic routines that should be periodically run according to the manufacturer's instructions, actual well temperatures should be checked against a set temperature using a thermocouple device; a record of the results must be kept as part of equipment quality control. Thermocouple[6] and temperature verification routines are provided by some manufacturers of temperature cyclers. It is also possible to accurately measure and record well temperatures relatively inexpensively using a type K thermocouple probe, minivolt converter, and chart recorder.[10] A simple check that should be performed each time a PCR cycling protocol is used is to note the beginning and ending time of each run. The actual run time is derived by subtracting the elapsed time shown on the thermocycler display (time elapsed during a programmed temperature hold cycle, which begins after completion of the last thermocycle) from the value obtained by subtracting the run start time from the time the specimens are removed. Generally, the run times should not vary more than a few minutes from run to run. Observed increases in run times represent a warning signal that adjustments to the machine are necessary to restore it to the proper operating condition.

Pipettors and pipette tips are a common source of errors and contamination in PCR testing laboratories. Pipettors should be decontaminated after use by a wipe with 10% bleach followed by a wipe with 70% ethanol. Calibration of pipettors should be performed on a periodic basis (e.g., quarterly or semiannually) as part of routine preventive maintenance. Calibration of micropipettors can by acheived by gravimetric methods (i.e., weight of distilled water) when an analytical balance is available or by pipetting radioactive solutions of defined specific activity. Pipette tips are an important component of both accuracy and prevention of carryover contamination. Pipette tips containing aerosol-resistant barriers are available from several commercial suppliers. It is important to verify the quality and performance of a selected brand of plugged tips using the micropipettors with which the tips are to be used. A truly airtight (aerosol-resistant) tip can be tested by pipetting a colored liquid containing 2% glycerol. Intentionally set the pipettor to deliver at least 10% more than the indicated maximum volume of the pipette tip and then pipette the solution. If the colored liquid rises above the plug inside the tip, the barrier is not airtight. Hydrophobic membranes bonded to the tip during manufacturing, in contrast to a plug that is inserted into the tip after molding, have proved to be the most reliable aerosol-resistant tips in our laboratory. A second, more reliable test that should be performed to document aerosol resistance of a pipette tip is to pipette a high copy number positive control (>1000 copies) followed by multiple negative controls.[11]

1. Pipette the positive control specimen up and down 10 times using the same pipette tip to simulate pipetting of successive positive specimens.
2. Add the required volume of a positive specimen to an amplification reaction.
3. Discard the used pipette tip and replace it with a fresh tip.
4. Using the same or fresh tips, pipette 10 negative control specimens consecutively into separate amplification reactions.
5. Amplify and detect as per the standard PCR procedure used in the laboratory.

The positive specimen should yield a clearly detectable signal following amplification and detection. None of the negative reactions should yield a detectable amplification product if the pipette tip is truly aerosol-resistant and therefore prevents conatmination of the bore of the micropipettor.

Decontamination of equipment and work surfaces can be accomplished by wiping or swabbing with 10% bleach followed by 70% ethanol to remove the residual bleach. Some equipment, such as tube racks and removable plastic rotors for microfuges, can be soaked in 10% bleach for 30 minutes and then rinsed with distilled water and dried after each day's use. Cross-linking of amplified products using ultraviolet (UV) irradiation is an alternative method for decontaminating bleach sensitive equipment.[12] Biologic containment hoods and bench-top enclosures should be equipped with UV lights that are left on for a minimum of 1 hour after surface decontamination with bleach and ethanol.

Personnel and Technique

Other than the factors to be considered for satisfying the regulations put forth in CLIA 1988 (see Chap. 22 for a description of CLIA), there are no specific educational requirements for personnel who work in PCR testing facilities. A background in biochemistry and molecular biology is helpful; but for routine testing with optimized PCR procedures, high quality results can be reproducibly generated by any appropriately trained individual with a medical technologist or college level science background. The key to a successful training program is a well thought out and well written standard operating procedure manual, which trainees can follow as they learn with hands-on supervision.

The most important factors for PCR testing personnel to consider are modification of their behavior with regard to movement within the PCR testing areas and learning the method by which specimens and controls are handled. The unidirectional work flow concept described earlier in this chapter must be understood and followed by all individuals who work in the PCR testing areas. The daily work schedule should begin in the pre-PCR area and proceed to the post-PCR area throughout the day, with minimum movement if any in the reverse direction. Separate laboratory coats should be worn in the pre- and post-PCR areas and should be removed before leaving to move into the next area. Specimens must be carefully handled to avoid contamination of gloves and subsequent cross contamination of other specimens. The use of 1 × 1 inch squares of gauze to remove and recap tubes containing specimens and controls is useful to avoid contamination of glove fingertips. Whenever a gloved hand is suspected or observed to have been contaminated with a specimen or control, it should immediately be discarded and a new glove put on before proceeding. Pipetting should be done with care to avoid generation of aerosols and contamination of pipettors. Reagents, especially those requiring addition of enzymes in solutions containing glycerol or control plasmid DNAs, should be thoroughly mixed and carefully pipetted to ensure uniform distribution while avoiding cross contamination. During amplification setup, specimen tubes should be capped before the next specimen is pipetted. When strips of caps are used, such as with the Gene Amp 9600 system (Perkin-Elmer Corporation), specimens should be pipetted individually into each row and capped before adding specmens to the next row of tubes. Equipment should be cleaned and decontaminated after each use as part of a daily routine set forth in the laboratory's standard operating procedure manual. Each member of the PCR testing laboratory must understand that their individual work habits affect all other members of the group. Strict adherence to operating procedures go a long way toward ensuring success of the PCR procedures used in the laboratory.

Reagent Preparation

Quality assurance for each PCR test procedure begins with quality control of the reagents used during the performance of the procedure. The decision as to whether to synthesize reagent components in house (e.g., primers, probes) and mix working stocks of reagents (e.g., reaction mixes, hybridization buffers, lysing solutions) rests with the laboratory and is determined by volume and cost constraints. Regardless of the source (in-house or commercial vendor) all reagents used during the testing process should be tested to ensure potency (sensitivity) and purity (specificity). A separate laboratory within the facility dedicated to the production, aliquotting, and storage of bulk reagents is a must for laboratories performing routine PCR testing using in-house reagents (noncommercial kit-based testing). Certificates of analysis should be obtained from commercial suppliers of chemical components to document composition and purity. Chemical components made in house, such as primers and probes, should be carefully checked for purity and potency. Dedicated high-performance liquid chromatography (HPLC) columns are recommended for purification of individual primer and probe sequences, and chart recordings should be carefully checked for the presence of failure sequences or nonconjugated species in the case of primers or probes containing attached ligands. Absorbance (260/280 nm) ratios between 1.6 and 1.9 are acceptable as estimates of concentration and purity following synthesis of oligonucleotides. Oligonucleotides can also be analyzed in polyacrylamide gels to assess purity.

The preparation of master mixes for amplification reactions is encouraged whenever possible. The use of master mixes containing as many of the individual reaction components as possible (e.g., without *taq* or without Mg^{2+}) minimizes the number of pipetting steps required compared with individually constructed reaction mixes. Correspondingly, the potential for introducing contaminants is reduced, and the observed interassay variability attributable to the amplification reaction itself is improved. A reagent production worksheet (Fig. 6.7) should be followed during the production of each batch of a reagent. The name, source (including lot number), and quantity required for each component of the reagent should be listed on the production worksheet. By following the worksheet, adding components in the order listed, and initialing in the space provided, a documented record of the construction of each reagent batch is created. Both the technician who prepared the reagent and the reagent laboratory supervisor or director should sign the document, providing a permanent record of the production procedure used to construct the final reagent batch. Following the steps listed in the reagent production worksheet each time a batch of a particular reagent is made helps to ensure that consistent performance is observed among batches of a reagent. Any change in component source or lot number is reflected in the production record in the event a batch of a reagent fails to pass quality control. Provided that consistent performance is maintained, with few exceptions (e.g., certain enzymes, primers, probes, dNTPs) the source of chemical components is not critical. Reagents should be dispensed and stored in small-volume aliquots. In general, the smallest-volume aliquot that is cost effective, as determined by testing volume (e.g., batch size), should be used to minimize the potential for reagent contamination caused by repeated use. Some reagents, such as PCR reaction mixes, should always be dispensed in single-use aliquots.

The quality control of reagent batches should be performed by PCR laboratory testing personnel (not reagent laboratory personnel) using aliquots of the master lot. Quality control testing of reagents should consist in standardized experiments designed to ensure consistent performance at an established level of sensitivity and specificity. Experiments must be specifically designed to assay for the function(s) for which the

<table>
<tr><td colspan="2">**PCR MASTER MIX: HIV ASSAY**</td></tr>
<tr><td colspan="2">DATE MADE: MM-DD-YY LOT#: YYMMDD</td></tr>
<tr><td colspan="2">Total Volume: 100 ml Total Reactions: 2000
Volume/Reaction 0.050 ml</td></tr>
<tr><td colspan="2">COMPOSITION: 25.0 pmol each primer
and 80 nMol dNTP's per reaction</td></tr>
<tr><td colspan="2">Primer #1: BDB78 at 109.0 pmol/ul
Primer #2: BDB79 at 167.0 pmol/ul</td></tr>
<tr><td>20.000 ml 10X Taq Buffer</td><td>Lot#: 931217</td></tr>
<tr><td>0.459 ml Primer #1: BDB78</td><td>Lot#: 940102</td></tr>
<tr><td>0.299 ml Primer #2: BDB79</td><td>Lot#: 940102</td></tr>
<tr><td>1.600 ml 100 mM dNTPs</td><td>Lot#: 931215</td></tr>
<tr><td>1.000 ml Taq Polymerase 0.5 ul/rxn</td><td>Lot#: Perkin Elmer 6789</td></tr>
<tr><td>76.642 ml ddH2O</td><td>Lot#: Sigma 43H9407</td></tr>
<tr><td colspan="2">100.0 ml Total Volume of Batch</td></tr>
<tr><td colspan="2">Aliquot size: 1.0 ml Reactions per Aliquot: 20</td></tr>
<tr><td colspan="2">COMMENTS:</td></tr>
<tr><td colspan="2">Made By: On:</td></tr>
<tr><td colspan="2">QC Completed By: PASS FAIL
On:</td></tr>
<tr><td colspan="2">Reviewed & Released By: On:</td></tr>
</table>

FIGURE 6.7. Reagent production worksheet.

reagent is intended. For example, simply testing a cell lysis solution to be certain that it is not contaminated with target nucleic acid does not ensure that the reagent will sufficiently lyse a given number of cells and provide amplifiable nucleic acid. Consistent performance of a new lot of a reagent can be ensured by testing low copy number (<50 copies) positive controls and multiple negative controls in the same experiment with the currently proved lot of reagent tested in parallel in order to provide reference values. It is important to use multiple negative control reactions to detect any low level of contaminating target DNA, which may not be consistently detected owing to inherent sampling bias introduced when testing aliquots of a larger volume. Once consistent performance has been documented by detection of all positive controls and the absence of detectable products from the negative control reactions, the quality control test results should be provided to the reagent laboratory director for final review and release of the reagent lot.

Amplification Parameters

A well designed PCR protocol begins with an optimized specimen preparation, amplification, and detection procedure. A discussion of the variables that must be considered, such as lysis and digestion buffers, primer and probe design, thermocycling parameters, product detection, and inactivation methods, are beyond the scope of this chapter. A number of useful reviews that describe the variables to consider when designing and optimizing PCR reactions have been published.[13,14] However, the issue as to which and how many primer pairs are necessary to provide an acceptable level of sensitivity and specificity warrants some mention here. It is prudent to examine multiple candidate primer pairs using clean (i.e., purified plasmid or genomic) targets to determine their analytic sensitivity and specificity. Once candidate primer pairs have been selected, one must compare the data obtained after testing a variety of well characterized clinical specimens to determine their clinical sensitivity and specificity. Once a database has been created for the cadidate primer pairs, one can compare the data and determine which pair, or pairs, will provide the most accurate detection. Often mistakes are made when trying to determine which primer pairs provide the most reliable detection of a given organism from a variety of clinical specimen sources. Too often a decision regarding the need for a second or third primer pair is made based on comparison of data from nonequivalent amplification systems. Some studies have shown that discordant results observed between alternate primer pairs designed to amplify the same target do not necessarily indicate that sequence variability or mutation was responsible for the discrepancy.[15] Data from PCR assays that may not have equivalent analytic sensitivities, unoptimized specimen preparation, amplification, or detection procedures, differences in laboratory expertise, and factors including sample bias introduced when low copy number specimens are used for comparison can contribute to a lack of concordance when comparing PCR amplification systems that employ different primer pairs. Careful consideration when selecting target regions to amplify and innovative design of primers capable of tolerating mismatched bases can often eliminate the need for inclusion of a second primer set.[5] A well designed concordance study with paired (or split) specimens, including comparison with alternate laboratory test results and supporting clinical data, is the best way to determine concordance among primer pairs and whether more than one primer pair is necessary to achieve an acceptable level of sensitivity and specificity.[16,17]

Assay Controls

Even a well designed PCR procedure is only as good as the way in which it is performed, including the design and use of assay controls. A problem often observed but seldom given adequate attention is the occurrence of false-negative results due to PCR inhibitors present in the processed specimen. An important part of an optimized PCR protocol is a robust specimen preparation procedure capable of efficiently liberating sufficient amounts of amplifiable *target* DNA free of inhibitory substances. A number of substances in biologic specimens can interfere with the amplification reaction, resulting in partial or total inhibition of target-specific amplification. Substances such as heme and other porphyrin-containing compounds inhibit amplification reactions by chelating divalent ions (Mg^{2+} or Mn^{2+}) and directly interfering with template recognition by Taq or *T. thermophilus* DNA polymerase. A specimen competence control should be included in addition to the assay positive and negative controls as a check to ensure that processed specimens are not grossly inhibitory to the amplification reaction due to the failure of the specimen preparation procedure to remove PCR inhibitors. For some specimen types, such as human blood, an intrinsic control can serve this purpose. Control reactions involving amplification of the human β-globin or HLA DQα sequence can serve to determine if significant levels of inhibitory substances are present in the prepared specimen.[18] For some specimens (e.g., cerebrospinal fluid and plasma, which do not normally contain significant levels of cellular DNA), a spiked control can be added to an aliquot of the specimen and amplified. The spiked control may consist of a low copy number (<50 copies) of the assay positive control DNA, which is added in a parallel amplification reaction and amplified along with the unspiked specimen reactions. Unlike qualitative PCR procedures, which should be monitored for significant levels of inhibitors to guard against false-negative results, quantitative PCR procedures require internal controls with amplification characteristics that closely resemble the natural target in order to monitor target-specific amplification reaction efficiency accurately. Quantitative standards (internal controls) containing the target nucleic acid primer binding regions but modified probe recognition sites or target primer binding sites but different intervening sequences have been described.[19,20] When added to the unprocessed specimen, recovered and amplified along with the natural target, this type of internal control provides an accurate measure of the amount of nucleic acid recovered during the preparation procedure and serves to monitor target-specific amplification reaction efficiency accurately. By comparing the amount of amplified target-specific product with the amount of product generated from the known input amount of the quantitation standard, changes in reaction efficiency can be measured and adjustments included in order to accurately determine the amount of natural target nucleic acid present in the specimen.

In addition to a specimen competence control, a low copy number positive control should be included and processed along with specimens. In addition, a negative control consisting of a known negative specimen or a "mock" reaction tube in which only the specimen preparation reagents have been added (no nucleic acid) should be included during the specimen preparation procedure to control for specificity (i.e., introduction of contaminants) for this part of the PCR protocol. It is acceptable to include a single positive and negative control for each batch of specimens to be processed. It is a good idea to include additional negative controls for preparation of particularly large batches of specimens (>30 specimens).

After specimen preparation, the processed specimens and controls can be transferred to the PCR setup area. Reaction mixes, complete with added polymerase enzyme and so on can be added to the reaction tubes followed by careful pipetting of the processed specimens and controls using aerosol-resistant pipette tips and capping of individual tubes after each addition, as described earlier. An amplification negative control consisting of amplification reaction mix and glass-distilled water or specimen resuspension buffer instead of control nucleic acid should be pipetted last, immediately following the addition of the assay positive control. Pipetting the amplification negative control last controls for any contaminating nucleic acid introduced during the PCR setup part of the protocol and maximizes the overall ability to detect carryover contamination for the batch of specimens to be amplified.

Carryover Prevention

Several methods of carryover prevention based on modification of the amplified PCR product (amplicon) have been described. Enzymatic restriction of uracil-containing amplicons using uracil-*N*-glycosylase,[21] photochemical inactivation of PCR products containing intercalated isopsoralen compounds,[22] and covalent modification of amplified products using hydroxylamine[23] have all been used successfully to reduce significantly the potential for carryover contamination by drastically reducing the amount of product that is able to serve as template DNA for reamplification. The enzymatic method involves substituting dUTP for dTTP in the amplification reaction mixture, such that U is substituted for T by Taq DNA polymerase during the primer extension (elongation) cycle of amplification. The resulting amplicons contain deoxyuracil residues instead of the naturally occurring deoxythmidine, thereby distinguishing the amplified product from native template DNA (i.e., specimens and controls), which is devoid of deoxyuracil (Fig. 6.8). Uracil-*N*-glycosylase (UNG) removes uracil residues from a deoxyribonucleic acid chain, an enzymatic function first observed among many bacterial species as a component of DNA excision repair. By adding UNG to the amplification reaction mix before amplification of specimens and controls, any contaminating carryover amplicons present in the reaction tube are subject to restriction with UNG. The uracil residues within the carryover amplicons are removed by UNG, and the resulting abasic sites are hydrolyzed under the buffer conditions and elevated temperature optimal for Taq polymerase activity during amplification. The hydrolyzed amplicon frag-

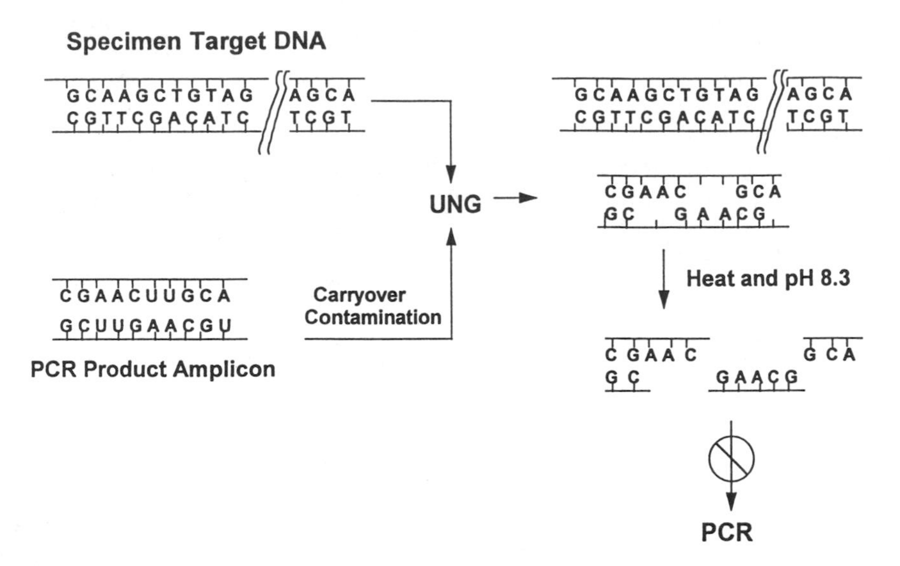

FIGURE 6.8. PCR carryover control with UNG.

ments do not serve as efficient templates for reamplification and thus do not contribute to positive signals for the specimens or controls. The naturally occurring template DNA from the specimens and controls is unaffected by UNG. Once the temperature has been raised above 60°C during the initial cycle of PCR amplification, the activity of the thermolabile UNG enzyme is destroyed, allowing newly synthesized uracil-containing amplified products to accumulate without being degraded. Therefore inclusion of dUTP and UNG enzyme in amplification reactions drastically reduces the potential for generating false-positive results by inactivating preexisting carryover amplicons and modifying newly synthesized product DNA amplicons so they can be degraded if carried over into subsequent PCR amplification reactions. Using UNG restriction, inactivation of 10^6 to 10^7 uracil-containing amplicons (aerosol range of carryover) is consistently achievable.

One important note to remember for users of dUTP and UNG is that UNG activity is restored upon cooling the reaction to below 55°C. To avoid degrading newly synthesized products prior to analysis, the reactions must either be frozen at −70°C, kept above 55°C, or the UNG activity permanently destroyed by adding NaOH (0.1–0.2 N final concentration). A separately recognized advantage of the UNG system of carryover control is that inclusion of UNG also improves amplification reaction sensitivity by degrading nonspecific extension products that result from low stringency primer binding, which occurs at suboptimal primer annealing temperatures (i.e., prior to the first cycle of PCR). In this manner, reaction sensitivity may be improved owing to the amplification and accumulation of only the desired target sequence, similar to the results observed using "hot start" PCR procedures.[24]

Each PCR procedure must be carefully optimized to determine the optimum amount of dUTP to include in the reaction mixture and the amount of UNG to include in each reaction. Some amplicons, such as those generated from G + C rich targets, may not be efficiently inactivated by UNG, and other product inactivation methods, such as those described above, may have to be considered.

Product Detection

The final step in the PCR test procedure is detection of the amplified products. Various product detection strategies have been used with success in different laboratories. Radioactive, chemiluminescent, and enzyme-labeled DNA probes are commonly employed to provide positive identification of amplified target sequences. DNA probe-based detection methods are recommended for diagnostic applications because they provide an additional level of specificity (if properly designed) when hybridized to the specific amplified target. Other product detection methods (e.g., restriction cleavage, visualization of proper size DNA fragments in stained gels, and nested PCR reactions) are also used but do not reliably provide the same level of sensitivity and specificity. Although nested PCR amplification can produce sensitive detection of low copy number target sequences, it is not recommended for clinical laboratory use. The potential to generate false-positive results is much greater, and product inactivation methods cannot be used because the first amplified product must serve as template DNA for the nested primer set.

Quality control for the postamplification process, like the pre-PCR procedures, must include daily decontamination and periodic scheduled preventive maintenance of all equipment used for product detection. Amplified products should be disposed of into 10% bleach or inactivated with UV irradiation or UV light depending on the system used by the laboratory.

Result reporting should follow a defined, unambiguous, reporting algorithm as outlined in the laboratory standard operating procedure manual. White et al.[25] provided useful suggestions for PCR clinical laboratory operating procedures. Assay positive (sensitivity) and negative (specificity) control reactions should be carefully reviewed to ensure that proper probe hybridization conditions have been established. Negative or weak reactions observed for any positive control should require that the run be repeated. Similarly, positive signals observed for any negative controls should require repeat testing of all specimens and controls in the batch. Repeatedly positive results for negative controls indicates that a reagent may be contaminated with amplified DNA products and necessitates that the reagents be checked before additional amplification reactions are performed. Any reagent found to be contaminated should be discarded and other aliqouts of the reagent batch checked for contamination. If the assay's positive and negative controls give the expected values, the specimen competence controls should be interpreted for each specimen. When it is determined that a prepared specimen is not inhibitory and all assay controls give expected values, a final result can be determined for each specimen. Specimens that yield discordant results should be retested. If results remain discordant, a second specimen should be requested.

Conclusion

Improvements in PCR technology and equipment have made implementation of PCR testing procedures with diagnostic applications feasible for most clinical laboratories. Facility design, personnel training, reagent production, testing procedure, and reporting algorithm must be considered during the development of a comprehensive quality assurance program for routine PCR testing. With modification of some laboratory procedures and strict adherence to the laboratory's standard operating procedures, a successful PCR testing program should not be beyond the reach of virtually all clinical testing laboratories.

References

1. Westgard JO, Barry PL. Washington, DC: American Association for Clinical Chemistry (AACC) Press, 1986:17
2. Kudlac J, Hanan S, McKee GI. Development of quality control procedures for the human immunodeficiency virus type 1 antibody enzyme-linked immunosorbent assay. J Clin Microbiol 1989;27:1303–1306
3. George JR, Palmer DF, Cavallaro JJ, Wagner WM. Principles of Radioimmunoassay. Immunology Series No. 12. Washington, DC: Public Health Service, US Department of Health and Human Services, 1984
4. Saiki RK, Scharf S, Faloona F, et al. Enzymatic amplification of β-globin genomic sequences and restriction site analysis for the diagnosis of sickle cell anemia. Science 1985;230:1350–1354
5. Mullis KB, Faloona F. Specific synthesis of DNA in vitro via a polymerase-catalyzed chain reaction. Methods Enzymol 1987;155:335–350
6. Saiki RK, Gelfand DH, Stoffel S, et al. Primer-directed enzymatic amplification of DNA with a thermostable DNA polymerase. Science 1988;239:487–491
7. Kwok S, Kellog DE, McKinney N, et al. Effects of primer-template mismatches on the polymerase chain reaction: human immunodeficiency virus type I model studies. Nucleic Acids Res 1990;18:999–1005
8. Kwok S, Higuchi R. Avoiding false positives with PCR. Nature 1989;339:237–238

9. McCreedy BJ, Callaway TH. Laboratory design and workflow. In Persing D, Smith T, Tenover F, White T (eds): Diagnostic Molecular Microbiology: Principles and Applications. Washington, DC: American Society for Microbiology, 1993:149–159
10. Louie P, Dela Rosa C, Madej R, Rodgers G. Evaluating the proficiency of PCR systems for clinical diagnostic testing [abstract] Clin Chem 1991;37:1059
11. Dragon AE, Spadoro JP, Madej R. Quality control of polymerase chain reaction. In Persing D, Smith T, Tenover F, White T (eds): Diagnostic Molecular Microbiology: Principles and Applications. Washington, DC: American Society for Microbiology, 1993:160–168
12. Ou C-Y, Moore JL, Schochetman G. Use of UV irradiation to reduce false positivity in polymerase chain reaction. Biotechniques 1991;10:442–446
13. Innis MA, Gelfand D-H. Optimization of PCR's. In Gelfand D, Sninsky J, White T (eds): PCR Protocols: A Guide to Methods and Applications. San Diego: Academic Press, 1990:3–12
14. Persing DH. Target selection and optimization of amplification reactions. In Persing D, Smith T, Tenover F, White T (eds): Diagnostic Molecular Microbiology: Principles and Application. Washington, DC: American Society for Microbiology, 1993:149–159
15. Schochetman G, Sninsky JJ. Direction detection of human immunodeficiency virus infection using the polymerase chain reaction. In Schochetman G, George JR (eds): AIDS Testing: Methodology and Management Issues. New York: Springer-Verlag, 1992:90–110
16. Butcher A, Salter L, Kinard S, et al. Evaluation of a single primer pair for the detection of HIV-1 in clinical specimens by PCR amplification: comparison of a non-radioactive microtiter plate assay with solution hybridization [abstract T-5]. In: Abstracts of the 92nd General Meeting of the American Society of Microbiology. Washington, DC: American Society for Microbiology, 1992:417
17. Lynch CE, Madej R, Louie P, Rodgers G. Detection of HIV-1 DNA by PCR: evaluation of primer pair concordance and sensitivity of a single primer pair. J Acquir Immune Defic Syndr 1992;5:433–440
18. Saiki RK, Bugawan TL, Horn GT, Mullis KB, Erlich HA. Analysis of enzymatically amplified β-globin and HLA DQ alpha with allele-specific oligonucleotide probes. Nature 1986;324:163–166
19. Mulder J, McKinney N, Christopherson C, et al. A rapid and simple PCR assay for quantitation for HIV RNA: application to acute retroviral infection J Clin Microsc (in press)
20. Vanden Heuvel JP, Tyson FL, Bell DA. Construction of recombinant RNA templates for use as internal standards in quantitative RT-PCR. Biotechniques 1993;14:395–398
21. Longo MC, Berninger MC, Hartley JL. Use of uracil DNA glycosylase to control carry-over contamination in polymerase chain reactions. Gene 1990; 93:125–128
22. Gimmino GD, Metchette KC, Tessman JW, Hearst JE, Isaacs ST. Post-PCR sterilization: a method to control carryover contamination for the polymerase chain reaction. Nucleic Acids Res 1991;19:99–107
23. Aslanzadeh J. Application of hydroxylamine hydrochloride for post-PCR sterilization. Ann Clin Lab Sci 1992;22:280

24. Chou Q, Russell M, Birch DE, Raymond J, Bloch W. Prevention of pre-PCR mispriming and primer dimerization improves low-copy-number amplifications. Nucleic Acids Res 1992;29:1717–1723
25. White TJ, Madej R, Persing D. The polymerase chain reaction: clinical applications. Adv Clin Chem 1991;29:161–196

7
HIV Culture

MARK A. RAYFIELD

Although serologic assays are capable of identifying prior exposure to human immunodeficiency virus (HIV), they cannot alone demonstrate whether an individual is currently harboring the virus. The first method used to ascertain if a blood specimen contained HIV was co-cultivation with stimulated primary human lymphocytes or continuous human T cell lines and monitoring the culture supernatants for the presence of reverse transcriptase. Although virus isolation has proved to be a poor diagnostic tool because of its relative insensitivity, high costs, and lengthy time requirements, culture has served as the standard by which all other diagnostic tests have been judged and established. Furthermore, virus culture remains the steadfast route by which new variants are identified, isolated, and initially characterized. The emphasis of this chapter is on delineation of the impact of a number of related fields of study—including epidemiology, applied serology, classic viral morphology, and molecular virology—on the culture and characterization of HIV specimens. In doing so we point out both advantages and disadvantage of the current approaches to virus isolation.

Evaluation and Selection of Culture Specimens

It is generally appreciated that the success rate of clinical cultures depends on identification of individuals appropriate for culture as well as the timing and method of sampling. In course each of these elements determines the quality of the specimen and thus the likelihood of success. When screening and selecting candidates on the basis of serologic data, it is important to remember that the positive predictive value (number of true positive tests/total number of positive tests) of the assay is proportional to the true prevalence of infection within the test population and as such is decreased among low prevalence populations. For example, the persistence of indeterminate or anomalous serologic findings in an individual at increased risk of HIV infection would warrant further analysis of virologic markers. Conversely, within a low prevalence population, similar findings are not singularly suggestive of HIV infection.[1,2] Furthermore, consideration should be given to questions related to the timing and mode of sampling. In the absence of overt evidence of increased viral replication, the choice should be tissue or fluid likely to contain cells that are actively expressing infectious virus. In perinatal studies these issues are compounded by reduced sample volumes and the difficulty of distinguishing residual maternal immunity from the infant's nascent serologic response. On such occasions direct detection by amplification methods such as the polymerase chain reaction (PCR) often best serve the immediate diagnostic needs for virologic markers.[3] Frank virus isolation is not always necessary for diagnostic purposes, but the choice of detection methods need not be mutually exclusive.

There is a strong correlation between the number of circulating cells carrying proviral DNA copies of HIV, which indicates infectious burden, and the frequency of isolation. In situ hybridization analysis of blood and lymphatic tissues indicates that on the order of only 1 in 10,000 to 1 in 100,000 peripheral blood mononuclear cells (PMBCs) are actively replicating virus.[4] Estimates of the number of PBMCs that might serve as infectious centers have also been derived from limiting dilution culture studies in vitro, using methods that detect active HIV replication such as p24 antigen capture, the reverse transcriptase assay, or RNA specific probing.[5–8] As the virus burden rises from 1 proviral DNA copy in 3000 PBMCs to 1 in 1000 or more, the frequency of isolation increases, generally following the rate of progression to Centers for Disease Control and Prevention (CDC) stage IV criteria within the patient.[8,9] This is not to say that culture is absolutely linked to disease progression, because successful isolations may result from PBMCs of asymptomatic, CDC stage II individuals.[10] The critical measure remains the number of cells carrying HIV proviral DNA copies that are replicating or that can be induced to replicate HIV as infectious centers in the culture inoculum. The cumulative findings from these studies indicate that not only does the number of HIV-infected PBMCs increase during disease progression, the percentage of these cells actively producing virus also rises sharply. These results provide a graphic explanation for the strong correlation between virus isolation rates and prognostic indicators of the clinical disease stages of acquired immunodeficiency syndrome (AIDS).

Culture and Virus Isolation

Reported rates of virus isolation from the blood of HIV-1 antibody-positive persons range from 65% to 100%.[11,12] HIV has also been occasionally isolated from lymphoid tissues, plasma, saliva, semen, tears, brain tissue, cerebrospinal fluid (CSF), breast milk, and urine.[13–17] Although HIV has been documented in saliva, tears, and urine, these findings are rare; and there is no epidemiologic evidence that contact with these secretions has resulted in infection. The more frequent recovery of HIV from the CSF of AIDS patients with neurologic disorders was predictive of a causal role.[18] Furthermore, the detection of viral antigens or isolation of HIV from plasma prior to complete seroconversion was initially the only means of demonstrating infection before the development of PCR technology. Therefore, although there are occasions when it may be valuable to attempt isolation from such fluids, the high isolation rates from PBMCs promote the use of whole blood as the principal source of diagnostic material.

Blood should be collected in the presence of an anticoagulant, usually heparin, sodium citrate, or EDTA, then stored at room temperature until the cells are separated. An effort should be made to collect 10 to 20 ml of blood from adults and 2 to 5 ml from infants or small children. A partially purified PBMC preparation, enriched for lymphocytes and adherent mononuclear cells, may be obtained by a single centrifugation through Ficol-Hypaque or similar gradient for lymphocyte separation. The resulting mixed population of cells is ideal for co-cultivation and virus isolation. Theoretically, it is best to culture the cells within a few hours of collection, but in practice there is little loss of infectivity in whole blood held as long as 24 hours before processing. The principal exception to this practice is the quantification of viral burden in infectious center assays, which require immediate sample preparation and co-cultivation. If virus cultivation is to be delayed longer than 24 hours, the PBMCs should be separated and the cells frozen. Cryogenic procedures that yield greater than 80% viability result in minimal loss of infectivity, and HIV may be isolated from specimens properly stored in this manner for periods in excess of 5 years.

Protocols for the isolation of HIV by co-culture with primary PBMCs generally require mitogen-stimulated normal donor cells as targets, cultivated at a concentration of 2×10^6 cells/ml. Donor PBMCs are separated by centrifugation over a Ficoll gradient; they are then precultured at 37°C for 72 hours in RPMI 1640 (GIBCO) media supplemented with 20 mM L-glutamine, 20% heat-inactivated fetal calf serum, phytohemagglutinin (PHA) 5 µg/ml, and antibiotics as needed. After activation the PHA is removed and media supplemented with anti-human interferon 100 to 200 U/ml, and interleukin-2 150 U/ml is added to the culture. PBMCs from the diagnostic specimen are added in an inoculum of 10^6 to 10^7 cells and the culture supplemented with fresh stimulated normal donor PBMCs at 3-day intervals. Optimal specimen/target cell ratios range from 1:1 to 3:1. Given these cell concentrations, comparable isolation rates may be obtained in culture volumes of 2 to 20 ml. Isolation rates may vary in culture volumes of less than 2 ml because of the reduced sample size.

Material withdrawn at the time of feeding is usually analyzed by the p24 antigen-capture assay, reverse transcriptase assay, or PCR for the presence of HIV. Repeatedly positive cultures with a rising titer in quantitative assays are considered indicative of infection. Cultures that remain negative are maintained and monitored for at least 4 weeks (Fig. 7.1).

These mixed cultures of immunocompetent cells promote the presentation of a variety of stimuli, including viral antigens as well as heterologous HLA and cell-surface markers, which in turn trigger a complex cascade of immune modulators from the participating cells. Among these are activation of the CD8 suppressor lymphocytes and potentially down-regulation of CD4 lymphocyte activity. In patients with elevated levels of suppressor-cytotoxic (CD8) T lymphocytes and a low frequency of HIV-expressing cells in the circulation, suppression in vitro may be sufficient to inhibit isolation. To avoid these difficulties two approaches are being developed: direct stimulation of selected T cell populations in mixed lym-

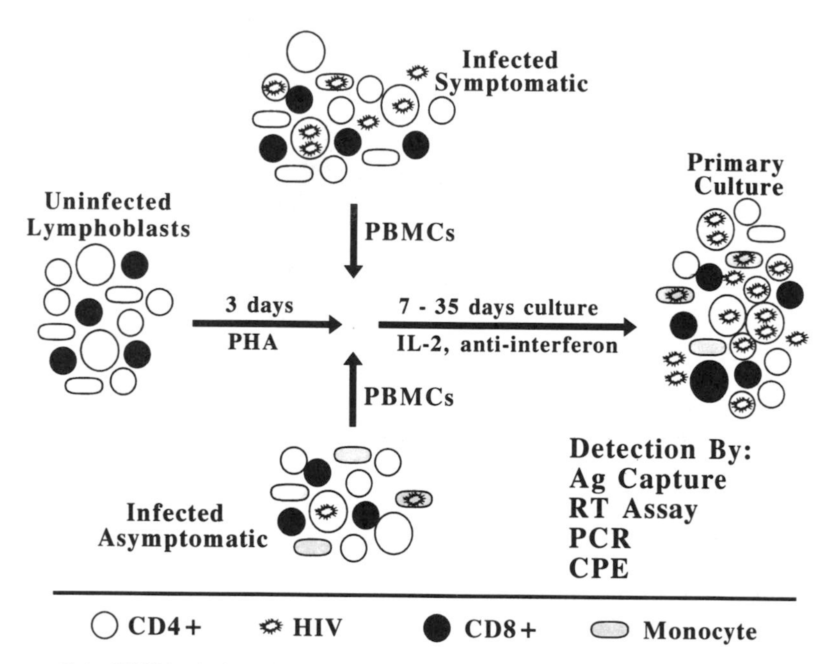

FIGURE 7.1. HIV isolation protocol. Peripheral blood mononuclear cells (PBMCs) from uninfected donors are pretreated with phytohemagglutinin (PHA) to enrich them for activated lymphoblasts, which serve as target cells for HIV isolation. Fresh PBMCs from asymptomatic or symptomattic HIV-infected persons are added at a specimen/target cell ratio greater than 1:1. The resulting mixed lymphocyte co-cultures are monitored regularly for evidence of HIV replication.

phocyte cultures[19] and, alternatively, the targeted culturing of only CD4 lymphocytes in the absence of CD8 cells.[20] Although culture methods that exclusively target CD4 lymphocytes have the disadvantage of loss of potentially productive synergy between heterologous cell types, cell sorting studies indicate that the preponderance of virus in the peripheral blood system is harbored in the CD4 lymphocyte subset.[21,22]

The initial propagation of HIV was achieved in neoplastic T cell lines, and much of the large scale preparation of virus continues to be done in continuous lymphoid cell lines.[23,24] Most such cell lines in use today are subclones of previously established lines that have been selected for favorable growth of HIV. Continuous cell lines have the advantages of consistency, availability, and lower maintenance costs over primary donor PBMC cultures. Derivations of the cutaneous human T cell lymphoma lines HUT-102 and HUT-78, including H9, are widely used. Cell lines that were derived from acute lymphoblastic leukemias include MT2, MT4, MOLT-3, MOLT-4, and CEM. These lines vary in their surface expression of CD4; and certain subclones, especially derivatives of CEM, are noted for their elevated expression of CD4, resulting in heightened HIV susceptibility. The H9, MOLT-3, MT4, and CEM cell lines are most frequently used in initial isolations, though they lack the sensitivity of primary PBMC cultures. Whereas these continuous lines are maintained as suspension cultures, the MT4 cell line and the CEM.SS derivative line of CEM have the advantage of adherence to poly-L-lysine-coated plastics, which is useful in quantal assays.[25] Finally, the human myeloid cell line U937 has proved useful in studies of host cell tropisms of HIV isolates because it expresses many of the characteristics of mature tissue macrophages. Studies suggest that isolates derived from monocytic cell lines or primary macrophage/monocyte cultures infect both monocytic and lymphocytic lines, whereas those isolated in lymphoblasts are restricted to lymphocytic lines.[26]

Detection and Characterization

A number of methodologies are available for the detection of HIV replication in vitro, including immunoadsorbance assays, electron microscopy, nucleic acid probes, and virus-specific enzymatic assays. Reverse transcriptase and antigen-capture assays can provide direct quantification of HIV in culture or, applied to infectious center studies in vitro, allow estimates of virus burden in the host. Limiting dilution analysis of syncytia formation and resulting CPE is useful in conjunction with neutralizing antisera to serotype isolates. Of these, the HIV-1 p24-based antigen-capture assay is today the most widely used method for monitoring diagnostic cultures.

Reverse Transcriptase Assay

The usual methods for measuring reverse transcriptase (RT) activity in culture are recent modifications of earlier methods for analyzing this distinctive enzyme of retroviruses.[27] Briefly, virus is pelleted by differential centrifugation from the supernatants of co-cultures and solubilized in the presence of a mild detergent. The viral lysate serves as a source of reverse transcriptase, which is detected by the incorporation of labeled nucleotides into acid-insoluble nucleic acid in the presence of Mg^{2+} and an RNA template primer. Duplicate samples are assayed per culture, and acid-insoluble label is collected on a filter and then measured in a beta scintillation counter or following autoradiography. In general, specimens with threefold increases over the values observed in negative control cultures are considered reactive, and rising titers should be repeatedly observed over a 3- to 6-day interval to confirm a positive culture. This procedure also serves as a quantitation of RT activity and is proportional to the original concentration of HIV in the co-culture. Obviously, analyses of HIV growth rates in culture are dependent on comparable RT activity between isolates, and isolates with nonlethal mutations in the *pol* gene could be distinct in their growth patterns.[28] Modifications to this technique include adaptation to a 96-well microtiter format, the use of semiautomatic cell harvesters, autoradiographic quantitation methods, and integration of nonisotopic labeling protocols.[29]

This assay is excellent for the detection of retroviruses, but it lacks specificity for HIV and requires 7 to 10 days of co-cultivation before detectable levels of RT activity are observed with most specimens. Although isolation of HTLV-I and HTLV-II has been reported from AIDS patients,[30,31] the frequency of transformation events leading to the successful propagation of these lymphoprolific viruses often requires cultivation periods of more than 4 weeks. Since these viruses also replicate via a reverse transcriptase, consideration should be given for them during the cultivation of specimens for extended periods. Finally, the spurious co-cultivation of agents such as the human herpesvirus type 6 group (HHV-6), which infect T lymphocytes, also competitively suppress HIV replication in vitro and inhibit any detection method that relies on active virus production.[32]

Viral Antigen

The methodology of the enzyme-linked immunoassay-based capture and detection of HIV p24 antigen has been discussed extensively elsewhere (see Chap. 5). This protocol is readily applicable to monitoring supernatants of infected cultures because it permits specific spectrometric quantitation of uncomplexed HIV core antigens. It has the further advantages of low cost, greater sensitivity (thus requiring less specimen volume), and greater specificity than the reverse transcriptase assay.[33] Although commercially available antigen capture assays have a sensitivity sufficient to detect 10 to 30 pg/ml, most laboratories consider the presence of 30 pg/ml or more indicative of a productive culture. Given a threshold reactivity of 30 pg/ml, repeatedly reactive specimens should show a minimal rise of 100 pg/ml in culture. In a direct comparison of RT activity versus antigen detection for monitoring cultured serial dilutions of HIV-1, the antigen capture method was found to be 100-fold more sensitive for detecting viral replication.[34] Because this assay incorporates antisera to HIV-1, it is highly specific and does not detect other human lymphotrophic viruses, such as HTLV-I or HTLV-II. Unfortunately, the sensitivity of current HIV-1-derived assays is to some extent strain-specific and is reduced against HIV-2 or distant variants from the prototype HIV-1, which express core antigens with altered epitopes or determinants.

Nucleic Acid Probe

Advances in molecular biology techniques have provided a number of methods of importance for detecting HIV replication. Principal among these are the polymerase chain reaction (PCR) for amplification of proviral DNA, in situ PCR,[35] and the more recent cRT-PCR assay for detecting viral RNA following reverse transcription and PCR amplification of cDNA transcripts.[36] Most hybridization assays using labeled nucleic acid probes can detect approximately 10 pg of targeted DNA or roughly one proviral DNA copy per 200 cells. Thus the sensitivity of Southern or dot blot hybridizations falls below that needed for direct detection of HIV in the peripheral blood and lymphatic tissues.[37]

As noted previously, PCR is the method of choice for direct analysis of infected tissues because of the high degree of amplification it affords. Conversely, each of these approaches is useful for monitoring HIV replication in co-cultures. Southern blot analysis can be used to fingerprint the endonuclease cleavage patterns of new isolates; in situ hybridization is useful for identifying specific cells replicating HIV; and RT-PCR can be used to screen for specific RNA transcripts in cultures. PCR often provides the greatest versatility in association with culture. Not all variants within any given isolation attempt propagate equally well in culture. PCR provides a means of analyzing minor species that are operationally lost to study as the more aggressive variants expand in vitro. To profit from such an approach, the PCR analysis must be initiated at the onset of culture or in conjunction with limiting dilution studies. PCR may also be used in a manner analogous to serotyping to characterize variants within a culture genotypically; or it can be used in conjunction with linker sequences to provide a rapid means of cloning and expressing the genomes of selected isolates.

Cytopathic Effects and Syncytia Formation

As can be inferred from the previous text, most primary HIV isolates exhibit a higher degree of replication and cytopathology in fresh PBMCs. However, since cytopathic effects (CPEs) and syncytia formation are host cell- and isolate-dependent, comparative studies are frequently done in continuous cell lines, which provide a stable background.[38] Given a stable culture environment, the rate of cell death, cytolysis, and fusion may vary greatly among primary isolates. Cell death may follow a general slowing of the mitotic cycle or be independent of cell division; and RT activity need not directly correlate with the extent of CPE present in the culture system.[28] Susceptibility to syncytial formation is also donor-dependent in primary PBMCs; and as with the continuous lines, most infected cells die prior to inclusion in giant cells.[39] For each of these reasons cytopathic assays that rely on the quantitation of infectious HIV are best done with isolates after establishment in a continuous cell line.

Microscopic Analysis

Initially used to demonstrate the presence of virus in diagnostic cultures, electron microscopy (EM) is generally supportive in nature and best suited for morphogenetic studies. Studies of HIV-1 in vitro replication first revealed many features in common with equine infectious anemia and the maedi and visna viruses of the Lentivirinae.[24] Now recognized as a member of the Lentivirus subfamily, the nascent HIV-1 virion resembles a type C retrovirus particle during the early stages of budding from the plasma membrane. However, the mature enveloped virus has a distinctive electron-dense nucleoid that is condensed into an eccentric, often bar-shaped structure that does not appear to contact the viral envelope.[40] Although its unique structure gives the HIV virion a readily identifiable character, significant viral replication must occur in the culture for EM analysis to be practical (Fig. 7.2).

Indirect immunofluorescence and similar enzyme-linked immunoassays are commonly used in conjunction with standardized antisera to detect HIV replication and define the subcellular loci of replicating viral elements. Culture samples are evaluated and quantified based on the percentage of labeled cells, intensity of the staining, and labeling pattern. Consequently, these techniques are frequently used to evaluate expansion of the initial isolate or determine the percentage of productive cells in large-scale culture systems. Cultures are best done in continuous cell lines, such as H9, HUT-78, or CEM cells, so the pooled sera or plasma might be standardized and preabsorbed to uninfected cells, thereby eliminating nonvirus-specific binding. Immunofluorescence assays and comparable immunoenzyme assays have the advantages of ease of use, as well as specificity and sensitivity, especially with respect to the cell-

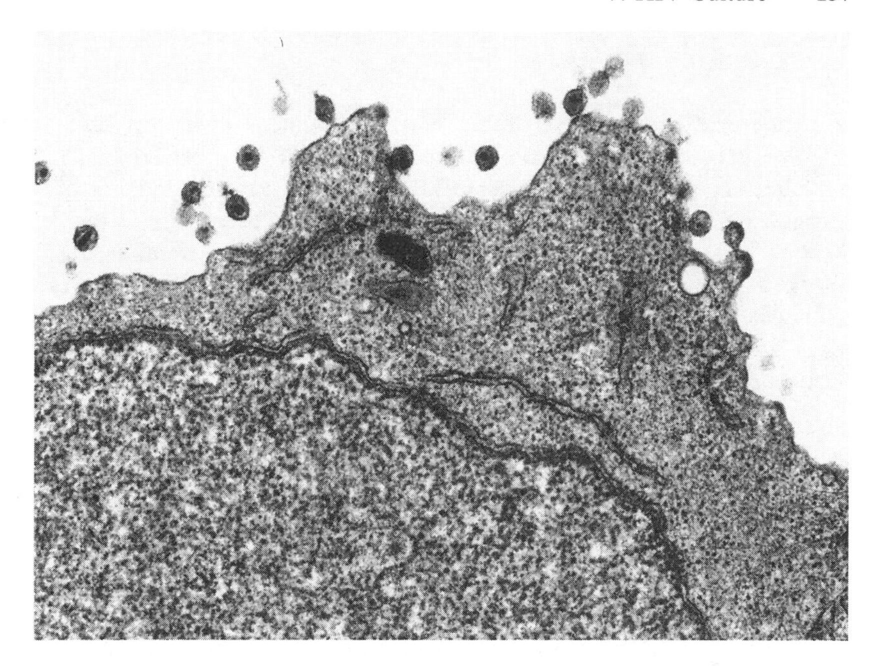

FIGURE 7.2. Electron micrograph of HIV particles in various stages of budding from an infected cell in a 10-day-old co-culture with primary human PBMCs. Mature infectious particles are visible in the extracellular medium as enveloped virions containing electron-dense, club-shaped nucleoid.

associated glycoproteins of the viral envelope.[41] The principal disadvantage of the microscopic immunoassays is that they are highly subjective, yet such methods provide a practical alternative to the radio immunoprecipitation assay (RIPA) for detection of the glycosylated HIV envelope proteins.

Western Blot and RIPA

Western blot and RIPA technology in conjunction with well characterized sera can be used to fingerprint the protein banding patterns of new viral isolates by polyacrylamide gel electrophoresis (PAGE). These assays are not quantitative, but the observed banding pattern is characteristic of the cultured isolate[42] and may be used to distinguish HIV variants. As is the case with their respective diagnostic applications, protocols for partial purification, reduction, and generation of the PAGE product have not been standardized and may vary greatly among laboratories. The reported migration patterns for a given isolate may therefore differ, especially in regard to the apparent molecular weights for glycosylated viral proteins. Distinctions are also noted in the expression and maturation of HIV proteins in differing continuous cell lines or host cells.

Conclusion

Although advances in serologic and biochemical detection techniques have largely supplanted virus culture as efficient diagnostic tools, related advances continue to be made in HIV isolation that extend its role as a prognostic standard. It is because of the strong correlation between isolation rates and circulating virus burden that culture remains a relvant marker for monitoring disease progression and therapeutic intervention. Regardless of its costs and stringent specimen requirements, culture will thus continue to serve the unique need for providing a definable infectious entity to the researcher for study.

References

1. Jackson JB, MacDonald KL, Cadwell J, et al. Absence of HIV infection in blood donors with indeterminate Western blot tests for antibody to HIV-1. N Engl J Med 1990;25:217–222
2. Dock NL, Kleinman SH, Rayfield MA, et al. Status of human immunodeficiency virus infection in individuals with persistently indeterminate Western blot patterns: prospective studies in a low prevalence population. Arch Intern Med 1991;151:525–530
3. Rogers M, Ou CY, Rayfield MA, et al. Polymerase chain reaction for early detection of HIV proviral sequences in infants born to HIV-seropositive mothers. N Engl J Med 1989;320:1649–1654
4. Harper ME, Marselle LM, Gallo RC, Wang-Staal F. Detection of lymphocytes expressing human T-lymphotropic virus type III in lymph nodes and peripheral blood from infected individuals by in situ hybridization. Proc Natl Acad Sci USA 1986;83:772–776
5. Richman DD, McCutchan JA, Spector SA. Detecting human immunodeficiency virus RNA in peripheral blood mononuclear cells by nucleic acid hybridization. J Infect Dis 1987;156:823–1828
6. Ulrich PP, Busch MP, El-Beik T, et al. Assessment of human immunodeficiency virus expression in co-cultures of peripheral blood mononuclear cells from healthy seropositive subjects. J Med Virol 1988;25:1–10
7. Ho DD, Moudgil T, Alam M. Quantitation of human immunodeficiency virus type 1 in the blood of infected persons. N Engl J Med 1989;321:1625–1626
8. Simmonds P, Balfe P, Peutherer JF, et al. Human immunodeficiency virus-infected individuals contain provirus in small numbers of peripheral mononuclear cells and at low copy numbers. J Virol 1990;64:864–872
9. Schnittman SM, Greenhouse JJ, Mitiades BS, et al. Increasing viral burden in $CD4^+$ T cells from patients with human immunodeficiency virus (HIV) infection reflects rapidly progressive immunosuppression and clinical disease. Ann Intern Med 1990;113:438–443
10. Jackson JB, Kwok SY, Sninsky JJ, et al. Human immunodeficiency virus type 1 detected in all seropositive symptomatic and asymptomatic individuals. J Clin Microbiol 1988;28:16–19

11. Francis DP, Jaffe HW, Fultz PN, et al. The natural history of infection with the lymphadenopathy associated human T lymphocytic virus type III. Ann Intern Med 1985;103:719–722
12. Jackson JB. Human immunodeficiency virus type 1 antigen and culture assays. Arch Pathol Lab Med 1990;114:249–254
13. Levy J, Kaminsky L, Morrow W, et al. Infection by the retrovirus associated with AIDS. Ann Intern Med 1985;103:694–699
14. Gaines H, Albert J, Von Sydow M, et al. HIV antigenemia and virus isolation from plasma during primary HIV infection. Lancet 1987;1:1317–1318
15. Ho D, Schooley R, Rota R, Kaplan J, Flynn T. HTLV-III in the semen and blood of a healthy homosexual man. Science 1984;226:451–453
16. Ho D, Rota R, Schooley R, et al. Isolation of HTLV-III from cerebrospinal fluid and neural tissues of patients with neurologic syndromes related to the acquired immunodeficiency syndrome. N Engl J Med 1985;313:1493–1497
17. Ho D, Byington R, Schooley R, et al. Isolation of HTLV-III from 83 saliva and 50 blood samples from 71 seropositive homosexual men. N Engl J Med 1985;313:1606
18. Koenig S, Gengelman HE, Orenstein JM, et al. Detection of AIDS virus macrophages in brain tissue from AIDS patients with encephalopathy. Science 1986;233:1089–1093
19. Walker C, Bettens F, Pichler WJ. Activation of T cells by cross-linking an anti-CD3 antibody with a second anti-T cell antibody: mechanism and subset-specific activation. Eur J Immunol 1987;17:873–880
20. Walker CM, Moody DJ, Stites DP, Levy JA. $CD8^+$ lymphocytes can control HIV infection in vitro by suppressing virus replication. Science 1986;234: 1563–1566
21. Schnittman SM, Psallidopoulos MC, Lane HC, et al. The reservoir for HIV-1 in human peripheral blood is a T cell that maintains expression of CD4. Science 1990;245:305–308
22. Spear GT, Ou CY, Kessler HA, et al. Analysis of lymphocytes, monocytes, and neutrophils from human immunodeficiency virus (HIV) infected persons for HIV DNA. J Infect Dis 1990;162:1239–1244
23. Barré-Sinoussi F, Chermann JC, Rey F, et al. Isolation of a T-lymphotropic retrovirus from a patient at risk for acquired immune deficiency syndrome (AIDS). Science 1983;220:868–871
24. Popovic M, Sarngadharan MG, Read E, Gallo RC. Detection, isolation, and continuous production of cytopathic retroviruses (HTLV-III) from patients with AIDS and pre-AIDS. Science 1984;224:497–500
25. Tateno M, Levy JA. MT-4 plaque formation can distinguish cytopathic subtypes of the human immunodeficiencey virus (HIV). Virology 1988;167:299–301
26. Gendelman HE, Baca LM, Husayni H, et al. Macrophage–HIV interaction: viral isolation and target cell tropism. AIDS 1990;4:221–228
27. Robert MS, Smith RG, Gallo RC, Sarin PS, Absell JW. Viral and cellular DNA polymerases: comparison of activities on synthetic and natural RNA templates. Science 1972;176:798–800
28. Cloyd M, Moore B. Spectrum of biological properties of human immunodeficiency virus (HIV-1) isolates. Virology 1990;174:103–116

29. Spira TJ, Bozeman LH, Holman RC, et al. Micromethod for assaying reverse transcriptase. J Clin Microbiol 1987;25:97–99
30. Hahn B, Popovic M, Kalyanaraman V, et al. Detection and characterization of HTLV-II provirus in patients with AIDS. In Gottlieb MS, Groopman JE (eds): Acquired Immune Deficiency Syndrome. New York: Alan R. Liss, 1984
31. Chorta T, Brynes R, Kalyanaraman V, et al. Transformed T-lymphocytes infected by a novel isolate of human T-cell leukemia virus type II. Blood 1985;66:1336–1342
32. Lopez C, Pellet P, Stewart J, et al. Characteristics of human herpesvirus-6. J Infect Dis 1988;157:1271–1273
33. Jackson JB, Sannerud K, Rhame FS, Balfour HH. Evaluation of two commercial tests for human immunodeficiency virus antigens in culture supernatant fluid. Am J Clin Pathol 1988;89:788–790
34. Feorino PM, Forrester B, Warfield DT, Schochetman G. Comparison of antigen assay and reverse transcriptase assay for detecting human immunodeficiency virus in culture. J Clin Microbiol 1987;25:2344–2346
35. Nuovo GJ. PCR In Situ Hybridization: Protocols and Applications. New York: Raven Press, 1992
36. Bagnarelli P, Menzo S, Valenza A, et al. Molecular profile of human immunodeficiency virus type 1 infection in symptomless and AIDS patients. J Virol 1992;66:7328–7335
37. Shaw GM, Hahn BH, Arya SK, et al. Molecular characterization of human T-cell leukemia (lymphotropic) virus type III in the acquired immune deficiency syndrome. Science 1984;226:1165–1171
38. World Health Organization Global Programme on AIDS. Report of a WHO workshop on the measurement and significance of neutralizing antibody to HIV and SIV. AIDS 1990;4:269–275
39. Somasundaran M, Robinson HI. A major mechanism of human immunodeficiency virus-induced cell killing does not involve cell fusion. J Virol 1987;61:3114–3119
40. Palmer E, Sporborg C, Harrison A, Martin ML, Feorino P. Morphology and immunoelectron microscopy of the AIDS virus. Arch Virol 1985;85:189–196
41. Sandstrom EG, Schooley RT, Ho DD, Byington R, Sarngadharan MG. Detection of human anti-HTLV-III antibodies by indirect immunofluorescence using fixed cells. Transfusion 1985;25:302–312
42. Kanki P. West African human retroviruses related to STLV-III. AIDS 1987;1:141–145

8 Direct Detection of HIV Infection Using Nucleic Amplification Techniques

GERALD SCHOCHETMAN and JOHN J. SNINSKY

Although infectious viral particles of human immunodeficiency virus (HIV) encapsidate single-stranded RNA (ssRNA) as the genetic information, the viral life cycle includes a compulsory conversion to double-stranded DNA (dsDNA), termed the provirus, which becomes integrated into the host cells' chromosomes (see Chap. 2 for more details of the virus life cycle). The integrated provirus remains associated with the cellular chromosomal DNA for the life of the infected cell. Furthermore, the integrated provirus can either actively transcribe the genes for the structural proteins of the virus, which results in the assembly and release of infectious virions, or by selective transcription of only the complex array of viral regulatory genes remain transcriptionally constrained and thereby not release viral particles. The latter condition is frequently referred to as the "latent state." Because proviral DNA is present regardless of the transcriptional state of the cell, early efforts targeted to direct detection of the virus used proviral DNA as a template.

Because of the low frequency of HIV-1-infected peripheral blood mononuclear cells (PBMCs) in a seropositive person,[1] conventional molecular biology techniques[2] were not sensitive enough to routinely detect and characterize HIV proviral DNA directly from patients' lymphocytes. Therefore HIV proviral DNA must first be amplified to detectable levels using the polymerase chain reaction (PCR). Prior to PCR, successful direct detection of HIV-1 required culturing the virus. The ability of PCR to amplify HIV sequences several orders of magnitude in vitro has obviated the need to propagate the virus for direct detection.

For the study of HIV infection and acquired immunodeficiency syndrome (AIDS), PCR has demonstrated both clinical and research utility for: (1) direct detection and quantitation of HIV DNA and RNA from cells of infected persons; (2) detecting infected persons during the window period (i.e., prior to the generation of HIV-specific antibodies); (3)

resolving the infection status of individuals with an indeterminate Western blot assay; (4) screening neonates for HIV infection; (5) distinguishing HIV-1 from HIV-2 infections; and (6) defining the patterns of transmission and evolution of the virus throughout the population.

PCR Methodology

The PCR process was originally developed as a technique for the in vitro amplification of targeted DNA sequences.[3–6] For PCR, sample preparation has employed separation of mononuclear from polymorphonuclear cells of the blood using a Ficoll-Hypaque gradient. After preparation of the PBMCs the cells are incubated with nonionic detergents and finally treated with proteinase K. Heat treatment (95°C for 15 minutes) is used to inactivate the proteinase K. The resulting DNA preparation is ready for PCR. Usually 1 μg of DNA (equivalent to about 150,000 mononuclear cells) (Table 8.1) is used per PCR reaction. PCR is a repetitive process consisting of three distinct steps (Fig. 8.1): (1) denaturation of dsDNA; (2) annealing of specific primers; and (3) extension of annealed primers. Because of the complementary and antiparallel nature of DNA, ssDNA can also serve as a template for amplification. When amplification of a specific RNA sequence is required, a DNA copy of the RNA sequence is produced using the enzyme reverse transcriptase prior to PCR amplification of the resulting DNA.[7,8] After the PCR process, a variety of techniques can be used to detect the amplified DNA sequences. Although the amplified DNA is of a defined size and can sometimes be visualized after gel electrophoresis, this method cannot provide definitive identification of the product. The confirmation of am-

TABLE 8.1. Numbers to consider for HIV infection.

General
Blood volume: 5.6 liters in a 70 kg person
White blood cells: 500,000/65 μl whole blood
White blood cells:
Polymorphonuclear cells (70%), 350,000
Mononuclear cells (30%), 150,000
Mononuclear cells
Monocytes (10%), 15,000
Large granular lymphocytes (LGLs) (10%), 15,000
B cells (10%), 15,000
T cells (70%), 105,000
T cells
T8 (suppressor) cells (35%), 37,000
T4 (helper) cells (65%), 68,000

plification of HIV DNA includes hybridization of a portion of the amplified DNA to a synthetic DNA probe that is complementary to a portion of the amplified DNA sequence. The probe can be labeled by a variety of means, isotopic (radioactive) or nonisotopic (colorimetric or chemiluminescent) (Figs. 8.2, 8.3, 8.4).

PCR Test System

A PCR system for qualitatively detecting HIV-1 has been developed as a simplified diagnostic test kit[9] that provides the inherent sensitivity and specificity of any research PCR assay in a user-friendly system format. The kit consists of three components: a specimen collection and preparation kit, an amplification kit, and a detection kit. All components are premixed for easy and reliable amplification and detection using the well known microwell plate colorimetric format. The test specimen for this kit is whole blood, and ambient temperature storage allows convenient shipping of clinical specimens. The whole blood lysis procedure is a simple method that requires only 0.5 ml of blood. A modified procedure using 0.1 ml of blood has also been developed for neonates, from whom virus culture requires too large a volume of blood. For the whole-blood lysis procedure red blood cells are selectively lysed while the leukocytes remain intact. After the leukocytes are pelleted and then washed several times, DNA is extracted from the final cell pellet. The entire procedure takes less than 2 hours. A 50 µl aliquot (approximately 200,000 cells) is used to amplify proviral HIV-1 DNA. Biotinylated primers used to amplify the HIV-1 DNA represent a 142 basepair sequence of the *gag* gene region. These primers have reportedly been tested on thousands of specimens and have been shown to amplify all samples efficiently, detecting virtually 100% of HIV-1 isolates worldwide.

During the early days of PCR, the potential to amplify minute amounts of contaminating material from previous reactions resulted in problems of false-positivity. Extremely small amounts of amplified target DNA or amplicon (10^{-8} µl) from a previous amplification contaminating a reaction tube may result in a false-positive reaction. To solve this problem, deoxyuridine 5′-triphosphate (dUTP) was incorporated into the amplification reaction, making it possible to distinguish amplicon DNA from native T-containing DNA found in specimen cellular DNA (Fig. 8.5). The active enzyme uracil-*N*-glycosylase (UNG) selectively recognizes and destroys the U-containing amplicons while leaving target DNA unaffected. Because all amplified products are synthesized with dUTP rather than deoxythymidine 5′-triphosphate (dTTP), they are susceptible to UNG activity in subsequent reactions. The development of this procedure has been critical to the successful transfer of PCR from the research laboratory to the clinical laboratory.

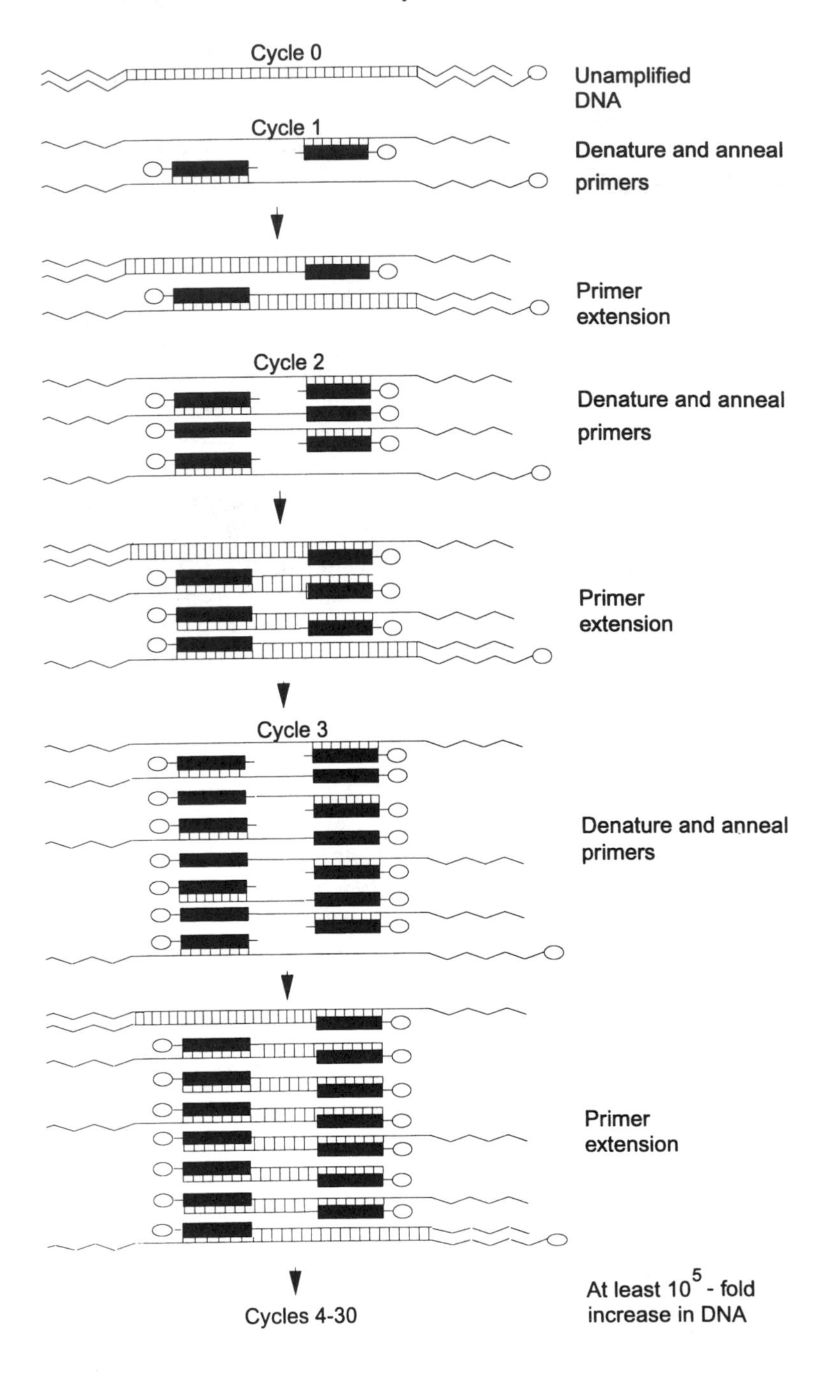
Cycle 0
Unamplified DNA
Cycle 1
Denature and anneal primers
Primer extension
Cycle 2
Denature and anneal primers
Primer extension
Cycle 3
Denature and anneal primers
Primer extension
Cycles 4-30
At least 10^5 - fold increase in DNA

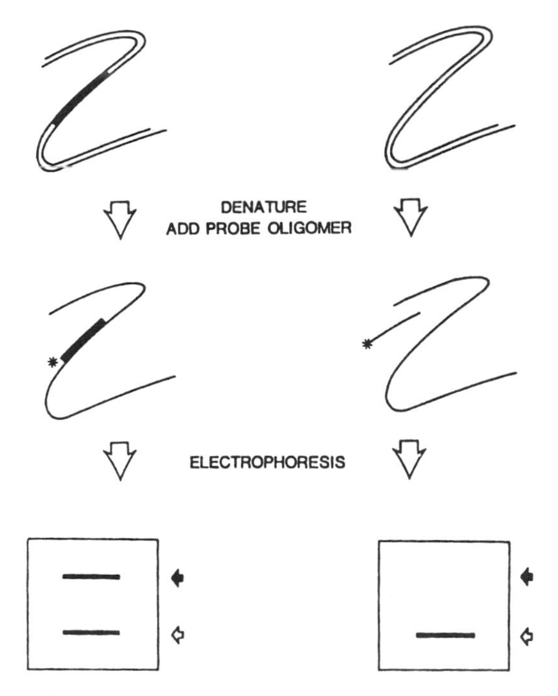

FIGURE 8.2. Detection of the amplified DNA involves hybridization of a portion of the amplified DNA product to a radioactive ^{32}P-labeled synthetic probe complementary to a portion of the amplified sequences followed by gel analysis and autoradiography.

All reagents necessary for amplification are provided in the kit; it is only necessary to add UNG to the premixed master mix vial. An aliquot of premixed master mix is then added to the appropriate number of amplification tubes. Subsequently, an aliquot of prepared specimen is added to the appropriate tube, the tube is capped, and the rack of reaction tubes is placed in the thermal cycler. At the end of the amplification cycle, the reaction tubes should be held at 72°C until the dena-

---

FIGURE 8.1. Polymerase chain reaction (PCR), which is a repetitive process that includes denaturation of double-stranded DNA (dsDNA), annealing of primers, and extension of bound primers. One PCR cycle usually takes about 3 minutes, and the cycle is repeated many times (usually 25–35 times). The dsDNA is first heated to 95° to 100°C to separate the strands of the duplex. During the subsequent annealing phase, oligonucleotide primers hybridize to the dissociated HIV DNA. Each primer is complementary to one of the original DNA strands, either the 5′ or the 3′ side of the sequence of interest. After annealing, a thermostable DNA polymerase from *T. aquaticus* (*Taq*) is used to catalyze the synthesis of new strands of DNA that are complementary to the intervening sequences primed by the opposing oligonucleotide primers.

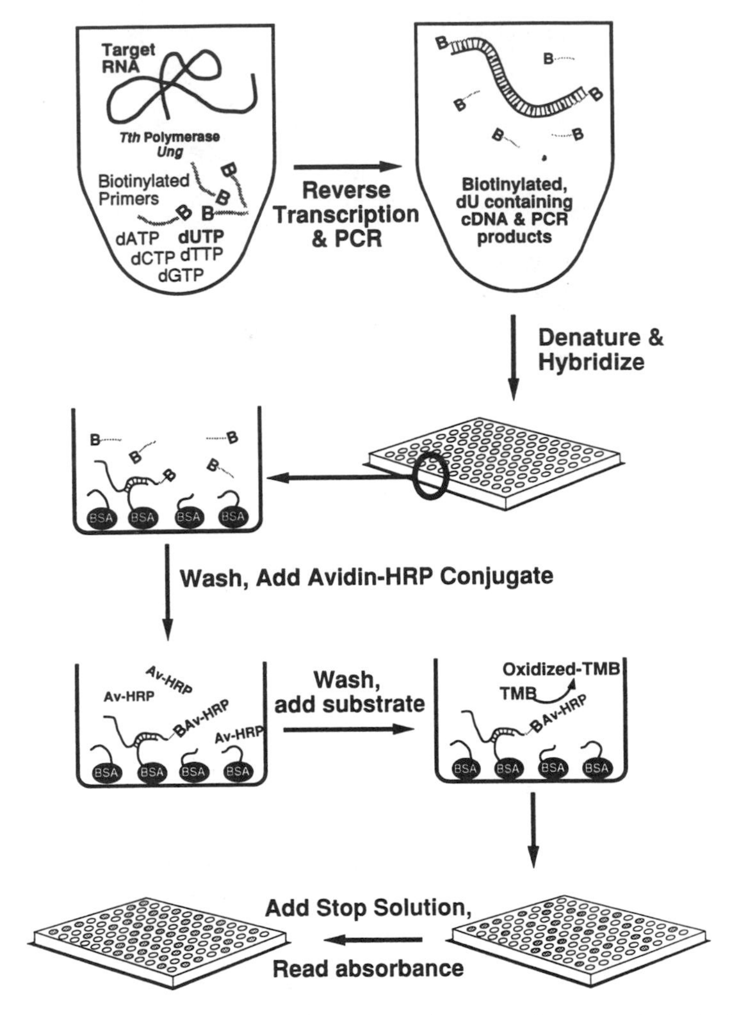

FIGURE 8.3. Detection of the amplified DNA involves the use of biotinylated primers to generate a tagged amplified DNA product, which is then hybridized to an immobilized probe complementary to a portion of the amplified sequences. This step is followed by incubation with avidin conjugated to horseradish peroxidase. This standard EIA format in a microplate yields a colorimetric readout for positive samples (C. Silver, M. Sulzinski, E. Dragon, and M. Longiaru, personal communication).

turation solution is added. Detection is based on the enzyme immunoassay (EIA)-like colorimetric microwell plate format (Fig. 8.3). Amplicons are captured by bovine serum albumin (BSA)-conjugated DNA probes specific for HIV that have been coated on the bottom of the well of a microwell plate. The specificity of the detection derives from the

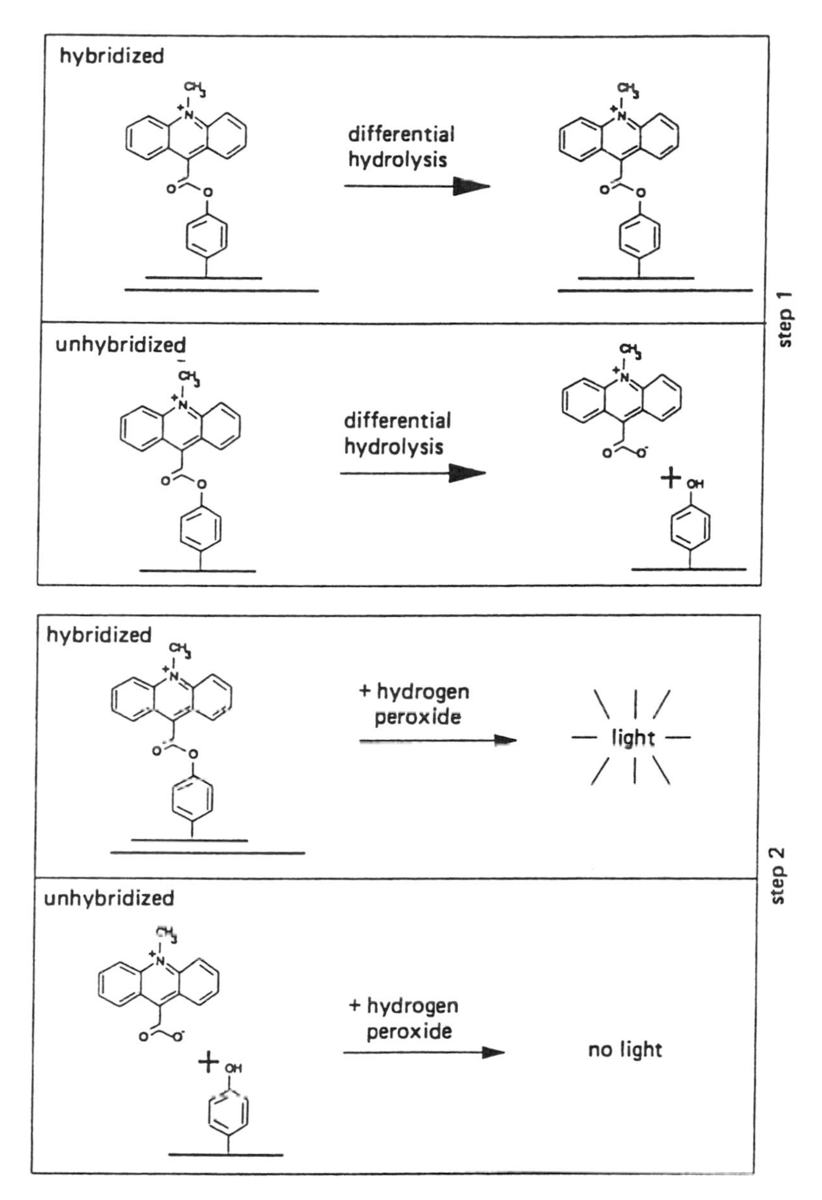

FIGURE 8.4. Detection of the amplified DNA involves hybridization of a portion of the amplified DNA product to an acridinium-labeled synthetic probe complementary to a portion of the amplified sequences followed by differential alkaline hydrolysis. The acridinium attached to the hybridized probe is relatively resistant to alkaline degradation and can chemiluminesce after oxidation by the addition of hydrogen peroxide. In contrast, the acridinium attached to the unhybridized probe is highly sensitive to alkaline degradation and loses its ability to chemiluminesce almost immediately.

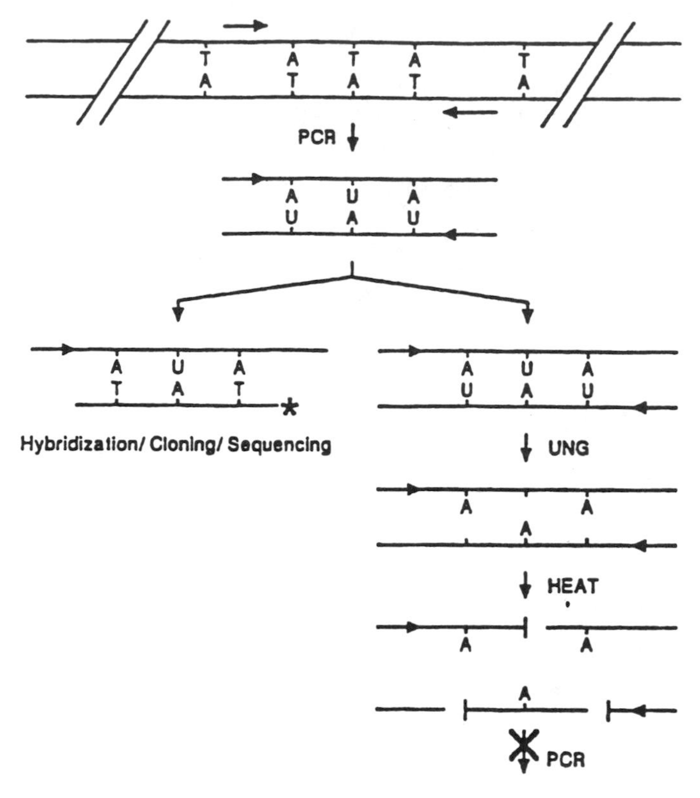

FIGURE 8.5. Use of dUTP and uracil DNA glycosylase to eliminate errant PCR products as templates for subsequent amplifications leading to false-positive results (see text for details).

BSA-conjugated DNA capture probe. Denatured amplification reaction is pipetted into the wells that contain the hybridization solution. After being covered and incubated for 1 hour at 37°C, the plate is washed, and avidin–horseradish peroxidase (HRP) conjugate is added to the wells and incubated for 15 minutes at 37°C. The plate is then washed again, and the substrate/chromogen is added. Color development is allowed to proceed for 10 minutes at room temperature in the dark. Acidic stop solution is added to the wells, and the resulting absorbances are read at 450 nm. The test kit should be rapid (results can be obtained within one working day), easy to perform, and amenable for use in the clinical laboratory. This system, which should provide a means of standardizing testing among laboratories, is being investigated at a number of designated clinical trial sites in the United States for Food and Drug Administration (FDA) licensure.

Direct Detection of HIV Proviral Sequences

The PCR has proved to be a powerful technique for diagnosing HIV infections. It has been used successfully for direct detection of HIV-1 proviral DNA sequences in PBMCs of seropositive persons[10–13] and to detect HIV-1 DNA in cells from seropositive subjects who were negative by virus co-culture.[10] Studies have also demonstrated that PCR has simplified the ability of researchers to directly clone and sequence HIV-1 DNA[14–16] and HIV-1 cellular RNA.[7] HIV sequences obtained by this method more accurately reflect the repertoire of viral sequences actually present in a patient.[15] It is possible because PCR does not require virus isolation, which leads to the selection of only a subset of the HIV strains that are present in any particular patient.[15,16]

The PCR has been used to establish that virtually all antibody-positive individuals are infected with HIV, lending further proof that HIV is the causative agent of AIDS.[17] This large-scale study involved testing PBMCs obtained from 409 individuals who were HIV antibody-positive. The group consisted of 56 individuals who had AIDS, 88 who had AIDS-related complex (ARC), and 265 who were asymptomatic. In addition, blood samples from 131 persons who were HIV antibody-negative were examined. All samples from the 56 AIDS patients, 87 of the 88 ARC patients (99%), and 259 of the 265 asymptomatic but HIV antibody-positive persons (98%) tested positive for virus by culture, PCR, or both analyses; in contrast, none of the 131 HIV-1 antibody-negative persons was positive for virus by culture or by PCR.

Quantitation of HIV Proviral Sequences

Because the amount of the original target DNA can be quantitated by PCR, the technique can be used to measure the number of infected cells[18–20] or to quantitate the amount of cell-free virus in the patient's plasma (viremia). A quantitative assay for proviral HIV-1 would be important for the evaluation of new drugs and vaccines or for monitoring disease progression. Quantitation of viral load in infected persons has been performed with varying success by limiting dilution culture methods in which culture supernatants were tested for the presence of HIV-1 *gag* protein (p24) or the presence of viral reverse transcriptase activity (or both).[21,22] Virus culture is expensive and time-consuming, requires handling large volumes of infectious material, and has not always been reliable. By amplifying a dilution series of known amounts of HIV DNA (e.g., a plasmid containing a full-length copy of HIV-1 DNA or DNA from a cell line containing one integrated copy of HIV-1 per cell), it is possible to quantify the virus burden in a person by determining the number of HIV proviral copies per given number of cells in a patient.[18–20,23–27] To ensure the biosafe use of a full-length proviral DNA standard, a replication-deficient HIV-1 proviral DNA has been developed.[28] A cell line, ACH2, which contains one integrated copy of HIV-1 and produces a noninfectious HIV, has also been developed.[29] Methods have also been developed for the quantitation of specific RNAs by the PCR technique.[30–32]

A rapid, quantitative detection procedure has been developed using a nonisotopic chemiluminescent DNA probe.[19] The total time for PCR amplification and DNA probing using this technique requires about 4 hours. Thus detection and quantitation of HIV DNA can be achieved within 1.0 to 1.5 days from the time of receipt of the blood sample.

The PCR has been used by a number of investigators to determine the relation between infected cell burden and immunologic status in persons with asymptomatic and symptomatic HIV infection.[20,25–27] In those studies, lysates of patient PBMCs were serially diluted, amplified, and detected with a radiolabeled probe. The signal intensity from each amplification was compared with the PCR performed on serial dilutions of the plasmid containing the HIV-1 genome or the ACH2 cell line that contains one integrated copy of HIV-1 per cell. The results of those studies demonstrated that there was a significant increase in viral burden per constant $CD4^+$ cells in patients as they progressed to clinical disease. There was also a concomitant quantitative depletion of $CD4^+$ cells. This finding contrasted with the stable viral burden and the maintenance of a relatively constant level of $CD4^+$ cells in patients who were clinically stable.[25] The number of HIV-infected cells can be calculated from the number of proviral copies, as it has been estimated that there is approximately one proviral copy per cell.[26] These results can be compared to the number of cells producing virus in asymptomatic versus symptomatic persons as determined by limiting dilution cultures.[21,22] From this type of analysis it can be estimated that approximately 10% of the infected cells in the blood of asymptomatic persons are actively expressing virus compared to about 100% of the infected cells in symptomatic persons. These results indicate that not only do the number of infected cells increase substantially, but the proportion of infected cells actively expressing HIV increases substantially as patients move from an asymptomatic to a symptomatic state. These results are consistent with a direct and probable causal relation between an increase in viral burden and immunosuppression and disease—presumably due to the increase in HIV expression leading to cell destruction.

Latent HIV-1 infection can be differentiated from active viral transcription[7,8,33] because HIV-specific RNA sequences can be detected in cells of infected persons by amplifying cDNA copies of reverse-transcribed cellular RNA. This assay was capable of detecting HIV RNA in one infected cell among 10^6 uninfected cells.[7] Direct comparison of the presence of detectable HIV serum antigen with HIV RNA expression in the same patients[7] demonstrated that RNA PCR was more sensitive than serum antigen detection (i.e., all patients who were antigen-positive were HIV RNA-positive, but all patients who were HIV RNA-positive were not antigen-positive).

Virus Infection and Seroconversion

Various reports have indicated that in some individuals HIV can be detected by virus isolation or by antigen detection prior to seroconversion.[34–38] Other studies have reported that HIV-1 proviral DNA can be detected in PBMCs before seroconversion.[39–42] Preliminary replication of one of these studies[42] has not confirmed the original results (unpublished data) and merits additional investigation. Those who have been followed prospectively after exposure to HIV-1 have generally seroconverted within 6 months.[43] However, cases of positive antigen reactions for as long as 14 months without detectable antibody have also been reported.[36,40] To define the length of time from infection to the development of detectable levels of HIV antibodies, a before and after seroconversion study was undertaken of 26 homosexual men and 11 men with hemophilia.[43] PBMCs from these men were analyzed for HIV-1 DNA by PCR using primers from two distinct regions of the viral genome. Using a Markov statistical model, the median time from infection with HIV-1 to seroconversion was estimated to be 3 months and that 95% of all persons who become infected would seroconvert within 6 months. Similar results were obtained studying infection in high risk seronegative prostitutes in Nairobi, Kenya,[44] where it was estimated that the interval between infection and seroconversion was 3 to 4 months. These results indicated that prolonged periods of latent infection without detectable antibody probably are rare.

Resolving Cases with Indeterminate Western Blot Assays

The PCR has been used to determine whether apparently healthy persons who have had repeatedly reactive EIAs and an indeterminate Western blot test for HIV antibody are infected with HIV-1.[45] A total of 99 volunteer blood donors in a low risk area of the United States with such a serologic outcome were coded and tested for the presence of HIV by culture and by PCR. Of the 99 blood donors, 98 had no reported risk factors for HIV-1 infection; 1 donor had used intravenous drugs. After a median 14 months from the time of the initial serologic tests, 65 donors (66%) were still repeatedly reactive for HIV-1 on at least one immunoassay. For 91 donors (92%) the Western blot results were still indeterminate. None of the 99 donors had evidence of either HIV infection as determined by culture or PCR. These results demonstrate that persons at low risk (e.g., volunteer blood donors) for HIV infection and who have persistent indeterminate HIV-1 Western blots are rarely infected with HIV-1.

HIV Typing

There is partial but significant serologic cross-reactivity between the *gag* (core) proteins of HIV-1 and HIV-2, whereas cross-reactive antibodies to the *env* (envelope) proteins are thought to be considerably less common.[46] There have been reports of individuals who possess antibodies reactive against the *gag*, *pol*, and *env* proteins of both HIV-1 and HIV-2.[46–50] Serologically, it has been difficult to determine whether this dual reactivity was due to a single HIV infection generating a broad immune response to determinants common to both viruses, an infection with a recombinant or third virus containing determinants of both HIV-1 and HIV-2, or a true mixed infection with both viruses in the same person. In regions where HIV-1 but not HIV-2 is highly endemic (e.g., the United States), it would be unlikely to find an individual infected with both viruses. However, in certain areas of West Africa where HIV-1 and HIV-2 are both prevalent, the probability of finding someone infected with both viruses is much higher.

The PCR has been used successfully as an adjunct to serologic testing to determine if a patient is infected with HIV-1 or HIV-2.[50] It has also been used to confirm the first case of HIV-2 infection in a person living in the United States[51] and to confirm the first case of a mixed HIV-1 and HIV-2 infection in the same individual.[50] That person was seroreactive by whole-virus EIAs, type-specific peptide EIAs, and Western blot assays for both viruses and contained proviral sequences of both HIV-1 and HIV-2 as determined by PCR.

HIV Infection in Newborns

The fact that about 13% to 40% of infants born to women with HIV-1 infection have acquired their infection from the mother together with the presence of maternal antibodies to HIV-1 in the newborn makes diagnosis of HIV-1 infection difficult. PCR has been used successfully to diagnose HIV-1 infection by detecting HIV-1 DNA during the neonatal period (first 28 days of life),[52,53] particularly in those infants born to HIV-seropositive mothers who develop a severe, rapid course of the disease.[53,54] The PCR has also detected virtually all HIV-infected children who are in the postneonatal period, usually a few months old.[53–55] Those HIV-infected infants born to seropositive mothers who develop a less severe, slow course of the disease become PCR HIV-positive by 4 to 6 months. However, diagnosis of HIV-1 infection during the neonatal period and assessment of disease outcome in seropositive infants[53,54] are essential for identifying the infants who might benefit from early therapeutic intervention[54,56] (see Chap. 16 for a more complete description).

Use of PCR to Monitor for Drug-Resistant HIV

The reverse transcriptase inhibitor 3′-azido-3′-deoxythymidine (AZT, or zidovudine) has demonstrated clinical utility for the treatment of AIDS and ARC. Specifically, this therapeutic strategy has been shown to extend life expectancy and to lower the frequency and severity of opportunistic infections. The initially reported isolates with reduced sensitivity in vitro to this nucleoside analogue,[57] cultured from patients receiving zidovudine, were subsequently found to harbor specific mutations within the coding sequence for the HIV reverse transcriptase.[58] Whereas the most resistant isolates have four amino acid substitutions (e.g., positions 67, 70, 215, and 219), isolates that have a subset of these four mutations are less resistant to the drug. Larder and his colleagues have identified six AZT-resistant mutations, including two at the same position, codon 215.[59] A "nested" or "double" PCR procedure was developed[60] to detect the common mutations found in residue 215 (for example, conversion of the threonine codon to one for either tyrosine or phenylalanine requires a two base change). High-level (>50-fold) AZT resistance required two or more of the mutations in the laboratory constructs examined. The mutations at codon 215 are the most important of the AZT resistance mutations, being almost invariably present in highly resistant isolates. Codon 215 mutations have been monitored recently in several clinical studies and have proved to be a good marker for imminent clinical or immunologic decline in AZT-treated persons from various cohort studies.[61] Under modified conditions,[62] PCR can also be used to selectively amplify sequences varying in a single nucleotide, and therefore the other codon changes should be amenable to similar analysis. The role these mutations play in the declining efficacy of zidovudine after protracted periods of treatment remains unclear, but rapid diagnostic procedures for their detection should assist in resolving their contribution.

The increasing use of ddI and ddC of course raises the threat of ddI and ddC drug resistance. In fact, such resistance occurs clinically, although reports thus far concern small numbers of patients. HIV-1 resistant to ddI was isolated[63] from three patients after 6 to 12 months on ddI; the resistance was due to a leucine 74 to valine mutation, which conferred cross-resistance to ddC. In another report[64] a methionine 184 to valine mutation, which has similar effects, has been detected in five ddI-treated patients. A resistance mutation has been detected in two ddC-treated patients, threonine 69 to aspartate.[65]

Analytic Sensitivity and Specificity

The PCR has the highest analytic sensitivity of any procedure used in the diagnostic arena. Single molecule detection has been reported by numerous laboratories. This exquisite analytic sensitivity does not necessarily translate into diagnostic sensitivity with the ultimate clinical utility. The procedure does not have an intrinsic analytic or diagnostic sensitivity and specificity. The diagnostic sensitivity and specificity are inextricably linked to the laboratories performing the procedure. As a result, the confidence in the reported results is directly proportional to the experience and critical interpretive criteria used by the laboratories performing the assay. Multiple parameters have been shown to dramatically affect the overall analytic sensitivity and, correspondingly, its diagnostic sensitivity. Factors that have been demonstrated to affect overall amplification efficiency beyond the obvious contribution of the selected primers and probes for amplification and detection, respectively, include the concentration of the various reagents and the thermocycling profiles used for amplification (Table 8.2). As a result, as with other diagnostic assays, the use of well characterized controls to monitor inter- and intraassay variability is essential. Introduction of the use of PCR for detecting HIV has resulted in several controversial reports that run counter to the experience of the remainder of the diagnostic community carrying out this procedure. PCR data in the absence of patient follow-up and supporting results from more established procedures, such as the FDA-approved EIAs or virus culture by an experienced laboratory, should be viewed with caution.

Similar to other diagnostic assays, the application of PCR to the detection of HIV proviral DNA has resulted in false positives and false negatives. False positives have been demonstrated to result from (1) cross-contamination of a negative sample from a positive sample; (2) contamination of clinical samples or the reagents for amplification with recombinant plasmids or phage harboring the entire HIV proviral genome or portions of it; and (3) "carryover" of PCR products from previous positive reactions. The latter is usually the reason for false positive because of the number of copies generated by the PCR (e.g., 10^6–10^{12}). Higuchi and Kwok[66] have recommended specific precautions to follow to minimize this type of contamination (see Carryover, below). In addition, two laboratories[67,68] have described the use of dUTP instead of dTTP and the other three conventional deoxynucleoside triphosphates in PCR as well as pretreatment of all reactions with uracil DNA glycosylase to eliminate or "sterilize" errant PCR products as templates for the amplification (Fig. 8.5). Just as PCR harnesses the replication capacity of cells, this procedure exploits the restriction/modification and excision/repair systems of cells. Because PCR products containing dU hybridize as efficiently as dT-containing PCR products and can be cloned and sequenced, this procedural modification promises to increase the reliability of positive results from a large number of laboratories.

Amplification of low copy numbers by PCR is vulnerable to interference by the amplified extension of primer pairs annealed to nontarget nucleic acid sequences in the test sample (termed mispriming) and by the amplified extension of two primers across one another's sequence without significant intervening sequence (termed primer dimerization). Primer dimers may experience amplified oligomerization during PCR to create complex mixtures of primer artifacts. The quantity of these primer artifacts often varies inversely with the yield of specific PCR product in low copy number amplifications. The resulting nonspecificity has numerous negative consequences for detection and quantitation of low copy number blood-borne infectious agents usually in the presence of high copy number host nucleic acid. PCR detection of HIV is typical of such an analysis, where amplification for 20 μl of blood containing about 1.6×10^5 diploid human genomes in 1 μg of DNA often generates an uninterpretable ethidium-stained amplified DNA pattern. It has been shown that, using PCR amplification of HIV-1 targets, most of the observed mispriming and primer dimerization arises during the customary and poorly controlled time interval (minutes) when reactants are mixed at room temperature before starting an amplification. To prevent this problem, a method called "hot start PCR" is employed. With this method some of the additions are delayed until all the reactants have been heated to a temperature that prevents primer annealing to nontarget sequences. Hot start PCR has been shown to increase amplification efficiency, specificity, and yield of low copy numbers of target sequences as with three gene targets of HIV-1.[69] An added benefit of the dUTP-UNG procedure is increased specificity and sensitivity of amplifications. By preincubating the amplification reactions at 50°C for 2 minutes prior to thermal cycling, non-specific extension products that were formed at ambient temperature are cleaved. Since the temperature of subsequent cycles is kept at or above the annealing temperature, non-specific extensions are reduced and an enzymatic hot start is achieved.[69a]

False-negative results have been attributed to compromised analytic sensitivity because of insufficient specificity either because of less than

TABLE 8.2. Reaction parameters to be evaluated for efficient PCR.

Annealing temperature and time
Denaturation temperature and time
Taq DNA polymerase addition at elevated temperature
Enzyme concentration
$MgCl_2$ concentration
Primer concentration
Co-solvents

optimal amplification conditions or the selection of primers and probes that do not readily recognize different sequence variants.[62]

As with all diagnostic assays, replication is an important factor for being confident that the results are reproducible. Samples that have disparate results in duplicate, not unlike the discordant EIA assays, can be caused by signals at the cutoff point for positivity or by sample mixup. Stochastically, a sample must contain five copies of HIV template to have a 99% likelihood of being reproducibly positive. If there are fewer than an average of five copies in a sample, the reactions may appear irreproducible due to sampling bias.

Previously, a multicenter, blinded proficiency trial was conducted using 105 HIV-1-seronegative, culture-negative samples from low risk blood donors and 99 HIV-1-seropositive and culture-positive samples.[70] The five laboratories participating in the study had significant experience with PCR, but the procedure and interpretive criteria varied somewhat, and only one of the multiple primer pairs was used in common. The average sensitivity for the laboratories was 99.0% and the average specificity 94.7%. One laboratory achieved 100% sensitivity and specificity. The overall false-positive, false-negative, and indeterminate rates were 1.8%, 0.8%, and 1.9%, respectively. This study demonstrated that PCR is a highly sensitive, specific assay for HIV-1 proviral DNA but that rigorous procedural and critical testing algorithms are required. Furthermore, the two primer pair systems targeted to the *gag* gene showed 100% sensitivity and specificity. This observation suggests that the inability to detect all samples known to contain HIV-1 proviral DNA at the requisite level with different primer pairs may be due to a laboratory's experimental performance rather than a viral sequence variant incapable of amplification.

The lack of concordance between duplicate samples may be due to sample mix-up. Resolution of sample mix-up when using serologic assays is difficult. Often the sample is either rerun, or another sample is obtained for analysis. However, PCR assays for HIV, particularly if amplification of the histocompatibility region is used as a control for the number of cells examined and amplification integrity of the sample, allows for simple resolution. The pioneering PCR studies of Erlich and colleagues[71] on HLA genotyping were later exploited for HIV[72] to demonstrate that due to the polymorphic nature of the region between the HLA DQα DNA primers used, samples from different individuals could be discerned because of the differential hybridization of HLA sequence-specific probes[73] (Fig. 8.6).

Quantitation of Plasma Viremia

Accurate determination of virus load, both cell-free virus and infected cell burden, is required for understanding the natural history of HIV-1

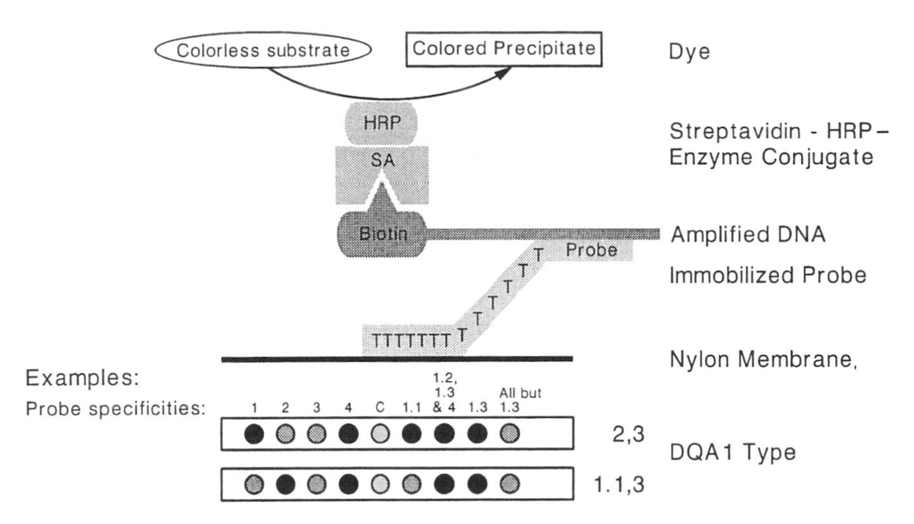

FIGURE 8.6. Immobilized oligonucleotide probe detection of amplified DNA. Reverse dot blot hybridization is used to rapidly and precisely type HLA DQα class II polymorphism. A sequence-specific probe is "tailed" with a dT homopolymer and immobilized on a solid support. The amplified PCR product, which has incorporated a biotinylated primer, hybridizes to the probe. After washing away unbound DNA and primers, the biotinylated, amplified DNA binds an avidin–horseradish peroxidase conjugate. The enzyme then converts a colorless dye to a colored precipitate. The format for detecting specific alleles in samples of amplified DQα DNA from heterozygous individuals is shown.

infection, predicting progression, and evaluating the efficacy of therapeutic drug regimens and vaccines. To date, CD4 T-helper lymphocytes together with elevated levels of β_2-microglobulin, neopterin, and interferon, delayed-type hypersensitivity, and clinical symptoms have served as surrogate markers for disease progression.[74,75] It is clear, however, that these markers represent indirect measures of the state of the viral infection and that virus load may provide better insight into the dynamics of the HIV-1 infection; moreover, virus load measurement is expected to complement the measurement of CD4 counts.

The ability to quantitate virus load in infected persons has been hampered by the low number of (1) circulating infected cells in the blood, (2) viral particles, and (3) expressed viral gene products. Viral cultures employing endpoint dilution have been used to quantify HIV-1 particles in plasma and in infected peripheral blood mononuclear cells[22,23] (see Chap. 7). This procedure is laborious, time-consuming, and expensive; and it requires handling large quantities of infectious virus. Direct detection of the HIV-1 p24 antigen in patient serum has been used, but the assay lacks sensitivity owing to its complexing with antigen-specific antibodies and to the low levels of the protein in the peripheral blood (see Chap. 6 for a detailed discussion). Acid dissociation of the immune complexes has been used to increase the sensitivity of p24 detection and has had some limited success.

The PCR technique for quantitation of RNA has been extensively reported and reviewed by Ferr[76] and Clementi et al.[77] The quantification of plasma HIV-1 RNA by PCR may prove to be a more important measure of infection status than culturable virus. RNA from both infectious and noninfectious particles can be detected by PCR; and given the fact that physical particles may be at a 10^4- to 10^7-fold excess over infectious particles,[78] PCR should be more sensitive that virus culture. The amount of viral RNA released into the plasma is an indirect measure of the transcriptional status of those infected cells that shed virus. The transcriptional activity of the virus is in turn associated with the expression of viral and cellular proteins that participate in viral pathogenesis. Most symptomatic patients with low CD4 counts have been shown to have higher viral titers than asymptomatic persons with high CD4 counts, although a wide range of viral titers have been reported at all stages of the infection cycle.[22,23] The variability of plasma virus titers among infected persons suggests that only by following plasma viremia in a single patient over time can the results assist in evaluating therapeutic efficacy. To quantitate HIV-1 RNA in patients encompassing the full spectrum of CD4 counts and disease states, an assay with the ability to detect as few as 200 copies/ml is required to avoid the need for excessively large volumes of plasma to monitor viremia.

The typical method of RNA PCR requires first the reverse transcription of the RNA (viral RNA in the case of HIV-1) to convert it to a DNA form, followed by standard PCR amplification. An assay has been described that has certain advantages over conventional reverse transcription coupled to PCR (RT-PCR) assays.[79] The sample preparation procedure requires only a single guanidium isothiocyanate (GuSCN) treatment of the plasma followed by an alcohol precipitation step. Instead of using commercially available retrovirus reverse transcriptases combined with *Taq* DNA polymerase, the new procedure uses a thermostable DNA polymerase that contains efficient RT and DNA polymerase activities.[80] The enzyme rTth DNA polymerase simplifies the two-enzyme systems and has been used successfully to detect hepatitis C virus.[81] The sensitivity of the amplification requires fewer cycles than are normally required. This assay incorporates an RNA quantitation standard that amplifies HIV-1 RNA without compromising amplification of the target. This point is critical for monitoring reaction variability. The new assay system utilizes the microwell detection assay, which provides a quantitative, colorimetric readout over a four-log dynamic range. Furthermore, the sensitivity of the amplification and detection system, in conjunction with the quantitation standard, provides quantification from a single amplification using only 50 μl of the sample being tested. This method contrasts with competitive PCR amplification systems, which require multiple amplifications.[82–84] The reported analytic sensitivity of this assay is 10 copies/ 50 µl, or 200 copies/ml of plasma. The effects of differential collection, storage, and processing of the patient's specimen have been studied. Studies have indicated that optimally sera or plasma should be separated from cells within 3 hours of collection and stored at −70°C.

Amplification procedures other than PCR have been developed, as well,[85–87] one of which is an RNA-based amplification system. It is an isothermal amplification technology capable of selective amplification of RNA in a large DNA background. The amplification is based on the simultaneous enzymatic activity of RNase H, AMV-reverse transcriptase (RT), and T7 RNA polymerase at 41°C. The isothermal reaction results in the accumulation of the target HIV sequence in single-stranded RNA, which allows direct, specific detection (Fig. 8.7). The method is based on the extension of primer 1 (which contains at T7 promoter) by AMV-RT on a (+) single-stranded HIV RNA, strand separation through RNA degradation by RNase H, synthesis of double-stranded DNA by AMV-RT, and large-scale (−) strand RNA synthesis by T7 RNA polymerase. The (−) strand RNA synthesis is the first step into the cyclic phase. The reported analytic sensitivity of detection is 10 to 100 copies of HIV-1 RNA. This technology has the capability of being adapted for quantitation of HIV RNA.[88]

Carryover

As mentioned above, considerable care must be taken to avoid carryover of DNA from one tube to another in order to prevent false positives.[66,89,90] Because amplified sequences are present in large numbers, carryover of minute quantities of amplified DNA can lead to significant false-positive problems. The following are procedures that should minimize carryover.

1. *Physical separation of pre- and post-PCR reactions.* To prevent carryover, a separate room or containment unit, such as a biosafety cabinet, should be used for setting up amplification reactions. A separate set of supplies and pipettes should be kept in this area and should be used only for setting up PCR reactions. Care must be taken to ensure that amplified DNA is not brought into this area. Reagents, devices, and supplies should never be taken and returned from an area where PCR analyses are being performed.

2. *Aliquot reagents.* Reagents should be aliquoted to minimize the number of repeated samplings. All reagents used in the PCR process must be prepared, aliquoted, and stored in an area that is free of PCR-amplified product. Similarly, oligonucleotides used for amplification should be synthesized and purified in an environment free of PCR product.

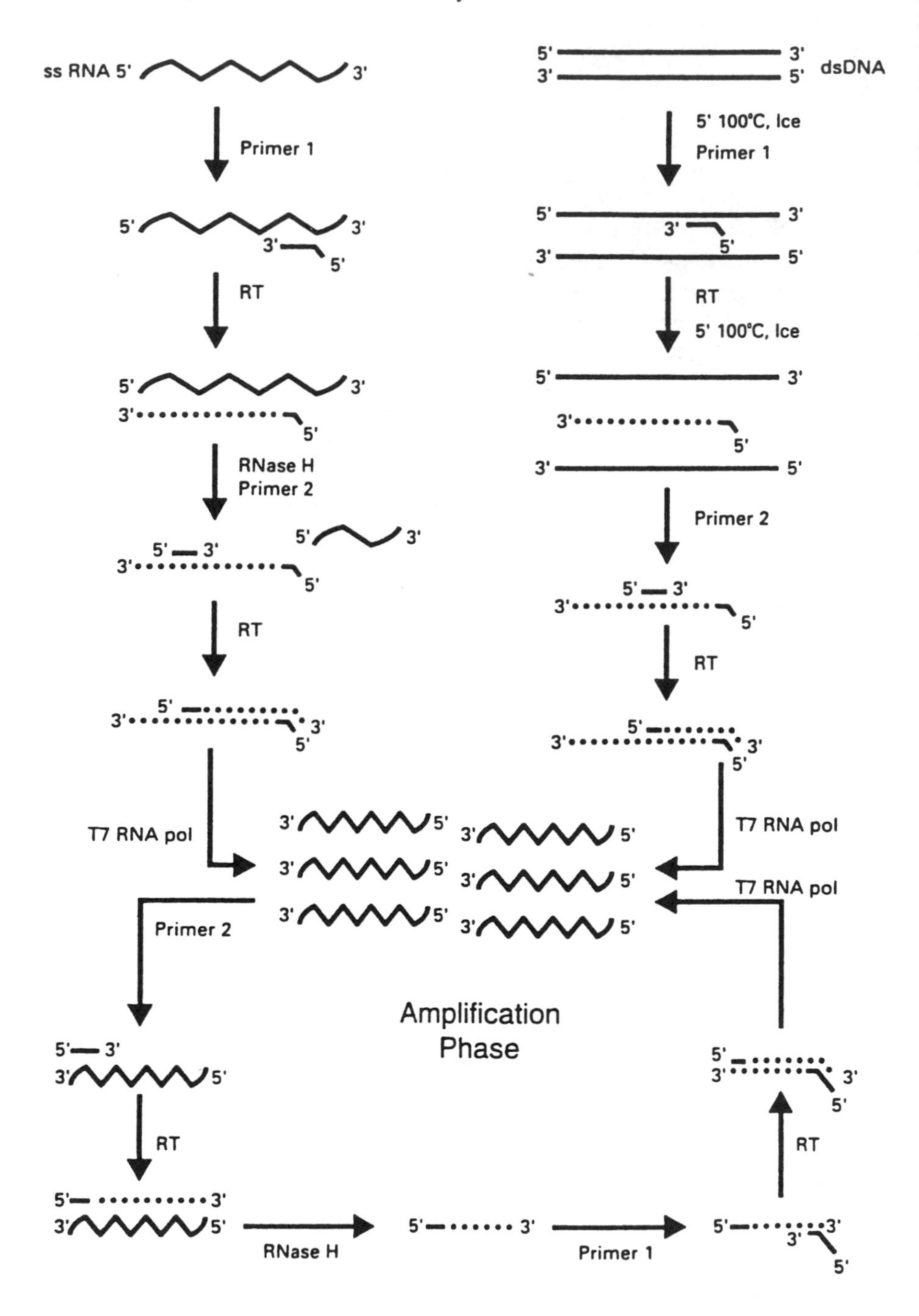

ss RNA 5'
3'
Primer 1
5'
3'
3'
5'
RT
RNase H
Primer 2
T7 RNA pol
Primer 2
RT
RNase H
Primer 1
Amplification
Phase
dsDNA
5' 100°C, Ice
Primer 1
RT
5' 100°C, Ice
Primer 2
RT
T7 RNA pol
T7 RNA pol
RT

3. *Positive displacement pipettes*. Contamination of pipetting devices can result in cross-contamination of samples. To eliminate cross-contamination of samples by pipetting devices, positive displacement pipettes with disposable tips or disposable tips with filters are recommended.

4. *Careful laboratory technique*. The following precautions should be taken during all aspects of PCR, from sample collection to PCR: (1) change gloves frequently; (2) uncap tubes carefully to prevent aerosols; (3) minimize sample handling; and (4) add nonsample components (mineral oil, dNTPs, primers, buffer, and enzyme) to the reaction mixture before adding the sample DNA. Cap each tube after the addition of DNA before proceeding to the next sample.

5. *Selection of controls*. For a positive control, select a sample that amplifies weakly but consistently. The use of a strong positive control results in the unnecessary generation of large amounts of amplified DNA sequences. Well characterized negative controls should also be used. The extreme sensitivity of the PCR process has the potential to amplify a nucleic acid sequence in a sample that is negative by all other criteria. Finally, multiple reagent controls should be included with each amplification because the presence of a small number of molecules of PCR product in the reagents may lead to sporadic positive results. The reagent controls should contain all the necessary components for the PCR process minus the template DNA.

In Situ PCR

The PCR has been successfully utilized to amplify DNA from formalin-fixed and even wax-embedded tissues. However, for the histopathologist, a limitation of PCR has been the inability to localize amplified DNA in

FIGURE 8.7. Isothermal amplification technology procedure applied to both single-stranded RNA (ssRNA) and double-stranded DNA (dsDNA). For RNA (top left) the technique is based on extension of primer 1 (which contains a T7 TNA polymerase promoter) by AMV-reverse transcriptase (RT) on the ssRNA template. RNase H degrades the RNA strand (or heat denaturation for the dsDNA) in the RNA:DNA hybrid that results from the AMV-RT enzyme activity. Subsequently, primer 2 can anneal to the resulting single-stranded cDNA, and the second DNA strand is synthesized by the DNA-dependent DNA polymerase activity of the AMV-RT, yielding a dsDNA molecule including a T7 RNA polymerase promoter sequence. The T7 RNA polymerase gives a 100- to 1000-fold increase in specific RNA. When DNA is used as input, two heat-denaturing steps (top right) are compulsory to obtain single-stranded DNA intermediates available for primer annealing. During the cyclic phase the events are the same as for RNA input, but primer 1 and primer 2 are incorporated in reverse order.

cells or tissue sections. This limitation has been overcome, however, with the reports of studies employing a combination of PCR with in situ hybridzation, termed in situ PCR. This technique allows localization of specific amplified DNA segments within isolated cells and tissue sections.

In situ PCR was first described by Haase and his colleagues,[91] who amplified a 1200 base pair *gag* gene segment of visna virus DNA from infected sheep choroid plexus cells followed by detection of the amplified DNA using in situ hybridization with a 150 basepair probe. The sensitivity of this in situ PCR was at least severalfold greater than in situ hybridization alone. The technique has subsequently been modified for the identification of different types of human papillomavirus in formalin-fixed, wax-embedded tissue sample.[92,93] In situ PCR has been used to detect low proviral copy HIV-1 in PBMCs when in situ hybridization alone could not detect HIV infection.[94,95] The usefulness of in situ PCR has also been demonstrated numerous times, including the identification of single-copy immunoglobulin gene arrangements in human B lymphocytes and the identification of various HLA DQ haplotypes in human PBMCs. The technique has also been successfully used to demonstrate latent and permissive HIV infection in routinely fixed and paraffin-embedded tissue.[96]

When compared with PCR performed in solution, DNA amplification from tissue sections or isolated cells is less efficient. It may be due in part to the relatively poor access of DNA primers, DNA polymerase, and other components to the target DNA. The most consistent in situ PCR results apparently were achieved using the "hot start" method of PCR in which the oligonucleotide primers and DNA polymerase are added only at high temperature and appeared to improve the DNA amplification efficiency.

Successful in situ PCR depends on the amplified DNA remaining localized. Why the amplified DNA does not diffuse during the procedure is unclear, but it may result from the network formation of the amplified DNA, which leads to the development of an insoluble high-molecular-weight DNA complex. Studies of in situ PCR have demonstrated that the length of the DNA fragment to be amplified did not alter the final localization of DNA after in situ hybridization.

After PCR, in situ hybridization is performed using standard methodology, including use of an oligonucleotide probe complementary to a portion of the amplified DNA sequences with the hybridization time kept to a minimum. Amplified DNA can be detected directly without performing in situ hybridization by the incorporation of labeled nucleotides or oligonucleotides in the PCR, which results in more rapid detection of the amplified DNA fragment.

To ensure that in situ PCR is amplifying and detecting a specific DNA sequence (e.g., from HIV proviral sequences) various controls must be included in the assay. The DNA amplification reaction must be shown to be enzyme-mediated, primer-dependent, and (if possible) of the appropriate molecular weight. Finally, omitting PCR should result in a reduced or absent in situ hybridization signal compared with in situ PCR.

Conclusion

As the AIDS epidemic continues to grow and spread, there is an increasing need for sensitive, quantitative assays for HIV. Quantitative tests for monitoring the infected cell load and cell-free particulate virus in infected persons are needed to monitor the *clinical* status of the patient and to evaluate the efficacy of new antiviral agents and potential vaccines.

The application of PCR technology to AIDS research opens up exciting possibilities for the sensitive, specific, direct detection and quantitation of HIV. PCR in its short existence has proved valuable for (1) detecting infection in seronegative persons, (2) quantifying the virus burden in a patient, (3) typing HIV infections, (4) measuring virus expression, (5) diagnosing perinatal transmission of HIV at an early stage, and (6) resolving indeterminate Western blottest results. As a research tool, the PCR technique is also proving useful for studying variant HIVs, distinguishing the important human retroviruses HTLV-I and HTLV-II, and discovering new pathogenic human retroviruses. Further simplification of PCR technology with the addition of sensitive nonisotopic detection systems requiring less than 1 hour for a quantitative readout should guarantee PCR a significant role in the diagnosis of HIV infection and AIDS.

References

1. Harper MH, Marselle LM, Gallo RC, Wong-Staal F. Proc Natl Acad Sci USA 1986;83:772–776
2. Shaw GM, Hahn BH, Arya SK, et al. Molecular characterization of human T-cell leukemia (lymphotropic) virus type III in the acquired immune deficiency syndrome. Science 1984;226:1165–1171
3. Mullis KB, Faloona FA. Specific synthesis of DNA in vitro via a polymerase-catalyzed chain reaction. Methods Enzymol 1987;155:335–350
4. Saiki PK, Scharf S, Faloona F, et al. Enzymatic amplification of β-globin genomic sequences and restriction site analysis for the diagnosis of sickle cell anemia. Science 1985;230:1350–1354
5. Saiki RK, Gelfand DH, Stoffel S, et al. Primer-directed enzymatic amplification of DNA with a thermostable DNA polymerase. Science 1988;239:487–491
6. Schochetman G, Ou C-Y, Jones W. Polymerase chain reaction. 1988;158:1154–1157
7. Hart C, Spira T, Moore J, et al. Direct detection of HIV RNA expression in seropositive subjects. Lancet 1988;2:596–599
8. Byrne BC, Li JJ, Sninsky J, Poiesz BJ. Detection of HIV-1 RNA sequences by in vitro DNA amplification. Nucleic Acids Res 1988;16:4165

9. Butcher A, Spadoro J. Using PCR for detection of HIV-1 infection. Clin Immunol Newslett 1992;12:73–76
10. Ou C-Y, Kwok S, Mitchell SW, et al. DNA amplification for direct detection of HIV-1 in DNA of peripheral blood mononuclear cells. Science 1988;239: 295–297
11. Kwok S, Mack DH, Mullis KB, et al. Identification of human immunodeficiency virus sequences by using in vitro enzymatic amplification and oligomer cleavage detection. J Virol 1987;61:1690–1694
12. Kwok S, Mack DH, Sninsky JJ, et al. Diagnosis of human immunodeficiency virus in seropositive individuals: viral sequences in peripheral blood mononuclear cells. In Luciw PA, Steimen KS (eds): HIV Detection by Genetic Engineering Methods. New York Marcel Dekker, 1989:243–255
13. Kellogg DE, Kwok S. Detection of human immunodeficiency virus. In Innis MA, Gelfand DH, Sninsky JJ, White TJ (eds): PCR Protocols. Orlando, FL: Academic Press, 1989:337–347
14. Ou C-Y, Schochetman G. Polymerase chain reaction in AIDS research. In Erlich HA, Gibbs R, Kazazian HH Jr (eds): Current Communications in Molecular Biology, Polymerase Chain Reaction. Cold Spring Harbor, NY: Cold Spring Harbor Laboratory Press, 1989:165–170
15. Meyerhans A, Cheynier R, Albert J, et al. Temporal fluctuations in HIV quasispecies in vivo are not reflected by sequential HIV isolations. Cell 1989;58:901–910
16. Goodenow M, Huet T, Saurin W, et al. HIV-1 isolates are rapidly evolving quasispecies: evidence for viral mixtures and preferred nucleotide substitutions. J Acquir Immune Defic Synd 1989;2:344–352
17. Jackson JB, Kwok SY, Sninsky JJ, et al. Human immunodeficiency virus type 1 detected in all seropositive symptomatic and asymptomatic individuals. J Clin Microbiol 1990;28:16–19
18. Kellogg DE, Sninsky JJ, Kwok S. Quantitation of HIV-1 proviral DNA relative to cellular DNA by the polymerase chain reaction. Anal Biochem 1990;189:202–208
19. Ou C-Y, McDonough SH, Cabanas D, et al. Rapid and quantitative detection of enzymatically amplified HIV-1 DNA using chemiluminescent oligonucleotide probes. AIDS Res Hum Retrovir 1990;6:1323–1329
20. Schnittman SM, Psallidopoulos MS, Lane HC, et al. The reservoir for HIV-1 in human peripheral blood is a T cell that maintains expression of CD4. Science 1990;245:305–308
21. Ho DD, Moudgil T, Alam M. Quantitation of human immunodeficiency virus type 1 in the blood of infected persons. N Engl J Med 1989;321:1626–1625
22. Coombs RW, Collier AC, Allain J-P, et al. Plasma viremia in human immunodeficiency virus infection. N Engl J Med 1989;321:1626–1631
23. Ratner L. Measurement of human immunodeficiency virus load and its relation to disease progression. AIDS Res Hum Retrovir 1989;5:115–119
24. Lion T, Nighet R, Hutchinson MA, Golomb HM, Brownstein BH. Rapid dot blot quantitation for viral DNA and amplified genes in less than 1000 cells. DNA 1989;8:361–367
25. Schnittman SM, Greenhouse JJ, Miltiades BS, et al. Increasing viral burden in $CD4^+$ T cells from patients with human immunodeficiency virus (HIV)

infection reflects rapidly progressive immunosuppression and clinical disease. Ann Intern Med 1990;113:438–443
26. Simmonds P, Balfe P, Peutherer JF, et al. Human immunodeficiency virus-infected individuals contain provirus in small numbers of peripheral mononuclear cell and at low copy numbers. J Virol 1990;64:864–872
27. Spear GT, Ou C-Y, Kessler HA, et al. Analysis of lymphocytes, monocytes, and neutrophils from human immunodeficiency virus (HIV) infected persons for HIV DNA. J Infect Dis 1990;162:1239–1244
28. Hart C, Chang S-Y, Kwok S, et al. A replication-deficient HIV-1 DNA used for quantitation of the polymerase reaction (PCR). Nucleic Acids Res 1990; 18:4029–4030
29. Clouse KA, Powell D, Washington I, et al. Monokine regulation of human immunodeficiency virus-1 expression in a chronically infected human T cell clone. J Immunol 1989;142:431–438
30. Wang AM, Doyle MV, Mark DF. Quantitation of mRNA by the polymerase chain reaction. Proc Natl Acad Sci USA 1989;86:9717–9721
31. Becker-Andre M, Hahlbrock K. Absolute mRNA quantification using the polymerase chain reaction (PCR): a novel approach by a PCR acided transcript titration assay (PATTY). Nucleic Acids Res 1989;17:9437–9446
32. Gilliland G, Perrin S, Blanchard K, Bunn HF. Analysis of cytokine mRNA and DNA: detection and quantitation by competitive polymerase chain reaction. Proc Natl Acad Sci USA 1990;87:2725–2729
33. Murakawa GJ, Zaia JA, Spallone PA, et al. Direct detection of HIV-1 RNA from AIDS and ARC patient samples. DNA 1988;7:287–295
34. Goudsmit J, Paul DA, Lange J, et al. Expression of human immunodeficiency virus antigen (HIV-Ag) in serum and cerebrospinal fluid during acute and chronic infection. Lancet 1986;2:177–180
35. Kessler H, Blaauw B, Spear J, et al. Diagnosis of human immunodeficiency virus infection in seronegative homosexuals presenting with an acute viral syndrome. JAMA 1987;258:1196–1199
36. Ranki A, Valle S, Krohn M. Long latency precedes overt seroconversion in sexually transmitted human immunodeficiency virus infection. Lancet 1987; 2:589–593
37. Simmonds P, Lainson FAL, Cuthbert R, et al. HIV infection and antibody detection: variable responses to infection in the Edinburgh haemophilic cohort. BMJ 1987;296:593–598
38. Ward JW, Schable C, Dickinson GM, et al. Acute human immunodeficiency virus (HIV) infection: antigen detection and seroconversion in immunosuppressed patients. Transplantation 1989;47:722–724
39. Loche M, Mach B. Identification of HIV-infected seronegative individuals by a direct diagnostic test based on hybridization to amplified viral DNA. Lancet 1988;2:418–421
40. Wolinsky SM, Rinaldo CR, Kwok S, et al. Human immunodeficiency virus type 1 (HIV-1) infection a median of 18 months before a diagnostic Western blot: evidence from a cohort of homosexual men. Ann Intern Med 1989; 111:961–972
41. Hewlett IK, Gregg RA, Mayner RE, et al. Detection of HIV proviral DNA and p24 antigen in plasma prior to seroconversion. IVth Int Conf AIDS 1988;1:137

42. Imagawa DT, Lee MH, Wolinsky SM, et al. Human immunodeficiency virus type 1 infection in homosexual men who remain seronegative for prolonged periods. N Engl J Med 1989;320:1458–1462
43. Horsburgh CR Jr, Ou C-Y, Jason J, et al. Duration of human immunodeficiency virus infection before detection of antibody. Lancet 1989;2:637–640
44. Farzadegan H, Vlahov D, Solomon L, et al. Detection of human immunodeficiency virus type 1 infection by polymerase chain reaction in a cohort of seronegative intravenous drug users. J Infect Dis 1993;168:327–331
45. Jackson JB, MacDonald KL, Cadwell J, et al. Absence of HIV infection in blood donors with indeterminate Western blot tests for antibody to HIV-1. N Engl J Med 1990;322:217–222
46. George JR, Rayfield MA, Phillips S, et al. Efficacies of U.S. Food and Drug Administration-licensed HIV-1 screening enzyme immunoassays for detecting antibodies to HIV-2. AIDS 1989;4:321–326
47. Rey F, Salaun D, Lesbordes JL, et al. HIV-1 and HIV-2 double infection in Central African Republic. Lancet 1986;2:1391–1392
48. Rey MA, Girard PM, Harzic M, et al. HIV-1 and HIV-2 double infection in French homosexual male with AIDS related complex. Lancet 1987;1:388–389
49. Foucault C, Lopez O, Jourdan G, et al. Double HIV-1 and HIV-2 seropositivity among blood donors. Lancet 1987;1:165–166
50. Rayfield M, DeCock K, Heyward WL, et al. Mixed human immunodeficiency virus (HIV) infection of an individual: demonstration of both HIV type 1 and HIV type 2 proviral sequences by polymerase chain reaction. J Infect Dis 1988;158:170–176
51. Centers for Disease Control. AIDS due to HIV-2 infection—New Jersey. MMWR 1988;37:33–35
52. Krivine A, Firtion G, Cao L, et al. HIV replication during the first weeks of life. Lancet 1992;339:1187–1189
53. Rogers MF, Ou C-Y, Rayfield M, et al. Use of the polymerase chain reaction for early detection of the proviral sequences of human immunodeficiency virus in infants born to seropositive mothers. N Engl J Med 1989;320:1649–1654
54. Rogers MF, Ou C-Y, Kilbourne B, Schochetman G. Advances and problems in the diagnosis of HIV infection in infants. In Pizzo PA, Wilfert CM (eds): Pediatric AIDS, The Challenge of HIV Infection in Infants, Children and Adolescents. Baltimore: Williams & Wilkins, 1990:159–174
55. Chadwick EG, Yogev R, Kwok S, et al. Enzymatic amplification of the human immunodeficiency virus in peripheral blood mononuclear cells from pediatric patients. J Infect Dis 1989;160:954–959
56. Pizzo PA, Eddy J, Falloon J, et al. Effects of continuous intravenous infusion of zidovudine (AZT) in children with symptomatic HIV infection. N Engl J Med 1988;319:889–896
57. Larder BA, Darby G, Richman DD. HIV with reduced sensitivity to zidovudine (AZT) isolated during prolonged therapy. Science 1988;246:1731–1734
58. Larder BA, Kemp SD. Multiple mutations in HIV-1 reverse transcriptase confer high-level resistance to zidovudine (AZT). Science 1989;246:1155–1158
59. Kellam P, Boucher CAB, Larder BA. Fifth mutation in human immunodeficiency virus type 1 reverse transcriptase contributes to the development

of high-level resistance to zidovudine. Proc Natl Acad Sci USA 1992;89: 1934–1938
60. Boucher CAB, Tersmette M, Lange JMA, et al. Zidovudine sensitivity of human immunodeficiency viruses from high-risk, symptom-free individuals during therapy. Lancet 1990;336:585–590
61. Kozal MJ, Shafer RW, Winters MA, Katzenstein DA, Merigan TC. A mutation in human immunodeficiency virus reverse transcriptase and decline in CD4 lymphocyte numbers in long-term zidovudine recipients. J Infect Dis 1993;167:526–532
62. Kwok S, Kellogg DE, McKinney N, et al. Effects of primer-template mismatches on the polymerase chain reaction: human immunodeficiency virus type 1 model studies. Nuclieic Acids Res 1990;18:999–1005
63. St Clair MH, Martin JL, Tudor-Williams G, et al. Resistance to ddI and sensitivity to AZT induced bya mutation in HIV-1 reverse transcriptase. Science 1991;235:1557–1559
64. Wainberg MA, Gu Z, Gao Q, et al. Clinical correlates and molecular basis of HIV drug resistance. J Acquir Immune Defic Syndr 1993;6(suppl 1): S536–546
65. Fitzgibbon JE, Howell RM, Haverzettl C, et al. Human immunodeficiency virus type 1 pol mutations which cause decreased susceptibility to 2′,3′-dideoxycytidine. Antimicrobial Agents Chemother 1992;36:153–157
66. Higuchi R, Kwok S. Avoiding false positives with PCR. Nature 1989;339: 237–238
67. Longo MC, Berninger MS, Hartley JL. Use of uracil DNA glycosylase to control carryover contamination in polymerase chain reactions. Gene 1990; 93:125–128
68. Sninsky JJ, Gates C, McKinney N, et al. dUTP and uracil-N-glycolase in the polymerase chain reaction: a resolution to carryover. In preparation
69. Chou Q, Russell M, Birch DE, Raymond J, Bloch W. Prevention of pre-PCR mispriming and primer dimerization improves low-copy-number amplifications. Nucleic Acids Res 1992;20:1717–1723
69a. Persing DH, Cimino GD. Amplification product inactivation methods in diagnostic molecular microbiology: principles and application. In Persing DH, Smith TF, Tenover FC, White TJH (eds): ASM, Washington, D.C. 1983; pp 1050–121.
70. Sheppard HA, Ascher MS, Busch MP, et al. A multicenter proficiency trial of gene amplification (PCR) for the detection of HIV-1. J Acquir Immune Defic Syndr 1991;4:277–283
71. Saiki RK, Bugawan TL, Horn GT, Mullis KB, Erlich HA. Analysis of enzymatically amplified beta-globin and HLA DQ alpha with allele-specific oligonucleotide probes. Nature 1986;324:163–165
72. Farzadeagan H, Polis MA, Wolinsky SM, et al. Loss of human immunodeficiency virus type 1 (HIV-1) antibodies with evidence of viral infection in asymptomatic men. Ann Intern Med 1988;108:785–790
73. Saiki R, Walsh R, Levenson CH, Erlich HA. Genetic analysis of amplified DNA with immobilized sequence-specific oligonucleotide probes. Proc Natl Acad Sci USA 1989;86:6230–6234
74. Anderson RE, Lang W, Shiboski S, et al. Use of β_2-microglobulin level and CD4 lymphocyte count to predict development of acquried immunodeficiency virus infection. Arch Intern Med 1990;15:73–77

75. Fahey JL, Taylor JMG, Detels R, et al. The prognostic value of cellular and serologic markers in infection with human immunodeficiency virus type 1. N Engl J Med 1990;322:166–172
76. Ferre F. Quantitative or semi-quantitative PCR: reality versus myth. PCR Methods Appl 1992;2:1–9
77. Clementi M, Menzo S, Bagnarelli P, et al. Quantitative PCR and RT-PCR in virology. PCR Methods Appl 1993;2:191–196
78. Layne SP, Merges MJ, Dembo M, et al. Factors underlying spontaneous inactivation and susceptibility to neutralization of human immunodeficiency virus. Virology 1992;189:695–714
79. Mulder J, McKinney N, Christopherson C, et al. A rapid and simple PCR assay for quantitation of HIV-1 RNA in plasma: application to acute retroviral infection. J Clin Microbiol 1994;32:292–300
80. Myers TW, Gelfand DH. Reverse transcription and DNA amplification by a Thermus thermophilis DNA polymerase. Biochemistry 1991;30:7661–7666
81. Young KKY, Resnick RM, Myers TW. Detection of hepatitis C virus RNA by a combined reverse transcription-polymerase chain reaction assay. J Clin Microbiol 1993;31:882–886
82. Becker-Andre M, Hahlbrock K. Absolute mRNA quantification using the polymerase chain reaction (PCR): a novel approach by a PCR aided transcript titration assay (PATTY). Nucleic Acids Res 1989;17:9437–9446
83. Gilliland G, Perrin S, Banchard K, Bunn HF. Analysis of cytokine mRNA and DNA: detection and quantitation by competitive polymerase chain reaction. Proc Natl Acad Sci USA 1990;87:2725–2729
84. Platak M, Saag MS, Yang LC, et al. High levels of HIV-1 in plasma during all stages of infection determined by competitive PCR. Science 1993;259: 1749–1754
85. Wu DY, Wallace RB. The ligation amplification reaction (LAR)—amplification of specific DNA sequences using sequential rounds of template-dependent ligation. Genomics 1989;4:560–569
86. Kwoh DY, Davis GR, Whitfield KM, et al. Transcription-based amplification system and detection of amplified human immunodeficiency virus type 1 with a bead-based sandwich hybridization format. Proc Natl Acad Sci USA 1989; 86:1173–1177
87. Kievits T, van Gemen B, van Strijp D, et al. NASBA isothermal enzymatic in vitro nucleic acid amplification optimized for the diagnosis of HIV-1 infection. J Virol Methods 1991;35:273–286
88. Van Gemen B, Kievits T, Schukkink R, et al. Quantification of HIV-1 RNA in plasma using NASBA during HIV-1 primary infection. J Virol Methods 1993;43:177–188
89. Dragon EA. Handling reagents in the PCR laboratory. PCR Methods Appl 1993;3:S8–S9
90. Hartley JL, Rashtchian A. Dealing with contamination: enzymatic control of carryover contamination in PCR. PCR Methods Appl 1993;3:S10–S14
91. Haase AT, Retzel EF, Staskus KA. Amplification and detection of lentiviral DNA inside cells. Proc Natl Acad Sci USA 1990;87:4971–4975
92. Nuovo GJ, MacConnell P, Forde A, Delvenne P. Detection of human papillomavirus DNA in formalin-fixed tissues by in situ hybridization after amplification by polymerase chain reaction. Am J Pathol 1991;139:847–854

93. Nuovo GJ, Becker J, Margiotta M, et al. Histological distribution of polymerase chain reaction-amplified human papillomavirus 6 and 11 DNA in penile lesions. Am J Surg Pathol 1992;16:269–275
94. Nuovo GJ, Margiotta M, Mac Connell P, Becker J. Rapid in situ detection of PCR-amplified HIV-1 DNA. Diagn Mol Pathol 1992;1:98–102
95. Bagasra O, Hauptman SP, Lischner HW, Sachs M, Pomerantz RJ. Detection of human immunodeficiency virus type-1 provirus in mononuclear cells by in situ polymerase chain reaction. N Engl J Med 1992;326:1385–1391
96. Embretson J, Zupancic M, Beneke J, et al. Analysis of human immunodeficiency virus-infected tissues by amplification and in situ hybridization reveals latent and permissive infections at single-cell resolution. Proc Natl Acad Sci USA 1993;90:357–361

9 Use of Flow Cytometry to Enumerate Lymphocyte Populations in HIV Disease

JANET K.A. NICHOLSON and ALAN L. LANDAY

Immunophenotyping combines the use of fluorochrome-labeled monoclonal antibodies (to help identify cell populations) and the flow cytometer (to evaluate these individual cells rapidly in a correlated multiparametric analysis).[1–3] Monoclonal antibodies for identifying human cell populations were first produced during the late 1970s[4–6] when cell sorters (large, complex flow cytometers that can physically sort individual cells) were available in limited numbers, mostly in research laboratories. Immunophenotyping using flow cytometry came into the forefront after the acquired immunodeficiency syndrome (AIDS) epidemic was recognized in mid-1981, and one of the laboratory findings was a decrease in the number of $CD4^+$ T cells.[7,8] Fortunately, at that time monoclonal antibodies to recognize human $CD4^+$ T cells had been developed, though many other antibodies to human antigens were not yet available. Since then, much simpler flow cytometers have been developed for the clinical laboratory; and many more monoclonal antibodies, recognizing a wide variety of functional and differentiation antigens of cell populations, have been developed.

The clinical flow cytometer is comprised of four basic units: (1) optical system; (2) fluidics system; (3) electronic system; and (4) analysis system, or computer (Fig. 9.1).[1–3] The optical system consists of a mercury arc lamp or, more commonly, a small air-cooled laser that is used to examine each cell and excite any fluorochromes that may be attached to the cell. To discriminate between various fluorochromes that may be excited on the cells, the emitted light passes through a series of optical filters to allow light of specific wavelengths into the fluorescence detectors. These wavelengths generally capture most of the emitted signal for a particular fluorochrome. In most clinical flow cytometers, there are three detectors capable of detecting three fluorochromes.

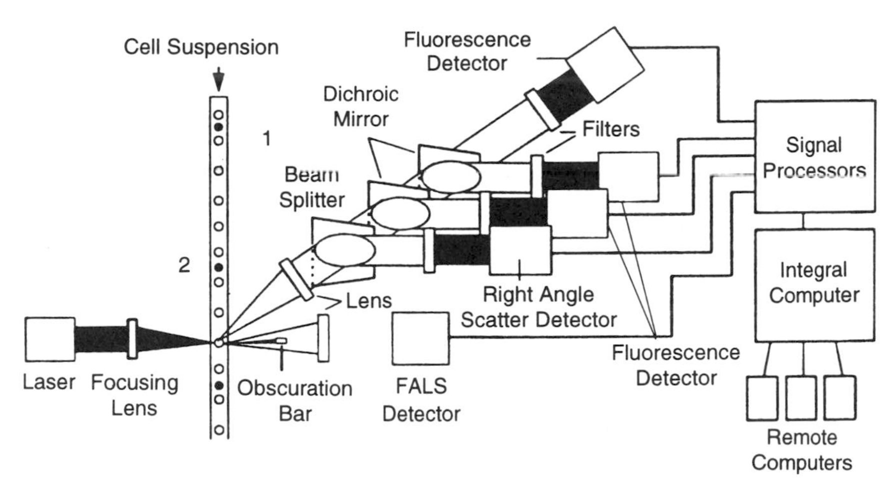

FIGURE 9.1. Schematic representation of modern flow cytometer suitable for clinical work. Stained cells enter flow chamber (1) and pass into center of stream of sheath fluid in single file (2). They are then struck by focused laser beam and emit scattered and fluorescent light, separated according to wavelength by appropriate mirrors and filters. An obscuration bar protects forward-angle light-scatter (FALS) detector from exposure to direct laser beam. Three fluorescence detectors are shown plus detector to measure laser light scattered perpendicular to laser beam by cells. Signals from detectors pass to amplifying processors and then to integral (onboard) computer, which digitizes, stores, and displays the signals. Detailed analysis of data is often most efficient with stand-alone computers.

The fluidics system allows the cells to pass individually through the flow cytometer, where they are interrogated by the laser light. This step is achieved by a process known as laminar flow or hydrodynamic focusing, whereby the sample containing individual cells is injected into a stream of saline (commonly called sheath fluid), which because of the hydrodynamic forces causes the sample stream and the cells in it to be centered in the fluid stream as it passes through a cuvette.

The electronics system collects light signals from the cells that have been interrogated by the laser. Detectors measure refracted light (forward-angle light scatter), reflected light (right-angle light scatter), and fluorochrome-excited light (commonly green, orange, and red). Because the fluorescence signals are of relatively low intensity, photomultiplier tubes (PMTs) are used to collect these signals. The light scatter detectors are commonly either photodiodes or PMTs that collect light scatter that has not been filtered through optical filters for fluorescence.

The analysis system consists of a computer that is used for data collection and analysis and for making adjustments to the electronics. A number of software programs for data analysis are available for on-line and off-line analysis. Data storage is usually on some sort of removable disk, as flow cytometry data can consume a significant number of bytes.

In general, the flow cytometer functions as follows. Cells in a single-cell suspension are introduced into the flow cytometer and pass single file through a cuvette. The laser light intersects the stream of cells; and as it strikes a cell, the cell scatters the light in all directions. This light is then collected in various detectors. Scattered light measured in the photodiode in the same plane as (and facing) the laser light is related to cell size. The remaining detectors measure scattered light at right angles to the laser light. This scattered light at 90 degrees is related to cell granularity. The remaining light, measured at 90 degrees, passes through optical filters and is collected in PMTs. Each PMT measures the fluorescence intensity at a wavelength specified by the filters placed in front of it. All these signals are collected and, after some conversions, are sent to the computer (Fig. 9.2). Thus as many as five parameters are measured on each individual cell. For each cell, the measurements take place within a few microseconds, thereby allowing the flow cytometer to analyze accurately thousands of cells in only a few minutes.

The flow cytometer is a powerful instrument with considerable flexibility, which is advantageous because there can be many applications using only one instrument. However, because of the flexibility, considerable expertise is needed to understand and use the flow cytometer optimally. Additionally, it is important to optimize the flow cytometer daily and ensure that it performs the same from day to day. This consistency is achieved through a comprehensive quality control (QC) program conducted on a daily basis.[9,10] Such a program consists of the following: (1) optimally aligning the flow cytometer so the laser beam intersects the center of the fluid stream (alignment); (2) adjusting the flow

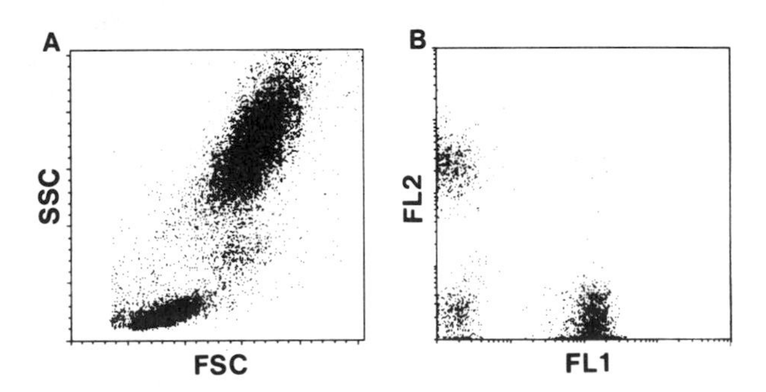

FIGURE 9.2. Light scatter and fluorescence histograms. Light scatter for whole blood (Panel A) shows linear forward scatter (FSC) on the x axis and linear side scatter (SSC) on the y axis. Fluorescence histogram for FL1 (FITC-CD3, in this case) and FL2 (PE-CD19, in this case) is in logarithmic units. Unlabeled cells (natural killer cells, in this case) are in the lower lefthand quadrant.

cytometer so a standard fluorescent particle falls in the same fluorescence channel every day (standardization); (3) resolving dimly labeled cells from negative cells (resolution); and (4) adjusting spectral compensation (color compensation).[10] (Color or spectral compensation is a way to subtract electronically the emitted fluorescence of one fluorochrome from the PMT measuring another fluorochrome.) A complete QC program consists of not only performing these functions but keeping logs of such activities and monitoring the instrument for any fluctuations in these parameters, which would indicate a mechanical problem.

Biosafety in the Flow Cytometry Laboratory

Specimens from patients infected with human immunodeficiency virus (HIV) and used for diagnostic purposes should be handled using biosafety level 2 procedures.[11] Universal precautions must be followed, and special care should be taken during the parts of this procedure that have the potential for creating aerosols or droplets (i.e., when unstoppering blood tubes, vortexing, and centrifuging). These procedures should be done in a biosafety cabinet (level I or II) or, in the case of centrifuging, in safety carriers.

Biosafety procedures unique to flow cytometers include procedures for decontamination. Household bleach is recommended for disinfecting the flow cytometer at the end of the day: A solution of 10% bleach (0.5% sodium hypochlorite) is introduced into the sample delivery system of the flow cytometer as one would enter a sample. This procedure is followed by flushing with water, as the bleach is corrosive. The wastes can be disinfected by putting undiluted bleach (5% sodium hypochlorite) in the waste receptacle to one tenth the volume of the receptacle. When the receptacle is full of waste, the final dilution of bleach is 10%.

Cell-free HIV has been shown to be inactivated using 1% paraformaldehyde, a fixative commonly used for immunophenotyping.[12] Some of the commercial reagents for lysing red blood cells (RBCs) and fixing leukocytes were tested for their effectiveness in reducing the viral load of HIV-infected cells in whole blood.[13] The reagents that inactivated cell-associated HIV 3 to 5 logs included FACS Lysing Solution, Coulter Immunolyse, Coulter Immunoprep (Q-Prep), and Gentrak Lysing Solution. Ammonium chloride, including Orthomune Lysing Reagent, is not effective in inactivating HIV-infected cells. Paraformaldehyde or formaldehyde fixation may be used to inactivate HIV-infected cells lysed with ammonium chloride. These fixatives should be used at 1% or 2% in buffered (pH 7.0–7.4) solution. An incubation at 4°C for 30 minutes has been shown to inactivate HIV-infected cells 3 to 4 logs.[13–15] Although these reagents, used as described, inactivated HIV-infected cells 3 to 5 logs, they did not totally inactivate the cells.[13] How much is enough is still open to question, as the frequency of HIV-infected cells in vivo in peripheral blood in various stages of HIV infection is not precisely known.

Methods for Immunophenotyping

Performing immunophenotypic analysis by flow cytometry begins with attention to sample preparation and staining.[9,10] Sample preparation for immunophenotyping is relatively straightforward. Whole anticoagulated blood is combined with fluorochrome-labeled monoclonal antibodies specific for cellular antigens of interest and incubated to allow antigen–antibody reactions on the cells to take place. The RBCs are lysed, and the specimens are then introduced into the flow cytometer and analyzed for fluorescence and light scatter patterns. Although this procedure is simple, there are a number of steps that require elaboration.

Specimen Collection

Because we are measuring cellular components of the blood, whole blood must be collected in an anticoagulant. The most commonly used anticoagulant for immunophenotyping is ethylenediamine tetraacetate (EDTA), either the disodium or the tripotassium salt. Because hematology measures may be done on the same tube of blood used for immunophenotyping, tripotassium EDTA (recommended for hematology) is often used. For immunophenotyping, the EDTA blood may be as old as 1 day and still yield results similar to those obtained with fresh blood. After this time, CD3 (T cell) percentages tend to increase, with a decrease in CD19 (B) cells.[16] In cases where specimens are older than 1 day but less than 2 days old when processed, heparin or acid-citrate-dextrose (ACD) may be used. Because specimens that are shipped cannot always have guaranteed arrival times, laboratories must strictly adhere to rejection criteria that include the age of the blood and the appropriate anticoagulant.

An absolute CD4 cell count is dependent on hematology measurements [white blood cell (WBC) count and leukocyte differential]; therefore care must be taken to ensure that specimens are collected and processed appropriately for hematology determinations. It is important that these measurements are made as soon after blood collection as possible, as the criteria used for distinguishing leukocyte populations may be sensitive to aged blood. When using automated hematology counters (the recommended procedure), specimens may be as old as 18 to 24 hours of age, depending on the instrumentation and specimen age limits provided by the manufacturer. If only normal data are provided by the manufacturer for these limits, it is necessary for laboratories testing specimens from HIV-infected individuals to establish blood age limits using blood samples from both HIV-positive and HIV-negative persons. If there are problems with blood integrity and stability, the specimen is likely from an HIV-infected individual.

Hematology tests should be done using automated hematology instruments. Although WBC counts are often done this way, leukocyte differential counts may not. It is important that automated differential counts be performed because it decreases the variability of the determination.[17] The hematology instrument can count thousands of cells accurately, whereas manual methods often count only 100 to 200. To approximate some confidence with the manual counts, at least 400 (preferably 1000) cells should be counted.[17] This procedure is time-consuming and labor-intensive, and smears may not have enough cells to enable counting 400 to 1000 cells. Automated hematology instruments flag specimens that are believed to have a problem—which is often in nonlymphocyte populations—and in these cases the automated lymphocyte differential is still acceptable. Only when the flag is for the lymphocyte population is a manual differential count, counting at least 400 cells, acceptable.

Transport temperatures have been a concern of laboratories since the reports of "refrigerator AIDS," a phenomenon where blood kept at 4°C and separated by density gradient centrifugation into lymphocyte preparations yielded a decreased percent of $CD4^+$ T cells.[18] Most laboratories that are immunophenotyping specimens no longer separate the lymphocytes before labeling them but, rather, use a whole-blood lysis technique. Cold temperatures do not have a deleterious effect on lymphocyte immunophenotyping by this method.[19] Nevertheless, most laboratories request that whole-blood specimens be held at room temperature until testing.

All specimens accepted in the clinical laboratory must be examined for specimen integrity, and each laboratory should have criteria for specimen acceptance and rejection. Such criteria pertain to hemolysis, clotting, temperature, age, and the anticoagulant. Specimens that do not appear in good condition (e.g., frozen, severely hemolyzed, clotted) should be rejected and new specimens requested. Under some conditions, for which the specimen integrity is questionable (e.g., hot to the touch), specimens may be accepted, but their condition should be noted on the laboratory worksheet. Effects of jostling or agitating specimens, as might result during shipping, have not been tested, though it might be surmised that it is not a problem, as most shipped specimens received in laboratories are of good quality. Any problems encountered during specimen processing or analysis may be a clue that the specimen is not acceptable. One particularly useful criterion for specimen integrity is the light scatter pattern (Fig. 9.3).

Specimen Processing

Whole blood is incubated with monoclonal antibodies that are conjugated with fluorochromes. The monoclonal antibodies commonly used for

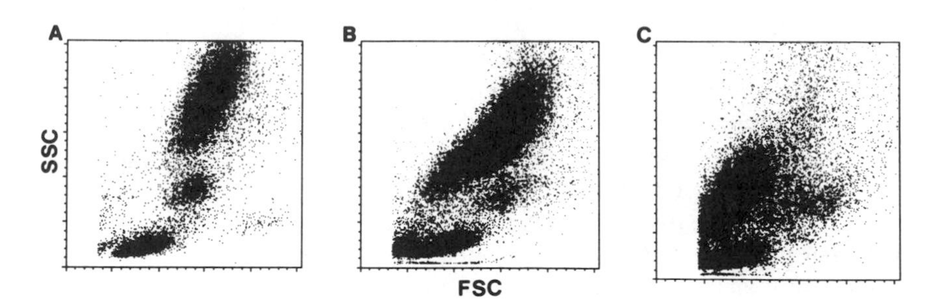

FIGURE 9.3. Light scatter patterns from poor specimens. Panel A shows normal light scatter. Panels B and C show abnormal light scatter (Panel B from blood heated to 47C for 30 minutes; Panel C from older blood from an AIDS patient).

routine immunophenotyping are lineage markers; that is, they identify cellular antigens associated with certain lineages of cells, such as T cells and B cells. In addition, monoclonal antibodies that recognize differentiation antigens, leukocyte adhesion molecules, activation antigens, and functional antigens may be used for studies that enumerate particular functional subsets of cells. Two-color immunofluorescence is most commonly used in laboratories that test specimens from HIV-infected patients. One advantage is that subsets of T cells, such as $CD4^+$ and $CD8^+$ T cells, can be more precisely identified by both the antigen of interest (e.g., CD4) and a marker for T cells (CD3). Another advantage is to eliminate unwanted cells from a particular analysis. For example, CD56 and CD16 are found on natural killer (NK) cells and some T cells (particularly CD56). Use of CD3 with CD56 or CD16 allows identification of non-T NK cells by enumerating only the $CD56^+$ or the $CD16^+$ cells that are $CD3^-$. Lastly, by using monoclonal antibodies that recognize two non-overlapping populations of cells, such as T cells and B cells, we can identify two populations in one tube, whereas we would use two tubes for single-color analysis.

The fluorochrome-labeled antibodies are incubated with whole blood at room temperature, commonly for 10 to 30 minutes, depending on the manufacturer's recommendation (Fig. 9.4). It is important that this incubation be done in the dark, as fluorochromes can be photobleached by exposure to light. An aliquot of blood (100 μl) is incubated with 10 to 20 μl antibody (usually 1–2 μg monoclonal antibody) in each tube. The antibodies should be used as premixed combinations provided by the manufacturer. If the laboratory elects to mix two single-color reagents, it is important to titrate these reagents properly with each other to determine the appropriate volumes of antibody needed.

The ideal processing methodology is one in which the cellular components of the blood are not altered; thus the cells sampled in the

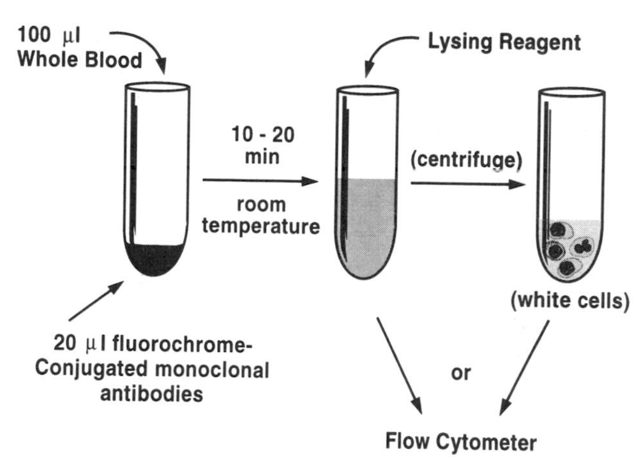

FIGURE 9.4. Scheme of immunophenotyping using the whole blood lysis technique.

specimen during data collection are representative of the cells found naturally in the specimen.[20] The best way to keep the cells similar to their natural state is to use a whole-blood lysis technique. After blood is incubated with the fluorochrome-labeled monoclonal antibodies, a lysing reagent is added. There are several commercial reagents that both lyse the red blood cells and fix the leukocytes. The mode of action of these reagents varies, depending on the vendor. In some cases but not others, volumes and incubation times must be exact.

Once the RBCs have been lysed, most protocols call for washing the specimens to remove cellular debris, although others do not. The need for this step depends on the reagents used and the flow cytometer on which the specimens will be analyzed. As with many commercial reagents, the protocols for their use are predicated on using reagents and instrumentation from one manufacturer, and laboratory protocols that intermix reagents (particularly lysing reagents) and instrumentation from different manufacturers must demonstrate that the results are comparable to those obtained when using products from the same manufacturer.

One particular problem with lysing RBCs is seen in pediatric specimens, which may contain nucleated RBCs that do not lyse with conventional lysing reagents. For these specimens, the RBCs must be removed from the specimen during analysis by choosing measures for the analysis that would remove them, such as adding CD45 (which does not label RBCs) or anti-glycophorin A (which does label RBCs) to the analysis tube.

The effects of centrifugation and vortexing are not clearly understood. All procedures require vortexing to mix the specimen properly with reagents and to keep cells from clumping. Some procedures do not require centrifugation, whereas others do. Centrifuging at high speeds and vortexing too much or too little may cause artifacts. The result is usually the appearance of cell aggregates, which because of their light scatter patterns fall outside the lymphocyte population and may be excluded from analysis.[21]

In addition to the isotype control, which is used to determine the level of negative fluorescence, two additional controls are needed. First, a control for the methodology must be included with each immunophenotyping run. Thus a whole-blood control must be processed at the same time the patient specimens are processed. This control specimen should be collected in the same anticoagulant and be of approximately the same age as the rest of the patient specimens. When there is a problem with labeling or lysis in the run, it becomes apparent in this specimen as well as the patient specimens. Biologic reasons for this specimen to fall outside normal limits for lymphocyte populations does not invalidate results from the run; only problems with specimen processing, which is evident in this tube and the patient tubes, can invalidate a run.

A second control is for the reagents. This control is used only to test the monoclonal antibodies, and then only when there is a question about labeling efficiency or when antibody lots or vendors change. Whole blood may be used for this purpose, although other materials may be used as well, including cryopreserved lymphocytes or commercial lymphocyte preparations specifically marketed for such purposes. The advantage of the commercial lymphocyte preparations is that they are normally provided with target values for the major subsets of lymphocytes that might be tested. When this reagent control is used, however, it is important to label not only with the antibody in question but also with the antibody currently used, whose performance characteristics are known.

Monoclonal Antibody Panels

A recommended panel of monoclonal antibodies is found in Table 9.1. This six-tube panel contains reagents to enumerate cells of particular interest ($CD4^+$ and $C8^+$ T cells) and ample quality control. CD14 and CD45 are used as gating reagents to identify leukocyte populations based

TABLE 9.1. Monoclonal antibody panel.

FITC reagent	PE reagent	Purpose
CD45	CD14	Gating reagent
Isotype Ig	Isotype Ig	Nonspecific + autofluorescence
CD3	CD4	$CD4^+$ T-cells; total T cells
CD3	CD8	$CD8^+$ T-cells; total T cells
CD3	CD19	B cells; total T cells
CD3	CD16 and/or CD56	Non-T NK cells; T cells

on their fluorescence patterns. The second tube contains an isotype control. Ideally, this tube serves as a control for both autofluorescence (which all cells have) and nonspecific binding of antibodies to the cells. Cursors, or integrators, are placed just beyond the fluorescence exhibited in this tube; and results from the other tubes are determined based on these cutoffs. In theory this technique works well, but in practice these controls may not be appropriate. At this time, use of isotype controls is still the standard of practice. The remaining four tubes contain CD3 combined with CD4, CD8, CD19, and CD16 or CD56 (or both). The first two of these four tubes are used to determine the proportion of T lymphocytes that are $CD4^+$ and $CD8^+$. The last two tubes serve primarily as quality control checks.

The antibody panel recommended by the National Institute of Allergy and Infectious Diseases Division of AIDS (NIAID DAIDS) for the AIDS Clinical Trials Group (ACTG) includes the following: CD14 and CD45; isotype control; CD3 and CD4; and CD3 and CD8.[22] CD19 is added for pediatric specimens in the clinical trials. For the ACTG, one marker for clinical outcome has been the change in CD4 T cells after therapy. In some trials CD8 T cells seem to have some utility. In the clinical trials B cells and NK cells are of less importance because these populations have had less prognostic significance in natural history studies.

The trend for the selection of monoclonal antibody panels is to increase the number of colors in one tube. Because most clinical flow cytometers have three fluorescence channels, many manufacturers are developing fluorochromes to be utilized in the third fluorescence channel, which until now has been used primarily by research laboratories. Fluorescein isothiocyanate (FITC) and phycoerythrin (PE) are the two fluorochromes that are commonly used in two-color combination. (FITC emits green light and PE orange light.) For the most part, the third color fluorochromes are tandem conjugates that combine two fluorochromes such as PE and Texas Red or Cy5 (emitting red light) where the emitted wavelength of the first is used to excite the second. The stability of these conjugates is a concern, and excellent quality control of the reagents by the manufacturers is a must. One naturally occurring fluorochrome, peridinin chlorophyll protein (PerCP), which is excited at 488 nm (the blue line of the argon laser in the clinical flow cytometers) and emits in the red spectrum, appears more stable than the tandem conjugates and is not as susceptible to many of the problems of the tandem conjugates.

There are several three-color combinations of monoclonal antibodies that are presently in validation studies. One panel includes an isotype control, CD3/CD4/CD8, and CD3/CD19/CD16 (Fig. 9.5). This combination would allow enumeration of T cells, B cells, NK cells, $CD4^+$ T cells, $CD8^+$ T cells, and cells that are $CD3^-CD4^+CD8^+$ (activated or immature T cells) and $CD3^+CD4^-CD8^-$ (presumably T γ/δ cells). There is no gating reagent in this panel, and the quality of the results should be

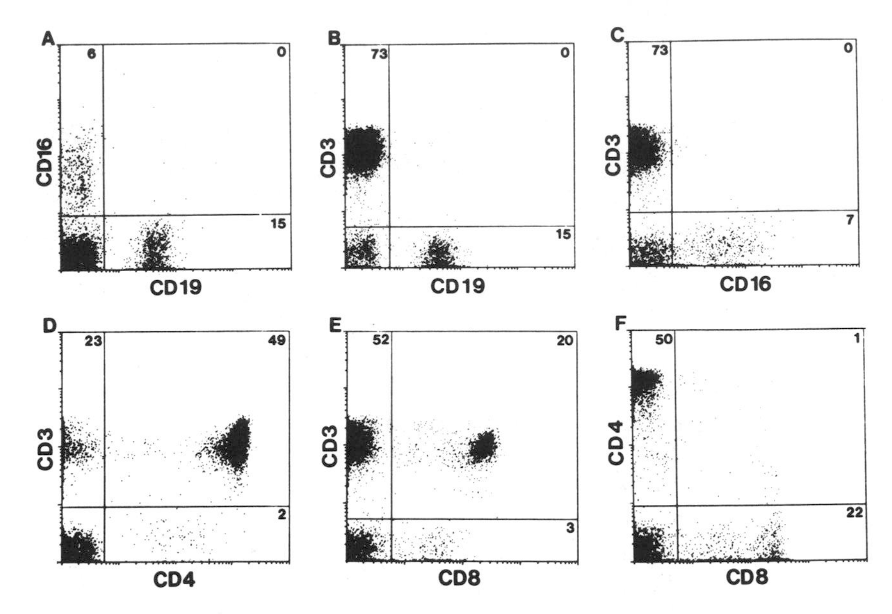

FIGURE 9.5. Fluorescence histograms for a three-color panel including CD3/CD4/CD8, CD3/CD19/CD16, and isotype control. Panels A, B, and C are from the CD3/CD16/CD19 tube; panels D, E, and F are from the CD3/CD4/CD8 tube.

deemed correct only when the CD3 determinations from each of these tubes are similar (within 2–3% of each other). Because a gating reagent is not included, the lymphocyte purity or recovery cannot be determined.

Another combination contains CD45 with any or all of the four last tubes in the six-tube panel (Fig. 9.6). Use of CD3/CD4/CD45 alone has shown some promise in the absence of additional tubes.[23] Because CD45 is used as one of the gating parameters, some assurance of the quality of

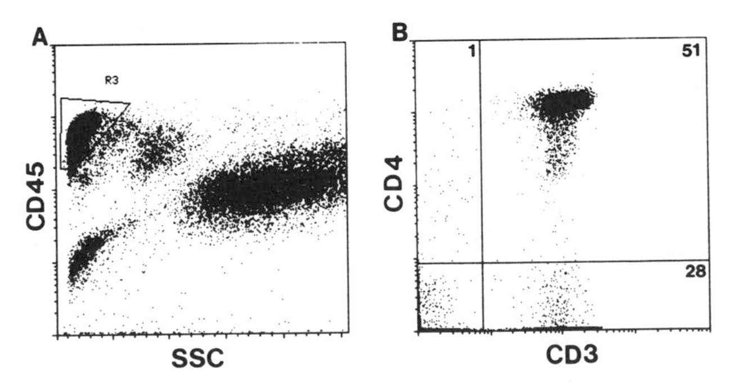

FIGURE 9.6. Scatter and fluorescence histograms for a three-color tube containing CD3/CD4/CD45. Panel A shows a gate set to include low side scatter and high CD45 fluorescence. Panel B shows a CD3/CD4 histogram of gated cells.

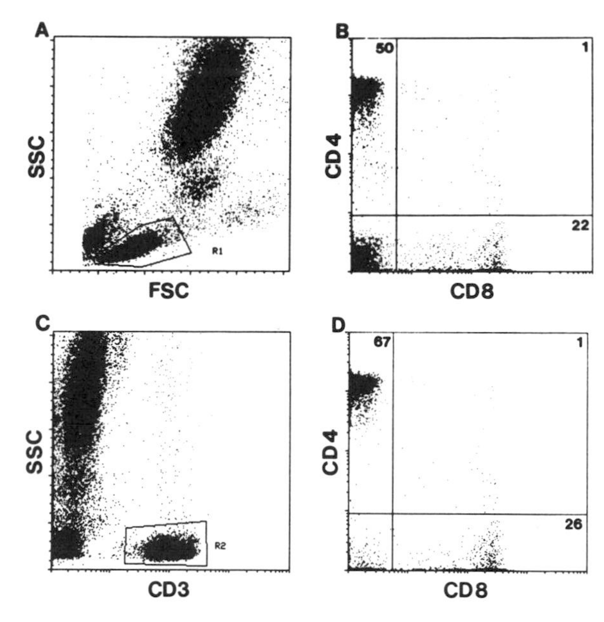

FIGURE 9.7. Scatter and fluorescence histograms for a three-color tube containing CD3/CD4/CD8. Panel A shows a light scatter gate with CD4 and CD8 analysis on the gated lymphocytes in Panel B. Panel C shows a gate set to include low side scatter and CD3-positive cells; panel D shows the CD4 and CD8 histogram of CD3-gated cells.

the lymphocyte gate could be achieved. If both CD4 and CD8 T cell determinations are needed, two tubes would be needed. One advantage of this combination is that the lymphocyte differential in an absolute CD4 count can be bypassed because the light scatter and CD45 reactivity can be used to identify not only lymphocytes but all leukocytes.

The last three-color panel consists of a single tube containing CD3, CD4, and CD8 (Fig. 9.7). CD3 and scatter are used to gate T cells, after which CD4 and CD8 determinations are made.[24] This analysis enables identification of $CD4^+$, $CD8^+$, $CD4^-CD8^-$, and $CD4^+CD8^+$ T cells. Unfortunately, an absolute number of CD4 or CD8 T cells cannot be directly derived from this analysis, and the subset results are in terms of T cells and not lymphocytes.

Sample Analysis

Samples are introduced into the flow cytometer, and either a set volume of sample is taken up and analyzed or a predetermined number of cells are counted. It is important that representative sampling takes place. That is, the cells must be equally dispersed in the tube so the sample

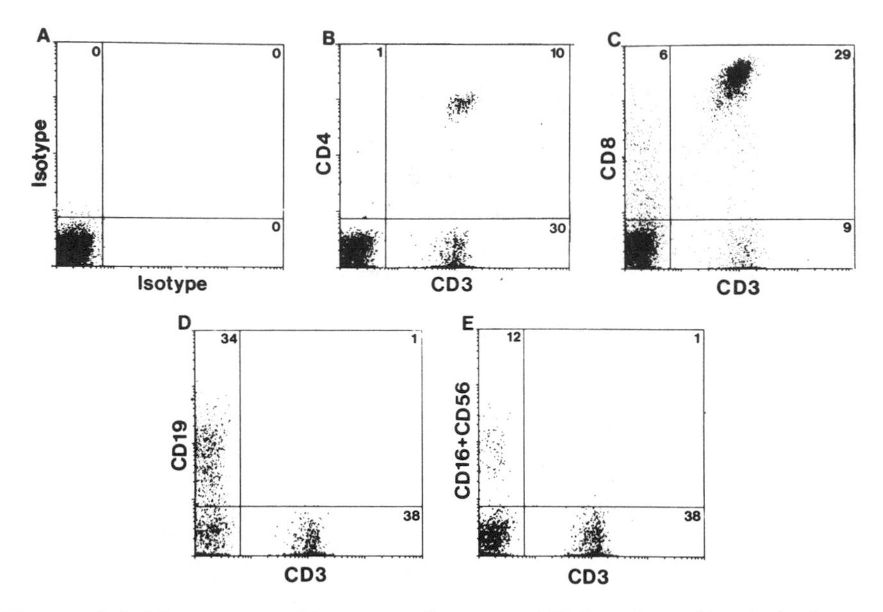

FIGURE 9.8. Fluorescence histograms from an AIDS patient. Panel A shows nonspecific binding of isotype controls, which is used to define where cursors or discriminators are placed to measure positive cells. Panels B through E show two-color fluorescence for the remaining 5 tubes of the six-tube panel.

taken into the flow cytometer is representative of the total population of cells. It assumes, of course, that specimen processing did not selectively remove cell populations. To help ensure representative sampling, it is important that the specimen be properly vortexed. Samples from the six-tube panel are analyzed in the following order: CD14 and CD45, isotype control, the remaining tubes (Fig. 9.8).

The number of cells to be analyzed depends on a number of factors. First, the accuracy of the results is related to the number of cells counted; therefore if approximate measures are all that are desired, fewer cells are needed than for more accurate measures. Second, the frequency of positive cells in the population and the level of variability acceptability are related, with greater variability found in populations that are closer to 50% positive and less variability in populations that are closer to 10% or 90% positive. Lastly, biases for the distribution of the cell population in the sample may contribute to the distribution of variability. In general terms, if 95% accuracy is expected, assuming a variability of 2% in the result, at least 2500 lymphocytes must be counted (binomial sampling model). Each laboratory should establish its own variability, which should be in the order of 2%. Hence replicate samples (at least eight) should be set up, processed, and analyzed. Assuming a normal distribution of the results, standard deviations can be determined. They should be approximately 1%, so that with a binomial distribution the range of variability is ±2%.

Data Analysis

The first step in data analysis is to set gates around the lymphocytes.[25] This technique is a means by which the computer analyzes only a portion of the total data collected. Because the population of cells of interest is lymphocytes, we must eliminate the remaining leukocytes (granulocytes and monocytes), platelets, and RBC debris from the analysis region (Fig. 9.9): We use CD14 and CD45 as well as light scatter parameters to identify lymphocytes with assurance. CD45 antigen is found on all leukocytes, though different cells express different amounts of the antigen and therefore have different fluorescence intensities. CD14 antigen is found on most mature monocytes. Using these two antibodies in combination, we can identify the lymphocytes as those cells that are the brightest for CD45 but negative for CD14. In addition, lymphocytes have low forward light scatter and low 90-degree light scatter. Taken together, the lymphocytes can be identified, and analysis gates are then placed around the light scatter pattern that corresponds to lymphocytes. The lymphocyte purity of the gated cells can be determined by evaluating the proportion of cells that are bright $CD45^+$ and $CD14^-$.

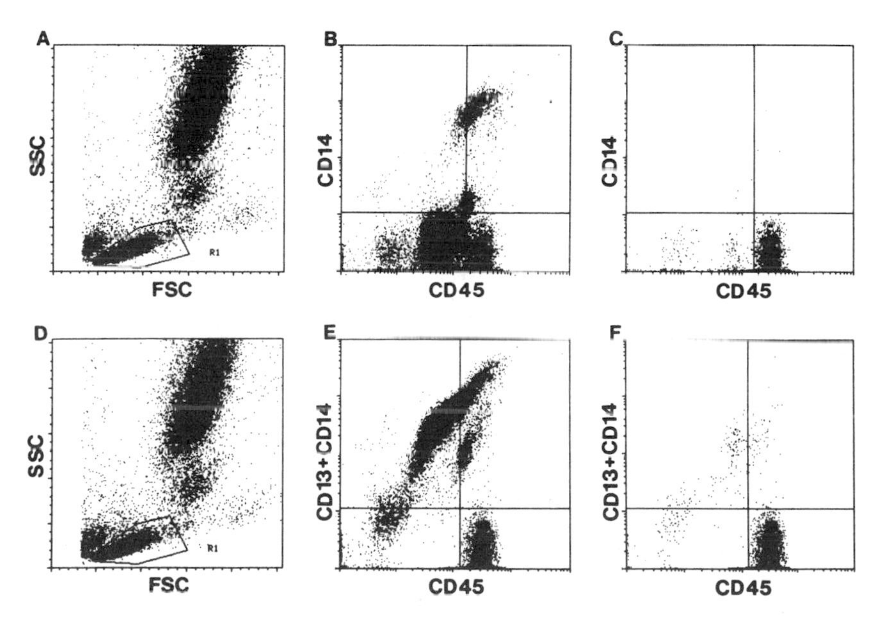

FIGURE 9.9. Light scatter and fluorescence for CD14/CD45 (panels A–C) and CD14+CD13/CD45 (panels D–F). Panels B and E are ungated, panels C and F are gated on the light scatter gate drawn in Panels A and D, respectively.

One problem with CD14 and CD45 as gating reagents is that in combination they do not necessarily identify definitive populations of cells. For example, CD14 does not identify all monocytes, and not all cells that exhibit light scatter of lymphocytes are bright $CD45^+$. It is not clear whether a granulocytic contaminant is found in the light scatter gate of lymphocytes or some lymphocytes exhibit CD45 reactivity similar to that of granulocytes. One potential solution to this dilemma would be the use of CD13 as a marker for both granulocytes and monocytes (Fig. 9.9). CD13 is found on neutrophils, monocytes, and basophils[26]; and combined with CD45 it might be useful for determining nonlymphoid contamination of lymphocyte light scatter gates. More exploration is needed to identify better gating reagents to help determine which cells are lymphocytes and which are not.

Ideally all the lymphocytes are included in the light scatter gate, but practically this recovery is usually slightly less than 100%. The lymphocyte purity in the gate should be as high as possible, but normally fewer than 100% of the gated cells are lymphocytes. At a minimum, the recovery should be more than 90% and the purity more than 85%.[10] The error, if any, should be on the side of purity because it is more important to include all the lymphocytes than to achieve a pure gate. Because some lymphocyte populations scatter light somewhat differently from others (note especially NK cells, which are slightly larger and more granular, and B cells, which are slightly smaller and less granular), to be sure all lymphocytes are included in the gate a rather generous gate should be drawn. One drawback to a large gate is the inclusion of nonlymphocytes, which may possess some of the same antigens as lymphocytes and be mistaken for lymphocytes in the analysis. The point is particularly true for granulocytes, which may possess some of the same antigens as NK cells.

The isotype control is commonly used to determine where nonspecific and autofluorescence end and specific fluorescence begins on the fluorescence scale. For the antibody combinations indicated in the six-tube panel, the need of isotype controls may not seem apparent. Most of the cell populations are brightly positive and discretely different from the negatively labeled cells in the histogram. Problems with the isotype control become readily apparent if the labeling in the remaining tubes is appropriate. One antibody combination for which the isotype control may be needed is CD3 and CD16, with or without CD56. The NK cells may appear dimly labeled with these antibodies, and use of isotype controls to determine positive cells is important.

Occasionally cells label as if they are uncompensated. Such labeling occurs relatively infrequently, though often enough to be a problem. In these cases nothing is usually wrong with color or spectral compensation, but there is an immunoglobulin in the plasma of some donors that allows nonspecific binding of antibodies to antibody-bound cells.[27] To eliminate this nonspecific binding, there are two easy remedies: (1) wash the plasma from the blood before labeling, or (2) add mouse immunoglobulin (10 μg) to the whole blood for 10 minutes before adding monoclonal antibodies (do not wash between these two additions).

Data should be reported as percent positive cells and corrected for lymphocyte purity.[10] This practice is recommended to help standardize results within and between laboratories. That is, the percent of the population of interest is divided by the lymphocyte purity and multiplied by 100. Correction is probably not as important for specimens in which more than 95% of the gated cells are lymphocytes and in which the populations of interest comprise only a small portion of all the lymphocytes. To be consistent, however, correcting for purity of less than 100% is recommended.

Quality Control in Data Analysis

The six-tube monoclonal antibody panel mentioned earlier contains additional tubes for quality control. It is important to recognize that when more than one tube is used for specimen analysis some assumptions are made. First, one assumes that procedures used for each tube do not vary from tube to tube. Second, the population of cells sampled in each tube is the same in each replicate tube. We know from simply performing this procedure that there are tube-to-tube variations in lysis because sometimes one tube does not lyse well when all the other replicate tubes do. Lysis is probably the most variable part of sample preparation, and reasons for this variability may be both biologic and procedural.

When one sets a light scatter gate using the CD14/CD45 tube, it is assumed that the light scatter and distribution of cells are the same in that tube as in the rest of the tubes from the same patient. As a check for any variability in the remaining tubes in terms of the lymphocyte distribution (and purity), CD3 (T cells), CD19 (B cells), and CD3 with CD16 or CD56, or both (NK cells), are included. The ability to sum T, B, and NK cells from these tubes ("lympho-sum") should help ensure that all the lymphocytes are samples appropriately in the tubes of the six-tube panel.[28] In a multicenter anticoagulant study, we found that it was possible to "lympho-sum" most specimens regardless of HIV status and blood age (Table 9.2). For this summation, it is important that the NK cells are enumerated with CD3 so cells positive for CD3 can be eliminated from the NK result.

Similarly, CD4 and CD8 should account for most all of the $CD3^+$ T cells, and addition of $CD4^+$ and $CD8^+$ T cells should approximate $CD3^+$ cells ("T-sum"). (In some cases T cells positive for the γ/δ receptor of the T cell–receptor complex may be numerous enough to invalidate this assumption.[29]) We found that criteria for "T-sum" acceptability cutoffs must be less stringent than for the "lympho-sum" cutoffs (Table 9.3), particularly for some of the HIV-positive specimens and for older specimens.

TABLE 9.2. Difference between the sum of %T, %B, % non-T NK ("Lymphosum") and lymphocyte purity.

Difference	Specimen type[a]					
	Fresh <6 hrs		24 hour		48 hour	
	HIV+	HIV−	HIV+	HIV−	HIV+	HIV−
<5%	12.7[b]	8.8	21.4	15.7	17	21.2
<10%	1.4	0	4.3	2.9	0	0

[a] Specimens from a multicenter study in eight laboratories.
[b] Percent of results that were beyond the different cutoffs for lymphosum differences.

Proficiency Testing

Few proficiency testing programs currently exist for immunophenotyping whereby laboratories are graded and remedial action is taken for poor performance. Most programs, whether they are performance evaluation (PE) or proficiency testing (PT), now send whole-blood specimens for the laboratories to process and report back immunophenotypes as percentages. This protocol is the best test of the laboratory's ability to carry out all the steps involved in specimen processing and analysis. Some programs also send out fixed specimens that need to be analyzed only on the flow cytometer, thereby controlling for flow cytometer performance. CD4 T cell results from these programs normally have a coefficient of variation (CV) of around 5% to 10%.[30–32] Most often, however, results from PE or PT programs are using standard deviation as better measures of CD4 variation over the range of CD4 results. In general, a range of ±5% is found,[32,33] which usually decreases as the CD4 results approach 1%.

Among the four largest and more comprehensive programs—the Centers for Disease Control and Prevention (CDC) Model Performance Evaluation Program (MPEP), the College of American Pathologists

TABLE 9.3. Difference between T cell determinations and sum of CD4 and CD8 T-cells ("T-sum").

Difference	Specimen type[a]					
	Fresh <6 hrs		24 hour		48 hour	
	HIV+	HIV−	HIV+	HIV−	HIV+	HIV−
<5%	19.5[b]	20	19.8	24.6	26.1	23.3
<10%	2.2	0	6.6	1.6	9.6	4.7

[a] Specimens from a multicenter study in nine laboratories.
[b] Percent of results that were beyond the different cutoffs for "T-sum" differences.

(CAP), the NIAID DAIDS program, and the Department of Defense (DOD)—there are similarities and differences (Table 9.4). Most of these programs do not pass or fail laboratories based on performance, although one does. Participation in the NIAID DAIDS program is required for laboratories that test specimens from AIDS clinical trials. As a result, the laboratories must perform acceptably to be able to test clinical specimens. Those laboratories that fail are put on probation; and if their performance does not improve they are barred from testing clinical specimens for the ACTG.

Ideally, the PT or PE specimens are received in the laboratory in the same manner as clinical specimens; they represent the type of specimens normally received in the laboratory; and they are handled the same way as clinical specimens. All of the current PT or PE programs are announced before shipment and are clearly labeled as such, with report forms included. Only the MPEP and NIAID DAIDS programs send blood from HIV-infected persons; the CAP manipulates the blood to deplete CD4 T cells or sends samples from B cell malignancies. In general, the manipulated blood is more variable in terms of specimen integrity and results obtained than is the nonmanipulated blood. How these specimens are handled once they are received in the laboratory is not known, as this type of information is not collected.

Although hematology results are required to calculate absolute CD4 T cell results, the only program that tests hematology parameters is the DOD program. The DOD laboratories all use similar hematology instruments; and after improving and standardizing their techniques, the variability in absolute CD4 results has decreased with time.[32] There are some reasons hematology has not been incorporated into other programs. First, it has been recommended that hematology measurements be done within 6 hours of the blood being obtained. Until recently, the transport of specimens has not permitted these samples to arrive in less than 6 hours. Moreover, hematology instrumentation has not been challenged to validate specimens older than 6 hours. Second, not all flow cytometry laboratories do hematology measurements. NIAID DAIDS tried to incorporate hematology into their program and abandoned it for many of the above reasons plus the poor quality of results they obtained. Now that many hematology instruments have been validated for older specimens, hematology measures will likely be incorporated into more of these programs.

As a result of the Clinical Laboratory Improvement Amendment of 1988 (CLIA '88), laboratories performing immunophenotyping must participate in proficiency testing programs when programs that meet the CLIA '88 requirements are identified and criteria for success/failure are determined. At this writing, PT programs for immunophenotyping have not been identified.

TABLE 9.4. Proficiency testing and performance evaluation programs for lymphocyte immunophenotyping.

Program	PE or PT	Type of samples	Sendouts per year	Specimens per sendout	Number of laboratories	Hematology	Cost
College of American Pathologists	PE	Whole blood, some CD4-depleted	4	3	600–700	No	$616
Model Performance Evalutation Program	PE	Whole blood, HIV+ and HIV−	About 2	5 plus 1 instrument performance control (IPC)	300–400	No	No cost
NIAID DAIDS	PT	Whole blood, HIV+ and HIV−	6, beginning 1/94	5, beginning 1/94	70–80	No	No direct cost
Department of Defense	PE	Whole blood	12	3 plus 1 IPC	13	Yes	No direct cost

Variability in Immunophenotyping Results

The analytic variability in immunophenotyping using flow cytometry is relatively small compared to the variability among hematology results. Calculating absolute lymphocyte population numbers combines the results from the WBC count, leukocyte differential, and flow cytometry; and the variability in the final number is a combination of the variability from each of these tests. For flow cytometry percent determinations, intralaboratory variability should be about ±2 percentage points. For the WBC count a 3% to 5% variability (CV) is normally expected. The differential has the largest variability, depending on the number of cells counted. For a manual differential count, 20% to 25% variability is found when counting 100 or 200 cells, and about 2% to 5% is found when counting 10,000 to 30,000 cells.[17] Combined, no less than about 8% variability is usually found, but this figure may depend on the laboratory. Thus a CD4 T cell count of 195 cells/μl could actually be less than 200 or more than 200 cells/μl.

Biologic variability can contribute significantly to CD4 T cell results. The WBC count changes diurnally, so specimens obtained in the morning have counts different from those obtained in the afternoon.[33] The differential count and percent CD4 T cells do not vary diurnally, as does the WBC count. For this reason, absolute $CD4^+$ T cells change with the time of day the specimen is drawn, but the percent CD4 T cells does not. Other factors that affect CD4 T cell results are found in Table 9.5.

Alternative CD4 Technologies

A number of manufacturers have developed assays for determining CD4 cell numbers directly from whole blood (Table 9.6). It should reduce the

TABLE 9.5. Factors that affect lymphocyte immunophenotyping results.

Factor	Effect	Result
Zidovudine (AZT, ZDU)	Increased granulocyte fragility; RBCs less susceptible to lysis	Poor light scattering resolution, thus more nonlymphocyte contamination in light scatter gates
Nicotine	Decreased lymphocyte counts	Lowered absoulte lymphocyte subset values
Corticosteroids	Decreased $CD4^+$ T-cell levels	Decreased $CD4^+$ T-cell percentage and absolute number
Diurnal	Changes in absolute lymphocyte count with time of day	Variable absolute lymphocyte subset values
Strenuous exercise	Decreased lymphocyte counts	Lowered absolute lymphocyte subset values

TABLE 9.6. Summary of alternative CD4 technologies.

Assay	Method	Specimen needed	Minimum volume needed	Assay time	Results readout
FACSCount by Becton Dickinson	Automated flow cytometry for CD3/CD4 and CD3/CD8 2-color fluorescence	Whole blood	200 μl per specimen	1½ hour incubation, 3–5 minute analysis per specimen, can batch for incubation	CD4 and CD8 percent of $CD3^+$ cells, absolute no. $CD3^+CD4^+$ and $CD3^+CD8^+$
Immuno VCS by Coulter Corporation	Flow cytometer (converted hematology instrument) using anti-CD4 microbeads	Whole blood	600 μl	2 min incubation plus analysis time (3–5 min per specimen)	CD4 absolute no. (flow cytometer does CD4 percent also)
Manual CD4 Count Kit by Coulter Corporation	Hemocytometer counting of cells bound to CD4-beads	Whole blood	150 μl	5 min incubation plus hemocytometer reading (5–10 min)	Absolute CD4 no. read off conversion chart
TRAx by T Cell Diagnostics	EIA on solubilized cells; assays soluble CD4	Whole blood	100 μl	4 hours (Can prepare lysates and store frozen); can batch ≤40 spceimens	Soluble CD4, with conversion to absolute CD4 numbers
Capsellia CD4/CD8 by Diagnositcs Pasteur	EIA for CD4 on captured T cells	Separated cells	500 μl for separated cells	1½ hours; can batch 8 per plate	pmol CD4 and CD8/L (converted to CD4 number)
Zymmune by Zynaxis	96-well plate format with CD4 or CD8 magnetic and fluorescent beads; assays for fluorescence associated with cells captured on the plate	Whole blood	150 μl	45 minutes for ≤12 specimens; can batch 12 per plate	Fluorescence converted no. CD4 and CD8

variability of the CD4 absolute number by eliminating the need for combining values from three separate technologies in the calculation. In addition, the price of these tests should be significantly reduced because they do not use clinical flow cytometers, which cost in the range of $100,000. Lastly, immunophenotyping using flow cytometry requires considerable expertise and training, which is not the case with the alternative CD4 technologies.

Four U.S. companies and one French company have developed assays that measure, in the end, absolute CD4 cells. Two are flow-based instruments: the FACSCount (Becton Dickinson Immunocytometry Systems) and the ImmunoVCS (Coulter Diagnostics). The FACSCount measures $CD3^+CD4^+$ and $CD3^+CD8^+$ cells by light scatter and fluorescence, using a calibrator bead to derive the number of positive cells. The ImmunoVCS measures cells that have bound beads conjugated to a CD4 antibody and calculates absolute cells from a WBC count performed with the same instrument. A version of the Coulter assay, the Manual CD4 Count Kit, uses a hemacytometer to count the CD4–bead-bound cells. TRAx CD4 (T Cell Diagnostics) is an enzyme immunoassay (EIA) that assays soluble CD4 in lysed whole blood, and absolute CD4 cells are extrapolated from a standard curve. Zymmune (Zynaxis) uses a combination of magnetic and fluorescent beads coated with CD4 or CD8 antibodies to capture CD4 or CD8 cells, and the fluorescence of the captured cells is read out and converted to the number of positive cells using a standard curve. Capsellia (Sanofi Diagnostics) captures T cells with CD2 antibodies bound to microtiter plates. The CD4 and CD8 T cells then are developed using EIA technology, with the captured cells serving as the solid phase in the assay. Absolute cell subset numbers are calculated from a standard curve. At the time of this writing, only the Manual CD4 Kit (Coulter) has been approved for diagnostic purposes. These tests will likely replace flow cytometry in some settings.

Use of Immunophenotyping Results in HIV Disease

The primary reason for immunophenotyping of patients with HIV disease is to enumerate the $CD4^+$ T lymphocytes. One of the first laboratory abnormalities found in what we now know as AIDS was a decrease in $CD4^+$ T cell percentages and absolute numbers.[7,8] This decrease is progressive throughout the infection.[34] $CD4^+$ T cells are important in both cellular and humoral immunity; they induce effector cells to perform their immune functions, which include antibody production, delayed-type hypersensitivity, and cell-mediated immunity.[35–37] A loss of these cells in HIV infection is correlated in vivo with decreased responses to recall antigens and increased susceptibility to opportunistic infections and Kaposi sarcoma.[38,39] In vitro, decreases in mitogen and antigen responses correlate with decreases in CD4$^+$ T cell numbers.[40,41] Cell-mediated cellular cytotoxicity (mediated by CD8$^+$ T cells) increases early in HIV infection, but during late-state disease these responses are obliterated as well.[42] Antibody production (by B cells) increases during early infection owing to polyclonal activation of B cells, but even this response wanes with disease progression.[41,43]

The CD8 T cells mediate major histocompatibility complex (MHC)-restricted cellular cytotoxic responses.[44] They also dampen other immune functions, such as antibody production and responses to antigen.[45,46] With viral infections such as those caused by Epstein-Barr virus (EBV) and cytomegalovirus (CMV), this cell population expands as a response to early infection.[47,48] HIV also induces an expansion of these cells; therefore during early HIV infection there is a much larger expansion of CD8 T cells than there is depletion of CD4$^+$ T cells.[7,8,49]

Early clinical studies focused heavily on the CD4/CD8 ratio, which is a number derived by dividing the CD4 percent by the CD8 percent. Because of the large influence of the increased CD8 cells on a decreased ratio early in infection, the use of this ratio has fallen out of favor. Most clinicians are more interested in CD4 cell percentages and numbers as an indication of immunosuppression. That is, the lower the CD4$^+$ T cell number, the greater are the chances that the patient will develop an opportunistic life-threatening infection. In many clinical trials, CD4 T cell numbers are used as criteria for entry into the trial and as an outcome parameter. CD8$^+$ T cells, on the other hand, may have a role in protection, at least early in infection. With late-stage AIDS, CD8$^+$ T cells decrease significantly.

Subsets of CD4$^+$ and CD8$^+$ T cells have been enumerated using two-, three-, and four-color immunophenotyping to help determine if there is a subset of these cells that are important in disease progression or in clinical outcome. B cells and NK cells, though being found to be decreased in HIV infection, have not been shown to have predictive or clinical value.

References

1. Loken MR, Stall AM. Flow cytometry as an analytical and preparative tool in immunology. J Immunol Methods 1982;50:85–112
2. Parks DR, Herzenberg LA. Fluorescence-activated cell sorting: theory, experimental optimization, and applications in lymphoid cell biology. Methods Enzymol 1984;108:197–241
3. Shapiro HM. Practical Flow Cytometry. New York: Alan R. Liss, 1988
4. Reinherz EL, Kung PC, Goldstein G, Schlossman SF. A monoclonal antibody with selective reactivity with functionally mature human thymocytes and all peripheral human T cells. J Immunol 1979;123:1312–1317
5. Kung PC, Goldstein G, Reinherz EL, Schlossman SF. Monoclonal antibodies defining distinctive human T cell surface antigens. Science 1979;206:347–349

6. Stashenko P, Nadler LM, Hardy R, Schlossman SF. Proc Natl Acad Sci USA 1981;78:3848–3852
7. Gottlieb MS, Schroff R, Schanker HM, et al. Pneumocystis carinii pneumonia and mucosal candidiasis in previously healthy homosexual men. N Engl J Med 1981;305:1425–1431
8. Masur H, Michelis MA, Greene J, et al. An outbreak of community-acquired Pneumocystis carinii pneumonia: initial manifestation of cellular immune dysfunction. N Engl J Med 1981;305:1431–1438
9. National Committee for Clinical Laboratory Standards. Clinical Applications of Flow Cytometry Quality Assurance and Immunophenotyping of Peripheral Blood Lymphocytes. NCCLS Publication H42-T. Villanova, PA: NCCLS, 1992
10. Centers for Disease Control. Guidelines for the performance of CD4$^+$ T-cell determinations in persons with human immunodeficiency virus infection. MMWR 1992;41(RR-8):1–17
11. Centers for Disease Control, National Institutes of Health. Biosafety in Microbiological and Biomedical Laboratories. Washington, DC: US Government Printing Office, 1993
12. Lanier LL, Warner NL. Paraformaldehyde fixation of hematopoietic cells for quantitative flow cytometry (FACS) analysis. J Immunol Methods 1981;47: 25–30
13. Nicholson JKA, Browning SW, Orloff SL, McDougal JS. Inactivation of HIV-infected H9 cells in whole blood preparations by lysing/fixing reagents used in flow cytometry. J Immunol Methods 1993;160:215–218
14. Lifson JD, Sasaki DT, Engleman EG. Utility of formaldehyde fixation for flow cytometry and inactivation of the AIDS associated retrovirus. J Immunol Methods 1986;86:143–149
15. Cory JM, Rapp F, Ohlsson-Wilhelm BM. Effects of cellular fixatives on human immunodeficiency virus production. Cytometry 1990;11:647–651
16. Nicholson JKA, Green TA, Collaborating Laboratories. Selection of anticoagulants for lymphocyte immunophenotyping: effect of specimen age on results. J Immunol Methods 1993;165:31–35
17. Koepke JA, Landay AL. Precision and accuracy of absolute lymphocyte counts. Clin Immunol Immunopathol 1989;52:19–27
18. Weiblen BJ, Debell K, Valeri CR. "Acquired immunodeficiency" of blood stored overnight. N Engl J Med 1983;309:793
19. Paxton H, Bendele T. Effect of time, temperature, and anticoagulant on flow cytometry and hematological values. Ann NY Acad Sci 1993;677:440–443
20. Hoffman RA, King PC, Hansen WP, Goldstein G. Simple and rapid measurement of human T lymphocytes and their subclasses in peripheral blood. Proc Natl Acad Sci USA 1980;77:4914–4917
21. Jackson AL. Basic phenotyping of lymphocytes: selection and testing of reagents and interpretation of data. Clin Immunol Newslett 1990;10:43–55
22. Calvelli T, Denny TN, Paxton H, Gelman R, Kagan J. Guidelines for flow cytometric immunophenotyping: a report from the National Institutes of Allergy and Infectious Diseases, Division of AIDS. Cytometry 1993;14: 702–715

23. Nicholson JKA, Jones BM, Hubbard M. CD4 T-lymphocyte determinations on whole blood specimens using a single-tube three-color assay. Cytometry 1993;14:685–689
24. Mandy FF, Bergeron M, Recktenwald D, Izaguirre CA. A simultaneous three-color T-cell subset analysis with single laser flow cytometers using T cell gating protocol. Comparison with conventional two-color immunophenotyping method. J Immunol Methods 1992;156:151–162
25. Loken MR, Bronman JM, Bach BA, Ault KA. Quality control in flow cytometry. I. Establishing optimal gates for immunophenotyping. Cytometry 1990;11:453–459
26. Griffin JD, Ritz J, Nadler LM, Schlossman SF. Expression of myeloid differentiation antigens on normal and malignant myeloid cells. J Clin Invest 1981;68:932–941
27. Ekong T, Gompels M, Clark C, Parkin J, Pinching A. Double-staining artefact observed in certain individuals during dual-colour immunophenotyping of lymphocytes by flow cytometry. Cytometry 1993;14:679–684
28. Schenker EL, Hultin LE, Bauer KD, et al. Evaluation of a dual-color flow cytometry immunophenotyping panel in a multicenter quality assurance program. Cytometry 1993;14:307–317
29. Margolick JB, Carey V, Munoz A, et al. Development of antibodies to HIV-1 is associated with an increase in circulating $CD3^+CD4^-CD8^-$ lymphocytes. Clin Immunol Immunopathol 1989;51:348–361
30. Homburger HA, Rosenstock W, Paxton H, Paton ML, Landay AL. Assessment of interlaboratory variability of immunophenotyping. Ann NY Acad Sci 1993;677:43–49
31. Gelman R, Cheng S-C, Kidd P, Waxdal M, Kagan J. Assessment of the effects of instrumentation, monoclonal antibody, and fluorochrome on flow cytometric immunophenotyping: a report based on 2 years of the NIAID DAIDS flow cytometry quality assessment program. Clin Immunol Immunopathol 1993;66:150–162
32. Rickman WJ, Waxdal MJ, Monical C, Damato JD, Burke DS. Department of the Army lymphocyte immunophenotyping quality assurance program. Clin Immunol Immunopathol 1989;52:85–95
33. Statland BE, Winkel P. Physiological variability of leukocytes in healthy subjects. In Koepke JA (ed): Differential leukocyte counting. Skokie, IL: College of American Pathology, 1978:23–38
34. Melbye M, Biggar RJ, Ebbesen P, et al. Long-term seropositivity for human T-lymphotropic virus type III in homosexual men without the acquired immunodeficiency syndrome: development of immunologic and clinical abnormalities. Ann Intern Med 1986;104:496–500
35. Mosier DE, Coppelson LW. A three-cell interaction required for the induction of the primary immune response in vitro. Proc Natl Acad Sci USA 1968;61:542–547
36. Cantor H, Asofsky R. Synergy among lymphoid cells mediating the graft-vs-host response. III. Evidence for interaction between two types of thymus-derived cells. J Exp Med 1971;135:764–779
37. Cher DJ, Mosmann TR. Two types of murine helper T cells clones. II. Delayed-type hypersensitivity is mediated by TH1 clones. J Immunol 1987;138:3688–3694

38. Roos MThL, Miedema F, Eeftink Schattenkerk JKM. Cellular and humoral immunity in various cohorts of male homosexuals in relation to infection with human immunodeficiency virus. Neth J Med 1989;34:132–141
39. Masur H, Ognibene FP, Yarchoan R, et al. CD4 counts as predictors of opportunistic pneumonias in human immunodeficiency virus (HIV) infection. Ann Intern Med 1989;111:223–231
40. Miedema F, Petit AJC, Terpstra FG, et al. Immunological abnormalities in human immunodeficiency virus (HIV)-infected asymptomatic homosexual men: HIV affects the immune system before $CD4^+$ T helper cell depletion occurs. J Clin Invest 1988;82:1908–1914
41. Nicholson JKA, McDougal JS, Spira TJ, et al. Immunoregulatory subsets of the T helper and T suppressor cell populations in homosexual men with chronic unexplained lymphadenopathy. J Clin Invest 1984;73:191–201
42. Grant MD, Smaill FM, Singhal DP, Rosenthal KL. The influence of lymphocyte counts and disease progression on circulating and inducible anti-HIV-1 cytotoxic T cell activity in HIV-1-infected subjects. AIDS 1992;6:1085–1094
43. Lane HC, Fauci AS. Immunological abnormalities in the acquired immunodeficiency syndrome. Annu Rev Immunol 1985;3:477–500
44. Shearer GM, Schmidt-Verhulst AM. Major histocompatibility comples restricted cell-mediated specificity. Adv Immunol 1977;25:55–91
45. Gershon RK, Kondo K. Cell interaction in the induction of tolerance: the role of thymic lymphocytes. Immunology 1970;18:723–737
46. Baker PJ, Stashak PW, Amsbaugh DF, Prescott B, Barth RF. Evidence for the existence of two functionally distinct types of cells which regulate the antibody response to type III pneumococcal polysaccharide. J Immunol 1970;105:1581–1586
47. Reinherz EL, O'Brien C, Rosenthal P, Schlossman SF. The cellular basis for viral-induced immunodeficiency: analysis by monoclonal antibodies. J Immunol 1980;125:1269–1274
48. Carney WP, Iacoviello V, Hirsch MS. Functional properties of T lymphocytes and their subsets in cytomegalovirus mononucleosis. J Immunol 1983;130:390–393
49. Nicholson JKA, McDougal JS, Jaffe HW, et al. Exposure to human T-lymphotropic virus type III/lymphadenopathy-associated virus and immunologic abnormalitics in asymptomatic homosexual men. Ann Intern Med 1985;103:37–41

10
Prognostic Indicators for Progression of HIV Disease

C. Robert Horsburgh Jr.

Infection with the human immunodeficiency virus (HIV) produces a prolonged, gradually progressive illness that eventually leads to opportunistic infections, malignancy, and death. The illness may last 10 years or longer and is usually asymptomatic for the first several years. Over time, the immune system gradually weakens, and clinical signs and symptoms become apparent. The course of the illness is variable among individuals in both its duration and its clinical and laboratory manifestations.

It would be useful to have prognostic indicators of the progression of the HIV disease to aid clinicians and patients in coping with the illness and to allow evaluation of the effect of therapeutic interventions. A number of markers of disease stage have been identified, including clinical syndromes and immunologic and virologic laboratory parameters. Most of these markers have been evaluated prospectively and were found to correlate with the risk of progression to clinical acquired immunodeficiency syndrome (AIDS), as defined in the 1987 U.S. Centers for Disease Control and Prevention (CDC) case definition. A few have been shown to predict survival as well.

Use of opportunistic infections and malignancies as endpoints for validating prognostic indicators for HIV disease introduces some difficulties, as the occurrence of an opportunistic infection or malignancy requires both exposure and susceptibility. Thus two individuals who are at the same stage of HIV disease may acquire AIDS at different times if one is exposed to an opportunistic pathogen and the other is not. Moreover, the 1987 AIDS case definition is heterogeneous; some AIDS-defining conditions occur at a time when immunologic impairment is minimal, whereas others occur only when impairment has become severe. Persons who present with Kaposi sarcoma have a significantly longer survival than those who present with *Pneumocystis carinii* pneumonia (PCP), confirming that these two AIDS-indicating diagnoses are not equivalent. Because studies of prognostic indicators with survival as the endpoint must wait additional years before analysis, most studies of prognostic markers of HIV disease have used the 1987 AIDS definition as the outcome variable. In 1993 the CDC amended the 1987 AIDS case definition by including persons with fewer than 200 $CD4^+$ cells per mm^3; however, this definition has not been used to evaluate prognostic indicators and is not further discussed.

The prolonged course of HIV disease has also made it difficult to identify early markers for progression. Even if AIDS rather than death is used as the endpoint, little information is available from persons who were infected with HIV during the early years of the HIV epidemic, and few persons who were in the early stages of HIV infection during the mid-1980s have progressed to AIDS; thus most of the markers that have been defined are those that indicate moderate to severe disease. It is hoped that earlier prognostic indicators will be defined in the coming years.

Lastly, although a large number of markers for disease progression have been identified, the prognostic value of multiple markers cannot be assumed to be additive. Thus the clinician must determine the markers that might be of value for an individual patient. A few studies have examined the interactions between various markers, but no study has examined the entire range of proposed indicators. The final section of this chapter examines some of these interactions and makes some general recommendations.

Clinical Markers

A number of clinical conditions have been shown to be associated with HIV infection. Among the first manifestations to appear after HIV infection are the acute retroviral syndrome, generalized lymphadenopathy, and a distinctive dermatitis resembling seborrheic dermatitis. Early reports had suggested that generalized lymphadenopathy was a significant marker of disease progression, but more recent studies have shown that this syndrome does not predict subsequent progression to AIDS.[1–3] Moreover, the resolution of such lymphadenopathy, which had been considered a grave sign, has not been confirmed as a prognostic indicator. Neither the presence nor absence of an acute retroviral syndrome or of dermatitis has been shown to relate to the eventual course of the disease. As noted above, this lack of association may be due to the difficulty of recognizing the value of such early signs at our present place in the epidemic. Hopefully, studies confirming or refuting the value of such signs in the prognosis of HIV infection will be forthcoming.

Several later clinical conditions have been recognized to be indicators of a more rapid progression to AIDS (Table 10.1). The first of these conditions to be recognized was the presence of oral mucocutaneous candidiasis (thrush). Several studies have confirmed that HIV-infected persons with thrush have a significantly increased risk of developing

TABLE 10.1. Parameters that may predict progression of HIV disease.

Clinical factors	Immunologic factors	Virologic factors
Thrush	$CD4^+$ cell counts or % β_2-microglobulin	p24 Antigen
Oral hairy leukoplakia		Quantitative plasma culture
Herpes zoster	Neopterin	Quantitative PCR
Bacterial pneumonia	Total IgA	Syncytium inducer phenotype
Pulmonary tuberculosis	p24 Antibody	Zidovudine resistance
	Skin testing	
Weight loss	$CD8^+$ and $CD38^+$ cell counts	

AIDS.[4,5] Although less common, an episode of herpes zoster also appears to herald a more rapid occurrence of AIDS.[6] Third, the presence of oral hairy leukoplakia (OHL), an infection of the tongue and oral cavity with Epstein-Barr virus, also is associated with subsequent development of AIDS.[7] Of these three conditions, thrush is the most useful indicator and has been evaluated in several of the combined systems (see Staging HIV Disease, below). OHL appears to be of limited value as a marker because it is reported infrequently in many populations, perhaps because it is often asymptomatic and therefore is not always recognized clinically.

Two other conditions that represent significant illness in HIV-infected individuals are bacterial pneumonia and pulmonary tuberculosis. Several authors have suggested that the occurrence of one of these illnesses may predict subsequent AIDS, but prospective studies to confirm these suspicions have not been reported.

Immunologic Markers

Early in the HIV epidemic it was recognized that HIV infection is characterized by progressive, irreversible decline of the helper-T lymphocyte population. These cells are essential for proper immunologic response to many pathogens, and their destruction is thought to be the explanation for the increased susceptibility of HIV-infected persons to fungal, viral, and parasitic pathogens. The cells are distinguished by the presence on their surface of an antigenic protein called CD4 (previously known as T4 or Leu3). Such cells are identified by staining with antibodies to CD4, which is usually performed with a cytofluorometer (see Chap. 9). The result is expressed as the percent of peripheral blood mononuclear cells (PBMCs) that are $CD4^+$. In the HIV-uninfected person, 20% to 40% of cells are $CD4^+$.

The percent of $CD4^+$ cells is commonly multiplied by the absolute lymphocyte count to produce a "helper-T cell count." This value is usually 500 to 1500 cells/mm^3 in the absence of HIV infection. The daily variability and interlaboratory variability of this parameter is considerable because of fluctuations in the white blood cell (WBC) count and the leukocyte differential count.[8] For this reason, some authors have advocated relying on helper-T cell percent rather than absolute number.[8-10] Although the helper-T cell percent is somewhat more accurate and reproducible, the two indices are clinically interchangeable, and the absolute $CD4^+$ cell count has more widespread usage.

The $CD4^+$ cell count has been shown in a number of studies to be the strongest prognostic indicator for both the appearance of AIDS and survival. Therefore, it is widely obtained and used in the clinical setting. It appears that the $CD4^+$ count declines gradually at a rate of about 80 cells/mm^3 per year from the time of HIV infection to a level of 400 to 500/mm^3.[11] At that point the decline may become more precipitous, although not always. Because zidovudine has been shown to decrease the occurrence of AIDS in persons with 500 or fewer $CD4^+$ cells/mm^3,[12,13] it is now recommended that zidovudine therapy be considered when the $CD4^+$ count falls below 500/mm^3. Although some persons show a transient rise in $CD4^+$ counts after the initiation of zidovudine therapy, $CD4^+$ counts continue to fall. Counts should be monitored every 6 months; and when they reach 200/mm^3, prophylactic therapy for PCP should be initiated.[14]

The suppressor-T lymphocyte population ($CD8^+$) is also affected by HIV infection. With acute HIV infection, $CD8^+$ cells increase and then return to normal numbers over a period of months. Then, as the HIV infection persists, $CD8^+$ cell numbers decline, though not as rapidly as $CD4^+$ cells. This decline is often expressed as a helper/suppressor ratio ($CD4^+/CD8^+$). Although this ratio is also a prognostic indicator for HIV disease, it is less specific than the $CD4^+$ cell count and has been largely replaced by it.[9] Evaluation of $CD8^+$ cell counts has not been shown to provide useful prognostic information, although one report has suggested that elevated levels of $CD38^+$ and $CD8^+$ cells add to the predictive value of decreased $CD4^+$ cells.[15]

Because enumeration of $CD4^+$ cells requires expensive laboratory equipment and specially trained technicians, efforts have been made to find serum markers of immunologic status that could substitute for the $CD4^+$ count. The most successful is the assay for β_2-microglobulin (β_2M). β_2M is a lymphocyte-surface protein that is released into the serum during destruction of cells. Thus as $CD4^+$ cell counts fall, β_2M levels rise. Normal β_2M values are less than 2 mg/liter, whereas persons with symptomatic HIV infection may have a β_2M level of 4 to 5 mg/liter. β_2M levels reflect turnover of all lymphocytes and are thus inherently less specific than $CD4^+$ cell counts. Moreover, the β_2M level rises with age in the normal population and may be increased by intravenous drug use, limiting its usefulness in some HIV-infected populations.

Nonetheless, as a predictor of prognosis of HIV disease, the β_2M assay is nearly as accurate as the $CD4^+$ count.[3,14] The combination of the two tests improves the prognostic accuracy of either test alone. At the present time, however, β_2M values have not replaced $CD4^+$ cell counts as the major predictor of disease progression, and the additional precision provided by combining the two tests does not appear to be worth the additional expense.

The serum neopterin level has also been evaluated as a prognostic indicator of HIV disease.[16] Neopterin, a metabolite of dihydroneopterin triphosphate, is produced from guanosine triphosphate (GTP) when cells of the immune system are activated. The average serum neopterin level in the HIV-uninfected individual is 5 nmol/liter, whereas in HIV-infected persons the average is nearly 14 nmol/liter. Neopterin levels are not as useful for prognosis as are $CD4^+$ counts, but as with β_2M the combination of assays for $CD4^+$ and neopterin is more accurate than either alone.[16]

Several other soluble proteins that are products of immune activation have been shown to have abnormal values in advanced HIV infection but have either not been confirmed as prognostic indicators or have not yet been prospectively studied. They include serum interferon-α, soluble CD4, soluble CD8, and soluble interleukin-2 receptor.

Total immunoglobulin levels, particularly IgA levels, rise precipitously after HIV infection, but their prognostic usefulness is unclear. A few studies have suggested that persons with high IgA levels had a rapid progression to AIDS,[16,17] but measurement of this parameter has not added significantly to the information provided by other indicators of disease progression.

Infection with HIV results in the production of antibodies to a number of HIV proteins, and several of these antibodies have been investigated as possible prognostic indicators. They include antibody to the p24 core protein, antibody to reverse transcriptase, antibody to the envelope protein gp120, and antibody to specific envelope regions such as the principal neutralizing domain or a portion of the gp41 moiety. The most widely investigated has been antibody to p24, which appears to decline prior to the development of AIDS.[17–19] The p24 antibody has been shown in some studies to have prognostic value when used in conjunction with the $CD4^+$ count.

In general, abnormalities of in vitro mitogen proliferation (e.g., phytohemagglutinin, concanavalin A, and pokeweed mitogen) are commonly reported to be abnormal in AIDS patients, but their usefulness as prognostic indicators has been limited. This lack is the result of the concordance, in most cases, of these defects with severely depressed $CD4^+$ cell counts. Thus in vitro mitogen stimulation assays do not usually give additional prognostic information.

On the other hand, HIV-infected persons may have defects in the ability of their lymphocytes to respond to antigen in vitro before significant decreases in $CD4^+$ cells occur. This situation has led several groups of investigators to evaluate such tests as prognostic indicators for

HIV disease.[20–23] Such tests include antigen-stimulated lymphocyte proliferation, antigen-stimulated lymphokine production, and lymphocyte proliferation to anti-CD3. Decreased responsiveness to these assays has correlated with an increased risk of progression to AIDS. It is not yet known if such in vitro tests can add significantly to the predictive value of $CD4^+$ cell counts. These tests are expensive, cumbersome, and time-consuming to perform, so their role in the clinical evaluation of HIV disease will probably remain limited.

Skin testing for determination of anergy has been proposed as an in vivo assessment of the ability of the immune system to respond to antigen stimulation and has the advantage of being inexpensive and simple to perform.[24] The test requires that patients return in 48 hours for reading, and the difficulty of accomplishing these return visits has led most investigators to omit this test. However, in situations where such testing can be accomplished, the results add to the predictive value of $CD4^+$ cell counts.[25]

Virologic Markers

Cross-sectional studies have suggested that the virus burden of the HIV-infected host increases over time. Thus laboratory quantitation of the amount of HIV present might be expected to be a useful prognostic indicator. Several methods are available for virus detection, including antigen detection, virus culture, and the polymerase chain reaction (PCR). The most extensively evaluated has been the p24 antigen assay.

Antigen to the core protein p24 can be detected and quantified using commercially available kits. Antigen can be detected before antibody appears in some patients and then may reappear with late stage disease; in rare patients it persists throughout the infection. The presence of antigen after detection of antibody correlates with more rapid progression to AIDS,[3,16,19] although no relation has been demonstrated between p24 antigen titer and disease progression. In one study where p24 antigen was evaluated along with $CD4^+$ counts and other indicators, antigen detection added little to the prognostic ability[16]; in another study, however, additional predictive value was noted.[3] The lack of utility of antigen detection appears to be related to the relative insensitivity of the test; high levels of antigen in serum are needed for a positive test result. Such high levels most commonly occur at late stages of disease; and at this point $CD4^+$ counts and other prognostic indicators are also markedly abnormal. More sensitive HIV antigen assays may increase the prognostic usefulness of antigen detection,[26,27] but such assays are not commercially available at the present time.

The amount of virus present in infected individuals can also be assessed by quantitative culture or DNA amplification. Quantitation of HIV in plasma has been reported to correlate with the stage of HIV disease.[28–30] However, such techniques have not been evaluated to determine if specific levels of plasma viremia have prognostic value.

Viral phenotype is also associated with progression of disease. HIV isolates that induce syncytic formation indicate a more rapid progression of disease than non-syncytium-inducing isolates.[31] The drawback to such assays is their expense and lack of general availability.

An additional virologic marker associated with poor prognosis is the appearance of zidovudine resistance (as assessed phenotypically or genotypically). This marker adds to the predictive value of the $CD4^+$ cell count,[32,33] but assays for it are not widely available.

Staging HIV Disease

Many clinical, immunologic, and virologic markers have prognostic value in HIV disease. The single most predictive marker is the $CD4^+$ cell count; the value of other markers must be evaluated in terms of what additional prognostic value they add to the $CD4^+$ count. Prospective studies of various laboratory markers have shown that β_2M, serum neopterin, p24 antigen, the presence of syncytial phenotype, and delayed hypersensitivity skin testing all add to the predictive value of $CD4^+$ counts, but the information gained for a particular patient is small. In most clinical situations, $CD4^+$ cell counts are the only laboratory marker routinely obtained. Of the clinical conditions, thrush is the most useful indicator that a patient needs more careful observation, even when $CD4^+$ counts are not severely depressed.

It would be optimal to combine the most promising clinical and laboratory markers into a staging system for HIV disease. An early staging system, the Walter Reed Classification System,[24] has had limited acceptance by many clinicians because it requires skin testing, which is not commonly performed. The CDC has established a staging system that has both a clinical and a laboratory component (Table 10.2).[34] With this system, clinical category A includes individuals with primary HIV infection, those without symptoms, or those with only persistent generalized lymphadeno-

TABLE 10.2. CDC classification system for HIV infection.

Laboratory axis: total $CD4^+$ count (/mm^3)	Clinical axis		
	Asymptomatic	Intermediate stage	Late stage (AIDS)
≥ 500	A1	B1	C1
200–499	A2	B2	C2
<200	A3	B3	C3

See text for explanation of categories A, B, and C and of 1, 2, and 3.

pathy. Clinical category B includes individuals with symptomatic conditions that are not AIDS-indicating diagnoses, such as mucocutaneous candidiasis, herpes zoster, or peripheral neuropathy. Clinical category C includes those with the AIDS-indicating conditions of the 1987 clinical AIDS case definition plus pulmonary tuberculosis, recurrent pneumonia, or invasive cervical cancer. In addition, each clinical category is subdivided into three immunologic categories: individuals with 500 or more $CD4^+$ cells/mm^3 are in immunologic category 1; those with 200 to 499 $CD4^+$ cells/mm^3 are in immunologic category 2; those with fewer than 200 $CD4^+$ cells/mm^3 are in immunologic category 3.

Several other systems have been proposed, but none is in widespread use. The major difficulty is that systems that require extensive laboratory evaluation soon become too complex to be clinically useful. $CD4^+$ cell counts are widely available in the United States and Europe and are an important parameter for management of HIV disease in those areas. It is important to emphasize, however, that even the $CD4^+$ count, by far the strongest predictor of the course of disease, can provide the clinician and patient with only a rough estimate of the risk of disease progression.

References

1. El-Sadr W, Marmor M, Zolla-Parner S, et al. Four-year prospective study of homosexual men: correlation of immunologic abnormalities, clinical status, and serology to human immunodeficiency virus. J Infect Dis 1987;155: 789–793
2. Lang W, Anderson RE, Perkins H, et al. Clinical, immunologic, and serologic findings in men at risk for acquired immunodeficiency syndrome. JAMA 1987;257:326–330
3. Moss AR, Bacchetti P, Osmond D, et al. Seropositivity for HIV and the development of AIDS or AIDS related condition: three year follow up of the San Francisco General Hospital cohort. BMJ 1988;296:745–750
4. Klein RS, Harris CA, Small CB, et al. Oral candidiasis in high-risk patients as the initial manifestation of the acquired immunodeficiency syndrome. N Engl J Med 1984;311:354–358
5. Phair J, Munoz A, Detels R, et al. The risk of Pneumocystis carinii pneumonia among men infected with human immunodeficiency virus type 1. N Engl J Med 1990;322:161–165
6. Melbye M, Grossman RJ, Goedert JJ, et al. Risk of AIDS after herpes zoster. Lancet 1987;1:728–731
7. Greenspan D, Greenspan JS, Hearst NG, et al. Relation of oral hairy leukoplakia to infection with the human immunodeficiency virus and the risk of developing AIDS. J Infect Dis 1987;155:475–481
8. Malone JL, Simms TE, Gray GC, et al. Sources of variability in repeated T-helper lymphocyte counts from human immunodeficiency virus type 1-infected patients: total lymphocyte count fluctuations and diurnal cycle are important. J Acquir Immune Defic Syndr 1990;3:144–151

9. Taylor JMG, Fahey JL, Detels R, et al. CD4 percentage, CD4 number, and CD4:CD8 ratio in HIV infection: which to choose and how to use. J Acquir Immune Defic Syndr 1990;2:114–124
10. Koepke JA, Landay AL. Precision and accuracy of absolute lymphocyte counts. Clin Immunol Immunopathol 1989;52:19–27
11. Lang W, Perkins H, Anderson RE, et al. Patterns of T lymphocyte changes with human immunodeficiency virus infection: from seroconversion to the development of AIDS. J Acquir Immune Defic Syndr 1990;2:63–69
12. Volberding PA, Lagakos SW, Koch MA, et al. Zidovudine in asymptomatic human immunodeficiency virus infection. N Engl J Med 1990;322:941–949
13. Fischl MA, Richman DD, Hansen N, et al. The safety and efficacy of zidovudine in the treatment of subjects with mildly symptomatic human immunodeficiency virus type 1 infection. Ann Intern Med 1990;112:727–737
14. Anonymous. Guidelines for prophylaxis against Pneumocystis carinii pneumonia for persons infected with human immunodeficiency virus. MMWR 1989;38:S1–9
15. Giorgi JV, Liu Z, Hultin LE, et al. Elevated levels of $CD38^+$ $CD8^+$ T cells in HIV infection add to the prognostic value of low $CD4^+$ T cell levels: results of 6 years of follow-up. J Acquir Immune Defic Syndr 1993;6:904–912
16. Fahey JL, Taylor JMG, Detels R, et al. The prognostic value of cellular and serologic markers in infection with human immunodeficiency virus type 1. N Engl J Med 1990;322:166–172
17. McDougal JS, Kennedy MS, Nicholson JKA, et al. Antibody response to human immunodeficiency virus in homosexual men: relation of antibody specificity, titer, and isotype to clinical status, severity of immunodeficiency, and disease progression. J Clin Invest 1987;80:316–324
18. Weber JN, Clapham PR, Weiss RA, et al. Human immunodeficiency virus infection in two cohorts of homosexual men: neutralising sera and association of anti-gag antibody with prognosis. Lancet 1987;1:119–122
19. Lange JMA, Paul DA, Huisman HG, et al. Persistent HIV antigenaemia and decline of HIV core antibodies associated with transition to AIDS. BMJ 1986;293:1459–1462
20. Murray HW, Hillman JK, Rubin BY, et al. Patients at risk for AIDS-related opportunistic infections. N Engl J Med 1985;313:1504–1510
21. Ballet J-J, Couderc L-J, Rabian-Herzog C, et al. Impaired T-lymphocyte-dependent immune responses to microbial antigens in patients with HIV-1-associated persistent generalized lymphadenopathy. AIDS 1988;2:291–297
22. Hoffman B, Bygbjerg I, Dickmess E, et al. Prognostic value of immunologic abnormalities and HIV antigenemia in asymptomatic HIV-infected individuals: proposal of immunologic staging. Scand J Infect Dis 1989;21:633–643
23. Schellekens PTA, Roos MTL, DeWolf F, et al. Low T-cell responsiveness to activation via CD3/TCR is a prognostic marker for acquired immunodeficiency syndrome in human immunodeficiency virus-1-infected men. J Clin Immunol 1990;2:121–127
24. Redfield RR, Wright DC, Tramont EC. The Walter Reed staging classification for HTLV-III/LAV infection. N Engl J Med 1986;314:131–132
25. Blatt SP, Hendrix CW, Butzin CA, et al. Delayed-type hypersensitivity skin testing predicts progression to AIDS in HIV-infected patients. Ann Intern Med 1993;119:177–184

26. Vasudevachari MB, Salzman NP, Woll DR, et al. Clinical utility of an enhanced human immunodeficiency virus type I p24 antigen capture assay. J Clin Immunol 1993;13:185–192
27. Miles SA, Balden E, Magpantay L, et al. Rapid serologic testing with immune-complex dissociated HIV p24 antigen for early detection of HIV infection in neonates. N Engl J Med 1993;328:297–302
28. Ho DD, Mougdil T, Alam M. Quantitation of human immunodeficiency virus type 1 in the blood of infected persons. N Engl J Med 1989;321:1621–1625
29. Piatak M, Saag MS, Yang LC, et al. High levels of HIV-1 in plasma during all stages of infection determined by competitive PCR. Science 1993;259: 1749–1754
30. Schnittman SM, Greenhouse JJ, Psallidopoulos MC, et al. Increasing viral burden in $CD4^+$ T cells from patients with human immunodeficiency virus infection reflects rapidly progressive immunosuppression and clinical disease. Ann Intern Med 1990;113:438–443
31. Koot M, Keet IPM, Vos AHV, et al. Prognostic value of HIV-1 syncytium-inducing phenotype for rate of $CD4^+$ cell depletion and progression to AIDS. Ann Intern Med 1993;118:681–688
32. St Clair MH, Hartigan PM, Andrews JC, et al. Zidovudine resistance, syncytium-inducing phenotype, and HIV disease progression in a case-control study. J Acquir Immune Defic Syndr 1993;6:891–897
33. Kozal MJ, Shafer RW, Winters MA, Katzenstein DA, Merigan TC. A mutation in human immunodeficiency virus reverse transcriptase and decline in CD4 lymphocyte numbers in long-term zidovudine recipients. J Infect Dis 1993;167:526–532
34. Centers for Disease Control. 1993 Revised classification system for HIV infection and expanded surveillance case definition for AIDS among adolescents and adults. MMWR 1992;41(RR-17):1–19

11
Testing for Other Human Retroviruses: HTLV-I and HTLV-II

RIMA F. KHABBAZ, WALID HENEINE, and JONATHAN E. KAPLAN

The first human retroviruses to be discovered, human T lymphotropic virus type I (HTLV-I) and human T lymphotropic virus type II (HTLV-II) were reported in 1980 and 1982, respectively.[1,2] HTLV-I was soon associated with adult T cell leukemia/lymphoma (ATL), a lymphoproliferative illness first reported in Japan in 1977. In 1985 a neurologic illness, now called HTLV-I-associated myelopathy/tropical spastic paraparesis (HAM/TSP), was also associated with HTLV-I. In contrast, HTLV-II, which was initially thought to be associated with hairy cell leukemia, has not been clearly associated with any disease. Screening of volunteer blood donors for HTLV-I began in 1986 in Japan and in 1988 in the United States; screening is also performed in Canada, France, the French West Indies, and Trinidad-Tobago.[3]

Virus Structure and Biology

These two viruses, HTLV-I and HTLV-II, belong to the oncovirus subfamily of retroviruses. They are distinct from the human immunodeficiency viruses, HIV-1 and HIV-2, which belong to the lentivirus subfamily of retroviruses and cause acquired immunodeficiency syndrome (AIDS). HTLVs characteristically cause cells to proliferate in culture, whereas lentiviruses cause cell death. Morphologically, HTLVs are classified as type C oncoviruses, referring to the electron-dense, centrally located nuclear core visible by electron microscopy.

The genomes of HTLV-I and HTLV-II are approximately 8.0 kilobase pairs (kbp) and 8.8 kbp in length, respectively, and they share approximately 60% of their nucleic acid sequences. These viruses share structural features with all retroviruses, including group-specific nucleoprotein antigens (*gag*), reverse transcriptase (*pol*), and transmembrane and external

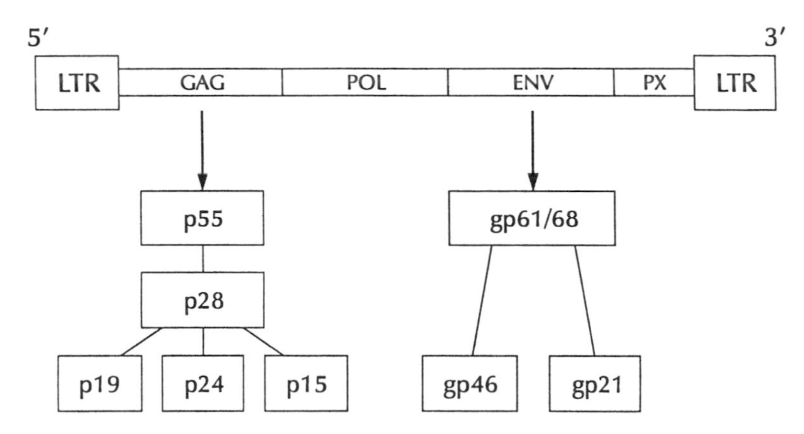

FIGURE 11.1. Genetic structure of human T lymphotropic virus type I (HTLV-I) and structural (*gag* and *env*) gene products.

envelope glycoproteins (*env*). In addition, HTLVs contain unique *px* genes at the 3′ end of the genome that encode for important regulatory proteins (*tax* and *rex*) (Fig. 11.1). The proteins and glycoproteins encoded by *gag* and *env* are highly immunogenic. Reactivities to the *gag* proteins p19, p24, p15, p28 (intermediate breakdown product), and p53 (precursor) and to the *env* glycoproteins gp46, gp21, and gp61/68 (precursor) form the basis for serodiagnosis of HTLV-I and HTLV-II. In contrast, the immunoreactivities of the *pol*, *tax*, and *rex* gene products have not been as well defined and are not generally incorporated into HTLV serologic testing algorithms.

Epidemiology

Infection with HTLV-I is highly endemic in southwestern Japan and in countries of the Caribbean basin.[3] Reported seroprevalence rates in the general populations of these areas are 5% to 20%. In both of these areas, HTLV-I seropositivity increases with age, and in older age groups rates are usually higher among women than among men. In the Caribbean, significantly higher HTLV-I seropositivity rates are found among Blacks than among other ethnic groups. HTLV-I infection is also endemic in areas of Melanesia, Africa, some Central and South American countries, and in isolated areas of Asia. In the United States, clusters of HTLV-I infections have been reported among African-Americans from the southeastern United States and among immigrants from HTLV-I-endemic areas. HTLV-I has also been found in persons who have had sexual contact with infected persons from the Caribbean basin or Japan. In Europe, HTLV-I is found primarily among West Indian immigrants. HTLV-I accounts for a small proportion of HTLV-I/II seropositivity among injecting drug users (IDUs).

The geographic distribution of HTLV-II is not well defined. HTLV-II infection is highly prevalent among IDUs in the United States and Europe. Otherwise, HTLV-II infection is endemic in several American Indian populations, including North American Indians in Florida and New Mexico, the Guaymi Indians in Panama, and the Cayapo and Kraho South American tribes of Brazil. HTLV-II infection has also been reported among Pygmies in Africa.

Approximately half of the U.S. volunteer blood donors seropositive for HTLV-I/II are infected with HTLV-II, and half are infected with HTLV-I.[4] Whereas blood donors infected with HTLV-II most often report either a history of drug injection or a history of sexual contact with an IDU, blood donors infected with HTLV-I most often report being born in the Caribbean or Japan or having sexual contact with persons from these areas. Smaller percentages of each group report a history of blood transfusion.

The HTLV-I is transmitted by sexual contact, blood transfusion, or sharing contaminated needles; it is also spread from mother to child.[3] Mother-to-child transmission occurs primarily through breast-feeding; in HTLV-I-endemic areas, approximately 25% of breast-fed infants born to mothers with HTLV-I acquire infection. Intrauterine or perinatal transmission of HTLV-I occurs less frequently; approximately 5% of children who are born to infected mothers but are not breast-fed acquire infection. Sexual transmission of HTLV-I appears to be more efficient from men to women than from women to men. In the United States approximately 25% to 30% of sexual partners of blood donors with HTLV-I/II are also seropositive. Transmission of HTLV-I occurs with transfusion of cellular blood products (whole blood, red blood cells, and platelets) but not with the plasma fraction or plasma derivatives from HTLV-I-infected blood. Seroconversion rates of 44% to 63% have been reported in recipients of HTLV-I-infected cellular components in HTLV-I endemic areas; lower rates (approximately 20%) have been reported in the United States.

The HTLV-II is believed to be transmitted the same way as HTLV-I.[3] Preliminary data from an HTLV-II-endemic population in which breast-feeding is the norm support mother-to-child transmission.[5] HTLV-II can also be transmitted sexually; the most commonly reported risk factor among U.S. female blood donors infected with HTLV-II is sexual contact with an IDU. HTLV-II can be transmitted by transfusion of cellular blood products, including whole blood, red blood cells, and platelets. The high prevalence of HTLV-II among IDUs is likely because of the sharing of blood-contaminated needles or other injection paraphernalia.

Diseases Caused by HTLV

The HTLV-I has been definitively associated with two diseases: ATL and HAM/TSP.[3] ATL is a malignancy of HTLV-I-infected CD4$^+$ T lymphocytes. The HTLV-I provirus is monoclonally integrated in the DNA of the malignant cells. Acute chronic, lymphomatous, and smoldering forms of ATL have been described. Acute ATL is characterized by infiltration of lymph nodes, viscera, and skin with malignant cells; this infiltration results in a constellation of clinical features, including lymphadenopathy, hepatosplenomegaly, skin lesions, lytic bone lesions, hypercalcemia, and abnormal liver function. Abnormal lymphocytes, called flower cells, are generally seen on the peripheral blood smear. Median survival for persons with acute ATL is 11 months from diagnosis. Conventional chemotherapy is not curative, and relapses often occur quickly, although prolonged survival has been reported. ATL has been estimated to occur among 2% to 4% of individuals infected with HTLV-I in regions where HTLV-I is endemic and where early childhood infection is common. The disease occurs most frequently in those 40 to 60 years of age, suggesting that a latent period of as long as a few decades is required for the disease to develop. Men and women are equally affected. One case of ATL in an immunocompromised patient has been reported in which infection appears to have been acquired by blood transfusion.[3]

A chronic, degenerative, neurologic disease, HAM/TSP is characterized by progressive and permanent lower extremity weakness, spasticity, hyperreflexia, sensory disturbances, and urinary incontinence. Unlike the symptoms of patients with multiple sclerosis, the signs and symptoms of HAM/TSP patients do not wax and wane; in these patients cranial nerves are not involved, and cognitive function is not affected. Disability increases with time, and patients may become confined to wheelchairs. HAM/TSP is believed to be immunologically mediated. Antibodies to HTLV-I are characteristically found in high titer in serum and cerebrospinal fluid, and treatment with corticosteroids has benefited some patients. Danazol, a synthetic androgen, reportedly improves symptoms including bladder dysfunction. Fewer than 1% of HTLV-I-infected persons develop HAM/TSP. The disease occurs most frequently in persons 40 to 60 years of age; women are affected more frequently than men. The latency period is shorter than that for ATL; cases of HAM/TSP have been associated with blood transfusion, with a median interval of 3.3 years between transfusion and the development of neurologic illness.

The full spectrum of HTLV-I-associated diseases may include other disorders.[3] Infective dermatitis, a chronic eczema associated with *Staphylococcus aureus* and β-hemolytic streptococcus has been reported in Jamaican children infected with HTLV-I. Cases of polymyositis, arthritis, pulmonary alveolitis, and uveitis also have been reported among HTLV-I-infected persons.

Infection with HTLV-II has not been clearly associated with any disease.[3] The virus was first isolated from two patients with hairy cell leukemia, but no evidence of HTLV-II infection was found among 21 additional patients with this disease. One study showed no increased rates of lymphoproliferative illnesses in New Mexico, where HTLV-II is present among American Indians. Cases of HAM/TSP-like neurologic illnesses and rare cases of mycosis fungoides and large granular lymphocyte leukemia have been reported among HTLV-II-infected persons. Cases of erythrodermatitis and bacterial skin infections have been reported among individuals infected with both HIV-I and HTLV-II.

Serologic Testing

Diagnosis of HTLV-I and HTLV-II infection relies primarily on serologic testing. The presence of antibodies indicates lifelong infection with the virus. Several assays are available for screening for HTLV-I/II, including the enzyme-linked immunosorbent assay (ELISA), particle agglutination, and immunofluorescence assay. All these assays have good sensitivity but are not specific for individual gene product reactivity.[4]

The ELISAs are rapid colorimetric tests that can be partially automated and can be used to test large numbers of samples. They have additional advantages that include low cost and ease of use. Samples (usually serum or plasma) to be tested are added to the solid-phase microtiter plates, which have been coated with antigen. After incubation, bound antibodies are detected by a form of anti-immunoglobulin–enzyme conjugate (e.g., peroxidase-conjugated anti-human immunoglobulin). An enzyme-reactant substrate is added, and the resulting color reaction is detected with a spectrophotometer. In the United States, ELISAs that use HTLV-I whole-virus lysate as antigen have been licensed by the U.S. Food and Drug Administration (FDA).[6] These tests vary in their sensitivity to detect antibodies to HTLV-II. An additional ELISA that contains HTLV-I lysate and recombinant *env*p21e appears to have better sensitivity for the detection of HTLV-II.[7]

Particle agglutination is the standard screening assay in Japan; it is a latex bead-based assay. The immunofluorescence assay has had limited use as a research screening test.

Specimens that are reactive by the screening assays should be confirmed by supplementary serologic tests, such as the Western blot assay and the radioimmunoprecipitation assay (RIPA), which can identify antibodies to specific HTLV gene products, such as *gag* and *env*.[6]

With the Western blot test, antibodies in the sample react with antigens on nitrocellulose or nylon strips. After incubation and washing, a horseradish peroxidase–labeled anti-antibody conjugate is applied, and the immune reactions are visualized with a substrate.

The RIPA is labor-intensive and requires handling HTLV-infected cell cultures and radionucleotides. Radiolabeled HTLV-infected cell lysates are reacted with serum or plasma and then precipitated with sepharose beads. The reaction products are electrophoresed on polyacrylamide gels and analyzed by autoradiography.

No supplementary tests have been licensed by the FDA, but Western blot strips are available to research institutions, blood banks, public health laboratories, and industry, and as in-house tests in some diagnostic laboratories. Western blot assays are sensitive for detecting *gag* products, and RIPA is often required to detect *env* reactivity.[8]

In 1988 a U.S. Public Health Service (PHS) working group recommended that specimens must demonstrate immunoreactivity to both *gag* p24 and *env* gp46 or gp61/68 by Western blot assay or RIPA to be considered seropositive for HTLV-I/II.[6] Reactive serum specimens not satisfying these criteria, but showing immunoreactivity to at least one suspected HTLV gene product, are designated "indeterminate." ELISA-reactive specimens with no immunoreactivity to any HTLV gene product are considered falsely positive. Several studies using provirus amplification have supported the accuracy of these diagnostic criteria; individuals whose specimens satisfy the criteria for seropositivity are virtually always infected with HTLV-I or HTLV-II.[9,10] In contrast, individuals whose specimens are "indeterminate" are rarely infected with either virus; in those who are found to be infected, repeat serologic testing frequently demonstrates seropositivity.[11] In rare instances, persons with reactivity to *gag* p19 and to *env* gp46 or gp61/68, but without reactivity to p24, have been found to be infected with HTLV-I/II.[12]

The development of the recombinant *env* protein p21e, which is highly sensitive for detecting both HTLV-I and HTLV-II infection, promises to simplify HTLV serologic testing algorithms.[7] HTLV-I Western blot assays that are spiked with *env* p21e demonstrate improved sensitivity over nonspiked Western blot tests for the detection of *env* (Fig. 11.2). However, the specificity of the current recombinant p21e appears to be only 97.5%.[13] Therefore, until a more specific p21e antigen is available, for purposes of notification and counseling it is prudent to confirm p21e reactivity by other serologic tests for *env* (gp46 or gp61/68 by Western blot or RIPA, peptide-based ELISA, type-specific recombinant protein) or by the polymerase chain reaction (PCR).

The supplementary serologic tests previously discussed cannot differentiate antibodies to HTLV-I and HTLV-II. The relative intensity of the reactivity to the *gag* proteins p19 and p24 on the Western blot assay has been used to differentiate HTLV-I from HTLV-II, but such differentiation may be unreliable.[14] Several synthetic peptides and recombinant proteins, all based on differences in the *env* region, have been developed for this purpose.[15–17] The synthetic peptides are available in an ELISA format; typing is achieved by testing the specimen in parallel against two peptides

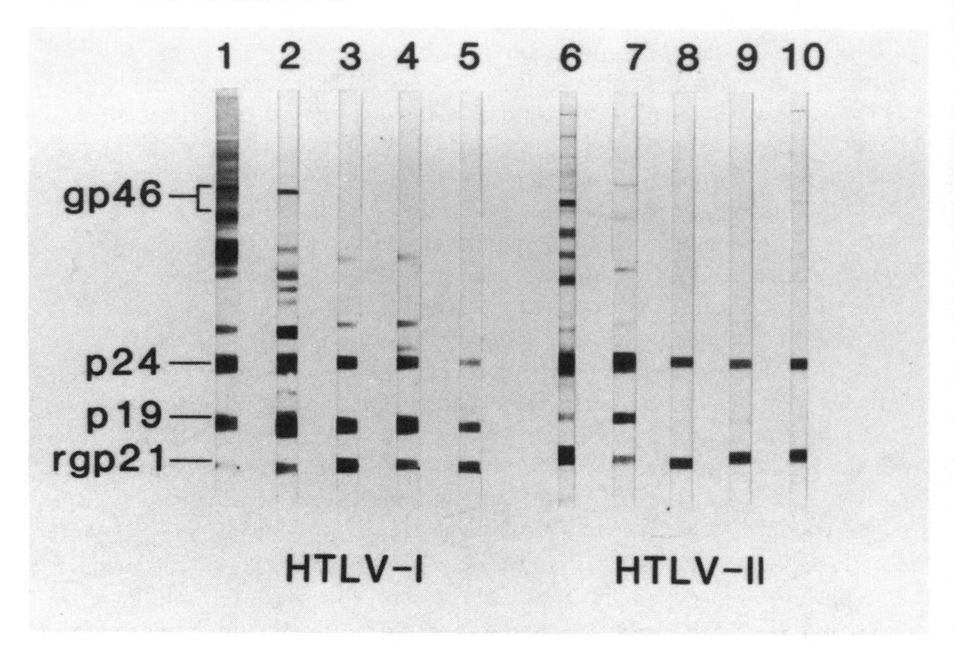

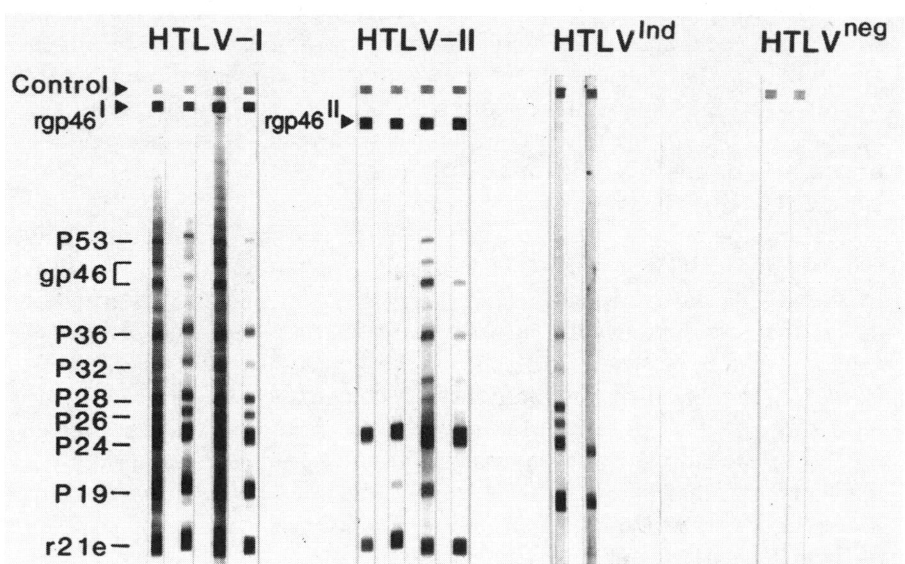

FIGURE 11.2. HTLV-I Western blot assays. (Top) HTLV-I- and HTLV-II-seropositive samples on p21e-spiked Western blot (Cambridge Biotech). Specimens demonstrating *gag* p24 and *env* p21e reactivity are considered seropositive for HTLV-I/II, although for notification and counseling purposes p21e reactivity should be confirmed by other evidence of *env* reactivity (see text). (Bottom) HTLV-I, HTLV-II, HTLV-indeterminate ($HTLV^{Ind}$), and HTLV-seronegative ($HTLV^{neg}$) serum samples on Western blot spiked with recombinant proteins p21e, gp46I, and gp46II (Genelabs Diagnostics).

(HTLV-I- and HTLV-II-specific peptides) and calculating the ratio of the respective optical densities. HTLV-I- and HTLV-II-specific recombinant proteins have been painted on an HTLV-I Western blot spiked with *env* p21e (Fig. 11.2). Like the previously discussed supplementary tests, all of these tests are available for research only. These assays appear to have excellent specificity (approximately 100%) for differentiating antibodies to HTLV-I and HTLV-II. However, not all HTLV-I/II-seropositive specimens can be typed as HTLV-I or HTLV-II using these tests (sensitivity range 76–100%). In such cases, more sophisticated methods, such as provirus amplification or virus isolation, may be needed to differentiate HTLV-I from HTLV-II infection.

Figure 11.3 shows a model algorithm for confirming and typing HTLV-I/II serologic reactivity. Variations of this algorithm depend on the assays used.

Confirmatory testing algorithms in the United States continue to change since the initiation of blood donor screening. Until November 1993, the practice of the American Red Cross (ARC), which is responsible for approximately half of the U.S. blood supply, was to confirm ELISA

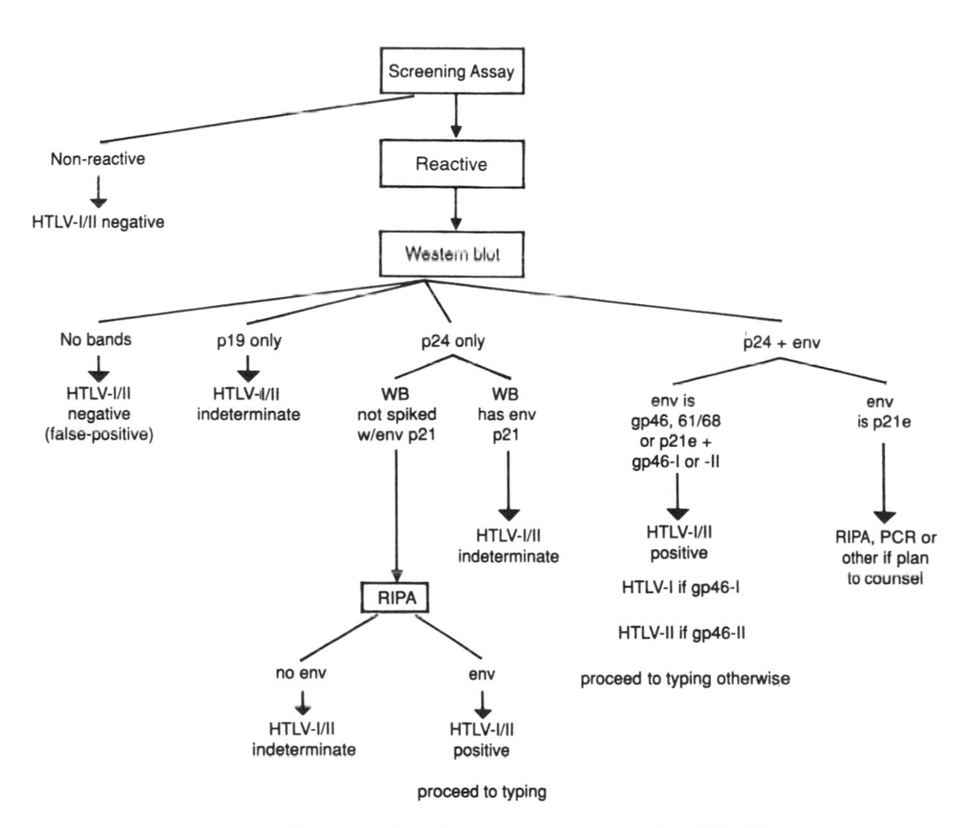

FIGURE 11.3. Algorithm for serologic testing for HTLV.

repeatedly reactive specimens by using an HTLV-I-based Western blot assay and a recombinant *env* p21e ELISA.[18] Since envp21e ELISA is no longer available, currently ARC is confirming ELISA repeat reactive specimens with a WB spiked with envp21e. Blood banks not affiliated with the ARC generally refer repeatedly reactive specimens to the ELISA manufacturer, where confirmation proceeds with the Western blot assay and RIPA. In some blood banks, seropositive specimens are typed serologically to differentiate HTLV-I and HTLV-II infections.

Molecular Testing

Like other retroviruses, HTLV-I and HTLV-II proviruses integrate into the DNA of the host cell after infection. Several molecular techniques can be used to detect the HTLV proviruses in infected cells.

PCR

Detection of proviral DNA by PCR amplification is a highly sensitive molecular technique for detecting HTLV-I and HTLV-II provirus in infected cells, such as peripheral blood mononuclear cells (PBMCs) from asymptomatic HTLV-infected persons. PCR has become a reference method for determining infection status, testing the validity of serologic assays, and distinguishing between HTLV-I and HTLV-II. Whereas HTLV-I/II seropositivity according to PHS criteria is a reliable indicator of infection, PCR is useful for confirming infection and studying seroindeterminate or seronegative persons at high risk for HTLV infection. PCR also is useful for testing infants, such as in mother-to-child transmission studies, in which the serostatus of the newborn may not be a reliable indication of infection because of passively acquired maternal antibodies. PCR also may be useful for detecting infection during the window period between exposure and seroconversion. Although HTLV-I and HTLV-II infections can now be differentiated with reasonable certainty by type-specific serology that uses peptides or recombinant proteins, PCR remains useful for confirming results and studying serologically untypable samples. Finally, PCR may be useful for studying in vivo virus load and tissue distribution.

Two strategies are used to confirm and differentiate between HTLV-I and HTLV-II infections by PCR. The first involves the use of HTLV consensus primers (e.g., *pol* SK110/SK111), which allow the amplification of both viruses. Typing is achieved by hybridizing the amplified product to an HTLV-I-specific probe (SK112) or an HTLV-II-specific probe (SK188) (Fig. 11.4).[19] The second strategy uses HTLV-I- and HTLV-II-specific primers and probes in separate amplifications (e.g., SK54/SK55, GAG49/GAG51 for HTLV-I; SK58/SK59, 2G1/2G4 for HTLV-II).[19] The first strategy is useful for the initial screening and typing, as it requires a single amplification reaction. Positive results can be further confirmed using type-specific amplification and probing. A sensitivity of 97% and a specificity of 100% have been achieved with PCR using the SK110/SK111 system.[19]

Proteinase K-digested PBMCs at a concentration of 6 million cells per milliliter can be used as templates for PCR. These lysates work as well as DNA extracted more laboriously by phenol/chloroform and ethanol precipitation. By minimizing sample handling, this method also can decrease the risk of contamination. Because hemoglobin can inhibit *Taq* polymerase, PBMC samples should be checked for red blood cell (RBC) contamination before digestion. Contaminating RBCs can be removed by lysing with ammonium chloride and subsequent washing. A simple, rapid alternative method for template preparation that obviates the need for separating PBMCs involves leukocyte nucleus preparations made by lysing 1 to 3 ml of noncoagulated blood in a sucrose-based lysis buffer.[19] After centrifugation, the pellet is washed once and then digested with proteinase K. This method requires a small volume of blood and is particularly suitable for studies involving pediatric patients. Storage of undigested nucleus preparations should be avoided, however, because of risks of DNA degradation, possibly by contaminating nucleases.

Aliquots of 25 μl of PBMC lysates (i.e., from 150,000 cells, ~1 μg DNA) are cycled 35 times at 94°C for 1 minute, 55°C for 1 minute, and

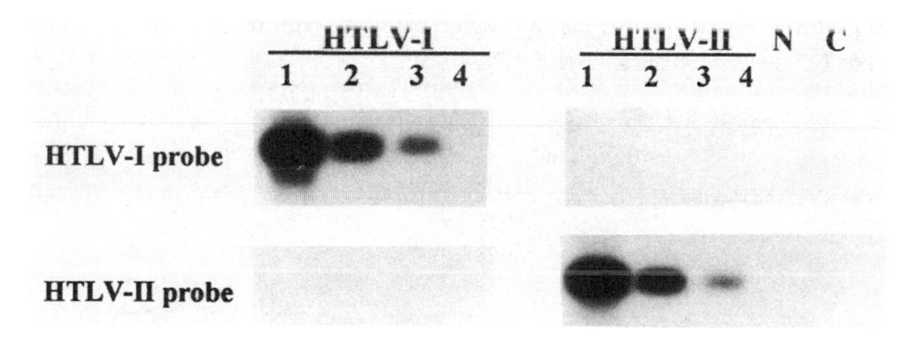

FIGURE 11.4. Generic PCR amplification of HTLV-I and HTLV-II by SK110/SK111 and differentiation by type-specific probing. Numbers 1, 2, 3, and 4 refer to lysates of uninfected Hut-78 cells containing 1500, 150, 2, and 0 HTLV-I-infected (MT-2) or HTLV-II-infected cells (Mo-T) cells, respectively. N = Hut-78; C = reagent cocktail control. The HTLV-I-specific probe is SK112, and the HTLV-II probe is SK188.

72°C for 1 minute in a 100-μl reaction volume.[19] The products are electrophoresed on 1.8% agarose gels and then hybridized by Southern blot to ^{32}P-end-labeled internal probes.[19] In addition to its ability to type and distinguish the HTLV PCR product from other nonspecifically amplified material, enhancing the specificity of the assay, probing also can increase the sensitivity of the assay by identifying signals that are too weak to be visualized on ethidium bromide-stained gels. Each experiment should be appropriately controlled. These controls should include both negative (uninfected) and positive (infected) cell controls, sensitivity controls (the highest dilution of infected cells that was prepared in uninfected cells and is PCR-positive), and reagent mixture control (no template). For an experiment to be acceptable, signals should be detected only in the positive and sensitivity controls (Fig. 11.4).

A system that obviates the need for electrophoresis for detecting the amplified DNA has become commercially available (Roche Laboratories, Piscataway, NJ). This system uses an ELISA plate in which the amplified product is hybridized to an oligonucleotide probe that is coated to the wells.

Occasionally, PBMC samples from HTLV-I/II-seropositive persons show a negative PCR result in an otherwise successful experiment. Although this falsely negative result rarely may be because of a low virus load, it is frequently the result of other template problems. For an initial assessment of the validity of the template, look in the agarose gel for any nonspecific DNA amplification. If amplification exists, the template is probably valid, and one of two methods can be attempted to improve the sensitivity. First, the sample can be retested by doubling the input DNA (i.e., 50 μl) and, second, if that reaction is still negative by applying nested PCR using 5 μl of the first reaction as template and (in the case of a *pol* amplification) using type-specific primers that are internal to SK110/SK111.[19] These approaches can boost the sensitivity of PCR to 100%.[19] If no amplifiable DNA is seen, the possibility of DNA degradation or nonspecific inhibition of the amplification reaction exists. These possibilities may be checked by PCR, amplifying an endogenous sequence, such as that from the β-globin gene. If this reaction is negative, another sample should be tested.

The high sensitivity of PCR does not come without problems. Reactions can easily be contaminated with target DNA from infected or previously amplified or cloned material, yielding a false-positive result. This problem merits constant attention; recommendations to prevent contamination should be strictly followed,[20] especially with nested PCR, with which the risks of contamination by amplified material from the first PCR run are high.

The PCR also can be used to detect viral transcripts by reverse transcriptase PCR (RT-PCR) in studies of virus expression in vivo[21,22] and in vitro.[23] In these assays, messenger RNA (mRNA) from infected

PBMCs or cultured cells is reverse-transcribed into DNA, and the viral transcripts are then PCR-amplified. To minimize the problems of contamination from the proviral DNA, RT-PCR assays have successfully used primers and probes designed on the basis of splice junctions of the viral mRNAs. Such assays detected the HTLV-I *tax/rex* transcript in the PBMCs of patients with HAM/TSP and ATL and of asymptomatic carriers.[21,22]

An in situ PCR method has been developed for HTLV-I in cell suspensions.[24] The method uses a system of multiplex primers for amplification and a digoxigenin colorimetric detection system. This technique combines the high sensitivity of PCR with the cellular localization ability of in situ hybridization. Therefore it should determine the percentage of infected cells in vivo, which is a better marker of the virus load than is the total number of proviral copies. However, the in situ PCR method currently used is complicated by frequent contamination; the amplified HTLV-I product from the infected cell, released by leakage or by cell lysis, contaminates the uninfected cells during the cycling process. Until this limitation is corrected, the technique should be considered qualitative rather than quantitative.

Other Molecular Techniques

Southern blot hybridization of enzyme-restricted DNA to radiolabeled HTLV-I or HTLV-II probes may identify viral sequences if a large proviral copy number is present in the sample. For example, Southern blotting may allow the detection of HTLV-I in lymphoma cells or in cultured cells; it has been used to demonstrate the monoclonal integration of HTLV-I in tumor cells from ATL patients.[25] Because of its low sensitivity in samples with low numbers of infected cells, however, Southern blot hybridization is not useful for detecting HTLV in PBMCs from asymptomatically infected persons.

In situ hybridization could theoretically be used for the cellular localization of HTLV-I DNA and RNA sequences in tissue, such as in biopsies of lymphoma. However, the successful application of this technique has been limited to the detection of HTLV-I RNA in PBMCs of patients with HAM/TSP.[26] Like the Southern blot, in situ hybridization is not sensitive enough to detect low copy numbers of HTLV-I or HTLV-II in asymptomatically infected persons.

Virus Isolation

Isolation of HTLV-I or HTLV-II from infected persons can be achieved by culturing PBMCs. The success of this technique depends on activation of T lymphocytes and on cell culture conditions that allow long-term growth of lymphocytes. The viruses are transmitted by cell-to-cell transfer. Because of these biologic factors, virus isolation from PBMCs normally involves an initial step of mitogen stimulation, after which cultures are monitored for HTLV production. The stimulated PBMCs also can be co-cultured with uninfected target cells (e.g., mitogen-stimulated PBMCs). Such co-cultivation techniques have been found to be more sensitive than primary cultures for isolating HTLV.[27]

Monitoring cell cultures for HTLV production can be achieved by detecting viral antigens or nucleic acids or by observing viral particles by electron microscopy. The antigen-capture assays use a monoclonal antibody directed against either an HTLV-I-specific epitope of p19 (Cellular Products, Buffalo, NY) or a common (HTLV-I/II) epitope of p24 (Coulter Immunology, Hialeah, FL). The bound antigen–antibody complex is detected colorimetrically. The presence of HTLV in cell cultures also can be detected by PCR amplification of proviral sequences in samples from the cultured cells. Screening cultures for reverse transcriptase by the current methods has little value because of the low levels produced by these viruses. Electron microscopy can be used to demonstrate the presence of type C retrovirus particles. HTLV virions, which are approximately 100 nm in diameter, are occasionally seen budding from the cell membrane. With all these detection techniques, multiple sampling of cultured cells early in the culture should be minimized to avoid depriving the culture of cells that harbor the virus and are needed to establish a spreading infection.

Attempts have been made to detect cell-free virus in the culture supernatant using RT-PCR.[28] This assay, which amplifies the particle-associated genome, is more sensitive than routine assays for early detection of the viruses in culture. It provides information on the release of packaged virus and eliminates the need for testing cultured cells.

Blood Screening

During 1989, the first full year of screening for HTLV-I in the United States, approximately 0.07% of volunteer blood donations were repeatedly reactive in the screening assay, and approximately 21% of these reactive specimens (0.014% of the total) were confirmed as HTLV-I/II-seropositive.[4] Limited testing suggested that approximately half of the seropositive donors were infected with HTLV-I and half with HTLV-II. Since then, HTLV-I/II-seropositive donors have been deferred from donating blood, and the current prevalence of HTLV-I/II seropositivity among blood donors has decreased to about 0.010%.[4]

To optimize the safety of the blood supply, all HTLV-I ELISA-reactive donations in the United States are not transfused, regardless of the results of confirmatory testing. In the ARC system, seropositive donors are indefinitely deferred from donating blood. Since October 1992, HTLV-indeterminate donors also have been deferred because such donors may be infected with HTLV-I or HTLV-II. Donors who are ELISA-reactive but confirmatory test-negative (false-positive screening test) may donate blood again, but they are indefinitely deferred if such reactivity occurs twice.

Counseling

In 1992 the U.S. Centers for Disease Control and Prevention (CDC) convened a PHS working group to develop guidelines for counseling HTLV-I/II-seropositive blood donors and other infected persons.[29] The following guidelines were proposed.

HTLV-I

Individuals found to be seropositive for HTLV-I/II, according to the PHS criteria, and positive for HTLV-I by additional testing should be informed that they are infected with HTLV-I. They should be told that HTLV-I is not the AIDS virus, that it does not cause AIDS, and that AIDS is caused by a different virus, called HIV. They should be told that HTLV-I is a lifelong infection and be given information regarding modes and efficiency of transmission of HTLV-I, disease associations, and the probability of developing disease.

Specifically, persons infected with HTLV-I should be advised to do the following.

1. Share the information with their physicians.
2. Not donate blood, semen, body organs, or other tissues.
3. Not share needles or syringes.
4. Not breast-feed infants.
5. Consider the use of latex condoms to prevent sexual transmission.

HTLV-II

Individuals shown to be seropositive for HTLV-I/II according to the PHS criteria and positive for HTLV-II by additional testing should be informed that they are infected with HTLV-II. They should be told that HTLV-II is not the AIDS virus, that it does not cause AIDS, and that AIDS is caused by a different virus, called HIV. They should be told that HTLV-II is a lifelong infection and be given information regarding possible modes of transmission of HTLV-II and the lack of firm disease associations.*

---

*Since publication of the guidelines,[29] additional cases of neurological disease in HTLV-II-infected persons have been reported, suggesting that HTLV-II may cause some neurologic illness. Persons infected with HTLV-II should be advised of this rare possibility.

Specifically, HTLV-II-infected persons should be advised to do the following.

1. Share the information with their physicians.
2. Not donate blood, semen, body organs, or other tissues.
3. Not share needles or syringes.
4. Not breast-feed infants (this recommendation is not as definite as for persons infected with HTLV-I).
5. Consider the use of barrier precautions to prevent sexual transmission (this recommendation also is not as definite as for persons infected with HTLV-I).

HTLV-I/II

Individuals found to be seropositive for HTLV-I/II according to the PHS criteria but without differentiation of the infection should be informed that they are positive for HTLV-I/II and that they are likely to be infected with either HTLV-I or HTLV-II. Because of the differences in epidemiologic and clinical correlates of HTLV-I and HTLV-II, an effort to type the infection should be made. If the efforts are unsuccessful, the person with HTLV-I/II should be given information regarding possible modes and efficiency of transmission of HTLV-I and HTLV-II, disease associations of HTLV-I, and the probability of developing disease. Specific counseling should be the same as for persons infected with HTLV-I.

HTLV-Indeterminate

Individuals with serum specimens that are HTLV-indeterminate on two occasions at least 3 months apart should be advised that their specimens were reactive in a screening test for HTLV-I but that their results could not be confirmed by a second, more specific test. They should be reassured that "indeterminate" test results are only rarely caused by HTLV-I or HTLV-II infections. Persons testing "indeterminate" for HTLV-I/II on only one occasion should be offered retesting to determine if they were recently infected with HTLV-I or HTLV-II and are in the process of seroconverting. If they have the same results, they should be reassured that they are unlikely to be infected with HTLV-I or HTLV-II.

False-Positive tests

Persons with serum specimens that are repeatedly reactive by HTLV-I ELISA but negative on the Western blot assay on two occasions should be advised that their HTLV-I screening test is falsely positive and that it could not be confirmed by a second, more specific test. They should be reassured that they are not infected with HTLV-I or HTLV-II.

Medical Follow-up

Medical evaluation of persons infected with HTLV-I or HTLV-I/II by a physician knowledgeable about these viruses is recommended. This evaluation might include a physical examination, including a neurologic examination, and a complete blood count with a peripheral blood smear examination. Medical evaluation of persons infected with HTLV-II should be optional.

Acknowledgments. We thank Dr. Renu Lal for providing Figure 11.2 and for critical review of the chapter, and John O'Connor for excellent editing of the chapter.

References

1. Poiesz BJ, Ruscetti FW, Gazdar AF, et al. Detection and isolation of type-c retrovirus particles from fresh and cultured lymphocytes of a patient with cutaneous T-cell lymphoma. Proc Natl Acad Sci USA 1980;77:7415–7419
2. Kalyanaraman VS, Sarngadharan MG, Robert-Guroff M, et al. A new subtype of human T-cell leukemia virus (HTLV-II) associated with a T-cell variant of hairy cell leukemia. Science 1982;218:571–573
3. Kaplan JE, Khabbaz RF. The epidemiology of human T-lymphotropic virus types I and II. Rev Med Virol 1993;3:137–148
4. Centers for Disease Control. Human T-lymphotropic virus type I screening in volunteer blood donors—United States, 1989. MMWR 1990;39:915, 921–924
5. Vitek CR, Gracia F, Fukuda K, et al. Evidence for sexual and mother-to-child transmission of HTLV-II among Guaymi Indians, Panama. Presented at the annual meeting of Laboratory of Tumor Cell Biology, Bethesda, August 1993
6. Centers for Disease Control. Licensure of screening tests for antibody to human T-lymphotropic virus type I. MMWR 1988;37:736–740, 745–747
7. Lillehoj EP, Alexander SS, Dubrule CJ, et al. Development and evaluation of human T-cell leukemia virus type I serologic confirmatory assay incorporating a recombinant envelope polypeptide. J Clin Microbiol 1990;28:2653–2658
8. Hartley TM, Khabbaz RF, Cannon RO, Kaplan JE, Lairmore MD. Immunoblot and immunoprecipitation tests for human T-cell lymphotropic virus types I/II: patterns of immune reactivity. J Clin Microbiol 1990;28:646–650
9. Kwok S, Lipka JJ, McKinney N, et al. Low incidence of HTLV infections in random blood donors with indeterminate Western blot patterns. Transfusion 1990;30:491–494
10. Busch MP, Kleinman S, Calabro M, Laycock M, Thompson R. Evaluation of anti-HTLV supplemental testing in U.S. blood donors [abstract]. Presented at the American Association of Blood Banks 45th Annual Meeting, San Francisco, November 1992

11. Khabbaz RF, Heneine W, Grindon A, et al. Indeterminate HTLV serologic results in U.S. blood donors: are they due to HTLV-I or HTLV-II? J Acquir Immune Defic Syndr 1992;5:400–404
12. Donegan E, Pell P, Lee H, et al. Transmission of human T-lymphotropic virus type I by blood components from a donor lacking anti-p24: a case report. Transfusion 1992;32:68–71
13. Kleinman SH, Kaplan JE, Khabbaz RF, et al. Evaluation of a p21e spiked Western blot in confirming HTLV-I/II infection in volunteer blood donors. J Clin Microbiol 1994;32:603–607
14. Wiktor SZ, Piot P, Mann JM, Nzilambi N, et al. Human T cell lymphotropic virus type I (HTLV-I) among female prostitutes in Kinshasa, Zaire. J Infect Dis 1990;161:1073–1077
15. Lal RB, Heneine W, Rudolph DL, et al. Synthetic peptide-based immunoassays for distinguishing between human T-cell lymphotropic virus type I and type II infections in seropositive individuals. J Clin Microbiol 1991;29:2253–2258
16. Lipka JJ, Miyoshi I, Hadlock KG, et al. Segregation of human T cell lymphotrophic virus type I and II infections by antibody reactivity to unique viral epitopes. J Infect Dis 1992;165:268–272
17. Chen Y-MA, Lee T-H, Wiktor SZ, et al. Type-specific antigens for serologic discrimination of HTLV-I and HTLV-II infection. Lancet 1990;336:1153–1155
18. Fang CT, Akins R, Sandler SG. An HTLV-I EIA using recombinant env-encoded peptides. Transfusion 1989;29(S):55f
19. Heneine W, Khabbaz RF, Lal RB, et al. Sensitive and specific polymerase chain reaction assays for diagnosis of human T-cell lymphotropic virus type I (HTLV-I) and HTLV-II-infections in HTLV-I/II seropositive individuals. J Clin Microbiol 1992;30:1605–1607
20. Kwok S, Higuchi R. Avoiding false positives with PCR. Nature 1989;339: 237–239
21. Gessain A, Louie A, Gout O, Gallo RC, Franchini G. Human T-cell leukemia-lymphoma virus type I (HTLV-I) expression in fresh peripheral blood mononuclear cells from patients with tropical spastic paraparesis/ HTLV-I-associated myelopathy. J Virol 1991;65:4398–4407
22. Kinishita T, Shimoyama M, Tobinai K, et al. Detection of m-RNA for the tax1/rex1 gene of human T-cell leukemia virus type I in fresh peripheral blood mononuclear cells of adult T-cell leukemia patients and viral carriers by the polymerase chain reaction. Proc Natl Acad Sci USA 1989;86:5620–5624
23. Kimata J, Ratner L. Temporal regulation of viral and cellular gene expression during human T-lymphotropic virus type I-mediated lymphocyte immortalization. J Virol 1991;65:1628–1633
24. Zaki SR, Heneine W, Coffield LM, Greer PW, Folks TM. Intracellular in situ PCR amplification and detection of human T-cell lymphotropic virus type I (HTLV-I) DNA. Lab Invest 1992;66:95A
25. Yamaguchi K, Seiki M, Yoshida M, et al. The detection of human T cell leukemia virus proviral DNA and its application for classification and diagnosis of T cell malignancy. Blood 1984;63:1235–1240
26. Bleike MA, Riding ID, Gravell N, et al. In situ hybridization detection of HTLV-I RNA in TSP/HAM patients and their spouses. J Med Virol 1991; 33:64–71

27. Kitamura K, Rudolph D, Goldsmith C, Folks TM, Lal RB. Isolation, characterization, and transmission of human T lymphotropic virus types I and II in culture. Curr Microbiol 1993;27:355–360
28. Kitamura K, Heneine W, Lal R, Folks T. Sensitive detection of HTLV-I and HTLV-II viral particles in culture supernatants using a reverse transcriptase-PCR assay. Presented at the American Society for Microbiology 92nd Annual meeting, New Orleans, 1992
29. Centers for Disease Control and Prevention, USPHS Working Group. Guidelines for counseling persons infected with human T-lymphotropic virus type I (HTLV-I) and type II (HTLV-II). Ann Intern Med 1993;118:448–454

12
HIV Testing in Blood Banks

MICHAEL P. BUSCH

Infectious disease testing presents special challenges in the blood bank setting, all of which are heightened with an agent as clinically serious and as politically and emotionally charged as the human immunodeficiency virus (HIV). Technical performance requirements for assays used in blood banks are particularly stringent, with maximum sensitivity critical to protect recipients; and high specificity is vital to prevent inaccurate notification and deferral of blood donors, high rates of indeterminate reactivity, and excessive wastage of blood and its derivatives. Because approximately 12 million blood donations are screened annually in the United States for each of seven infectious disease markers, issues such as automated sample handling, rapid and high through-put instrumentation, and data transfer–computer interfacing are major considerations. The performance of infectious disease testing in blood banks is closely monitored by the U.S. Food and Drug Administration (FDA), which has imposed increasingly stringent standards of "good manufacturing practices," originally developed for pharmaceutical companies.[1,2] As a result, levels of training, procedural controls, documentation, and quality assurance are mandated that may not be generally required in other HIV testing settings.

This chapter reviews HIV testing protocols employed in blood banks and considers the special requirements imposed on assays used for donor screening. The performance of anti-HIV screening and supplemental tests used in blood banks since 1985 are reviewed, and data on the current risk of transmission of HIV-1 and HIV-2 from screened blood transfusions are summarized. The potential value of performing additional tests directed at viral antigen or nucleic acids is then considered.

HIV-1 Testing Algorithm Used in Blood Banks

Routine screening of blood donors is carried out according to the scheme shown in Figure 12.1. As discussed below, the FDA now requires that all units of blood and components be screened using licensed enzyme

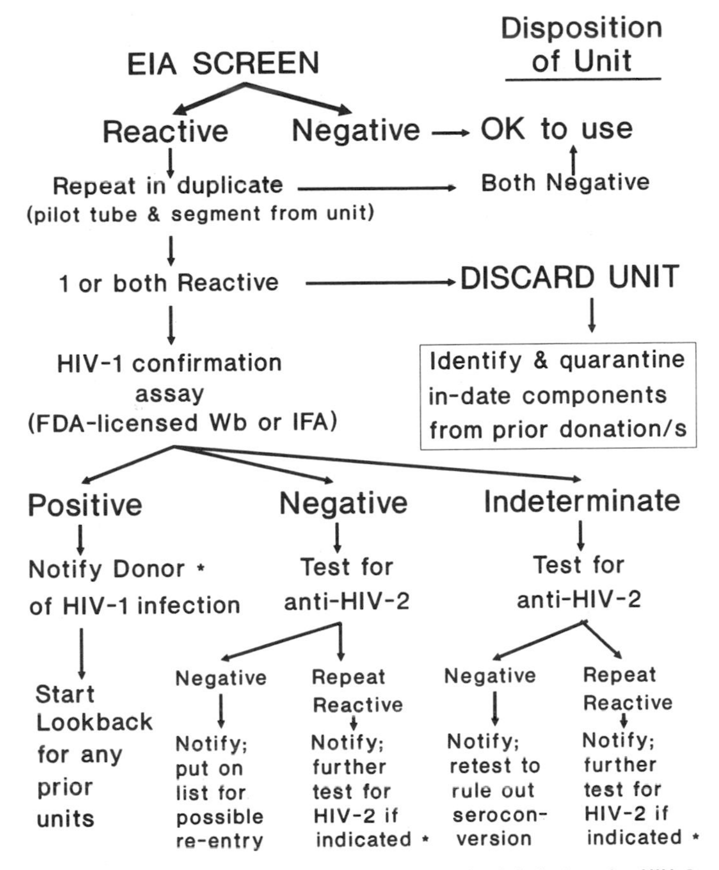

FIGURE 12.1. Algorithm for current routine screening of blood donors.

immunoassays that are sensitive to both anti-HIV-1 and anti-HIV-2.[3,4] Initial testing is performed on serum from "pilot" tubes collected at the time of donation. If this test is reactive, duplicate repeat testing is generally performed on both the pilot tube serum and an aliquot of plasma obtained from segmented tubing attached to the blood components. This element of the algorithm is designed as an additional safeguard to identify and resolve possible specimen-labeling errors. If one or both repeat tests are reactive, the unit is designated "repeat reactive" and discarded. A process is also initiated to identify and quarantine all in-date components from any prior donation(s) from that donor.[5]

Supplemental testing must utilize FDA-licensed reagents and rule out both HIV-1 and HIV-2 infection. Although combination HIV-1/HIV-2 supplemental assays using recombinant DNA-derived or synthetic peptide assays have been developed that accurately detect and discriminate anti-

HIV-1 and anti-HIV-2, they are not yet licensed by the FDA. Therefore current algorithms in U.S. blood banks employ the HIV-1 viral lysate-based Western blot assay or immunofluorescence assays, in combination with a licensed anti-HIV-2 enzyme immunoassay (EIA) and unlicensed HIV-2 supplemental assays.[3,4] All repeat reactive donors are deferred from further blood donations and notified of their screening and supplemental test results. For confirmed positive donors, recipients of prior donations are traced in a process called "lookback."[5,6] Donors whose supplemental test results are completely negative are eligible for possible reentry into the donor pool according to an FDA-specified protocol.[3] Critical aspects of the reentry protocol include a 6-month interval between the initial repeat reactive donation and a reentry sample, both of which must test negative on reentry testing using two licensed EIAs (at least one of which is viral lysate-based) and a licensed supplemental assay. Although donors with indeterminate Western blot patterns are rarely infected with HIV,[7,8] they are currently ineligible for reentry.

Special Considerations for Blood Donor Screening Assays

Table 12.1 lists a number of considerations that apply to test selection in blood banks. They include such issues as compatibility with test formats used for other donor screening assays, automation and computer interfacing, turnaround time and through-put, and cost. Because of space limitations, we focus our discussion on the technical performance characteristics of assays.

Sensitivity to Early HIV-1 Seroconversion

Safeguarding recipients from transfusion-transmitted infections is the fundamental goal of blood donor screening. Given that more than 90% of anti-HIV-positive transfusions result in recipient infection,[9] the cost of

TABLE 12.1. Considerations when selecting HIV tests employed in blood banks.

1. Maximum sensitivity to prevent infection of recipients.
2. Capacity to detect variant strains of HIV (e.g., HIV-2).
3. Optimal specificity to minimize false-positive reactivity.
4. Supplemental assays (algorithm) must be FDA-approved and capable of detecting anti-HIV-1 and anti-HIV-2 with sensitivity equal to that of EIAs.
5. Compatibility with automation to allow high through-put testing.
6. Objective results allowing automated reading and direct data transfer to computer.
7. Rapid processing time enabling rapid release of blood components.
8. Standard format: similar to that used for other donor screening assays.
9. Economical, given that large numbers of donations are tested for a variety of agents.

false-negative EIA results in the donor setting is enormous. For this reason, blood bank infectious disease specialists constantly evaluate new assay developments to ensure that they are employing the most sensitive tests available.

Developments in test methodology have occurred at a rapid rate, and blood banks have implemented improved assays as soon as possible after FDA licensure. The evolution of antibody assays has centered around improvements in the antigens on the solid phase and in assay formats. For crude viral lysates, production moved to viral lysate assays spiked with purified natural antigens, and then to use of cloned [recombinant DNA (rDNA)-derived] and synthetic peptide antigens. Although use of the rDNA and peptide antigens initially caused concerns within the FDA regarding sensitivity to immunologically variant strains, it is now well established that these assays employ highly selected antigenic regions of HIV that are capable of excellent sensitivity to infected subjects from around the world. Consequently, recombinant DNA-derived and synthetic peptide antigen-based assays are now in wide use in U.S. blood banks. Test manufacturers have also developed assay formats with increased capacity to detect low-titer IgM antibody produced during early seroconversion. For example, the recombinant antigen sandwich EIA format employed in Abbott's anti-HIV-1/HIV-2 EIA allows use of lower dilutions of donor sera and detects IgM with improved sensitivity.[10] Progressive introduction of assays with increased sensitivity has led to significant narrowing of the seroconversion window (Fig. 12.2), with concomitant reductions in the risk of screened blood transfusions.[1,11,12]

Detection of Variant HIV Strains

Concern over sensitivity to variant viruses such as HIV-2 is heightened in the blood bank setting. The decision by the FDA and blood bank organizations to require implementation of combination assays for detection of anti-HIV-1 and anti-HIV-2 by mid-1992[3] illustrates the unique decision process in the donor-screening arena. Extensive data available at the time of that decision indicated that (1) the prevalence of HIV-2 in the general U.S. population was (and remains) low; (2) no HIV-2-infected blood donors were detected after active surveillance of the equivalent of more than 10 million blood donors; (3) no cases of HIV-2 transmission by transfusion have been detected in the United States; and (4) the existing HIV-1 tests detected most HIV-2-infected subjects (reviewed by O'Brien et al.[4]). On the other hand, it was clear that the prevalence of HIV-2 would likely increase, albeit slowly, in the United States and that anti-HIV-1 tests would fail to detect some of these cases. There was also concern that ongoing surveillance of blood donors for HIV-2 would be difficult to maintain, and that eventually a case of HIV-2 transmission might occur. Moreover, the earlier policy of geographically excluding

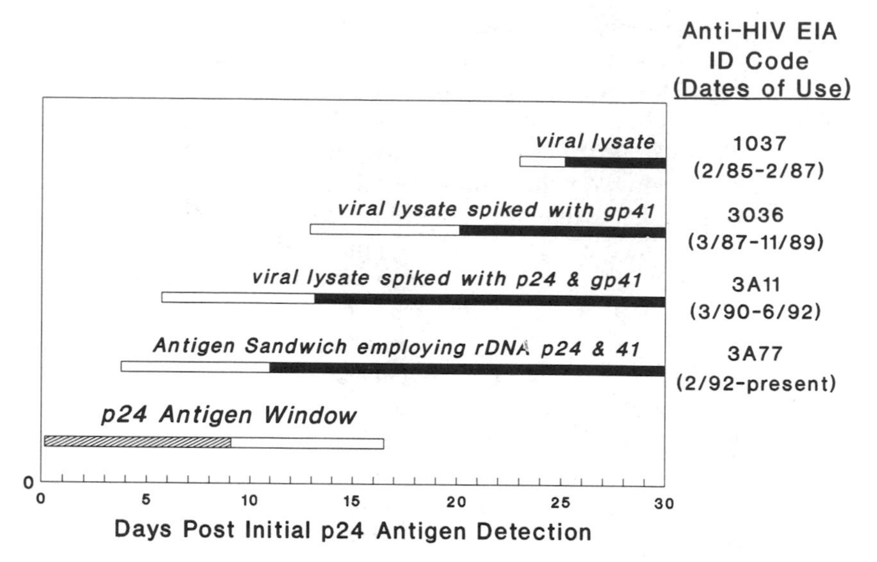

FIGURE 12.2. Progressive reduction in HIV seroconversion window due to improved sensitivity of anti-HIV assays. Serial plasmapheresis units from 25 source plasma donors later determined to have seroconverted to anti-HIV positivity were tested with four FDA-licensed anti-HIV assays and one FDA-licensed p24 antigen assay (Abbott Laboratories, Abbott Park, IL). On average 6.7 units collected over a 30-day period from each donor (average 4.5 days between collections) were available for testing. Data are plotted as days following earliest detectable p24 antigen reactivity. Filled bars indicate the time period during which all samples from all plasma donors were reactive on respective tests. Open bars represent periods during which variability in the serologic response between seroconverting individuals was observed. (Data courtesy of Drs. Susan Stramer and Bruce Phelps, Abbott Laboratories. Reprinted from Busch,[11] by permission.)

persons from sub-Saharan Africa was politically problematic. During the 24 months (as of this writing) since combination anti-HIV-1/HIV-2 tests were universally implemented, no HIV-2-infected donors have been identified by U.S. blood banks (personal communication, R.Y. Dodd, Head of Transmissible Diseases, American Red Cross Holland Laboratory, Rockville MD).

Optimal Specificity to Minimize False-Positive Reactivity

Blood donors are highly selected through predonation eligibility requirements and screening interviews designed to detect risk factors for bloodborne infectious agents.[13] Potential donors are given specific literature describing HIV risk factors and disease symptoms and are requested to defer from donation if they perceive themselves as at risk. They are then asked a series of direct and specific questions about HIV risk behavior in writing and in private interviews with trained medical historians. Although the brief physical examination given to donors is primarily intended to ensure that the donor is healthy enough to give a unit of blood, certain elements of it, such as examination of both arms for evidence of drug injection marks, serve to detect persons at risk for HIV. Only donors who pass each of these steps are allowed to donate.

As a result of these selection criteria, the prevalence of HIV-infected persons among volunteer blood donors is low, with the current rate at approximately 0.001% (i.e., one in 10,000 blood donations are confirmed as anti-HIV-positive). This rate is approximately 1/50 that estimated for the overall U.S. population.[14] This low prevalence (and even lower incidence) imposes stringent requirements for screening tests of high specificity and for supplemental tests that accurately discriminate true-positive from false-positive results. High specificity is particularly important in the blood donor setting because units with falsely reactive screening test results are discarded, and the false-positive donors must be notified of their results and deferred from subsequent donation. Although reinstatement of donors who have EIA-reactive and Western blot-negative results is possible, it requires extensive testing of both the index and follow-up samples, is of relatively low yield and high cost, and carries regulatory and legal liability risks.

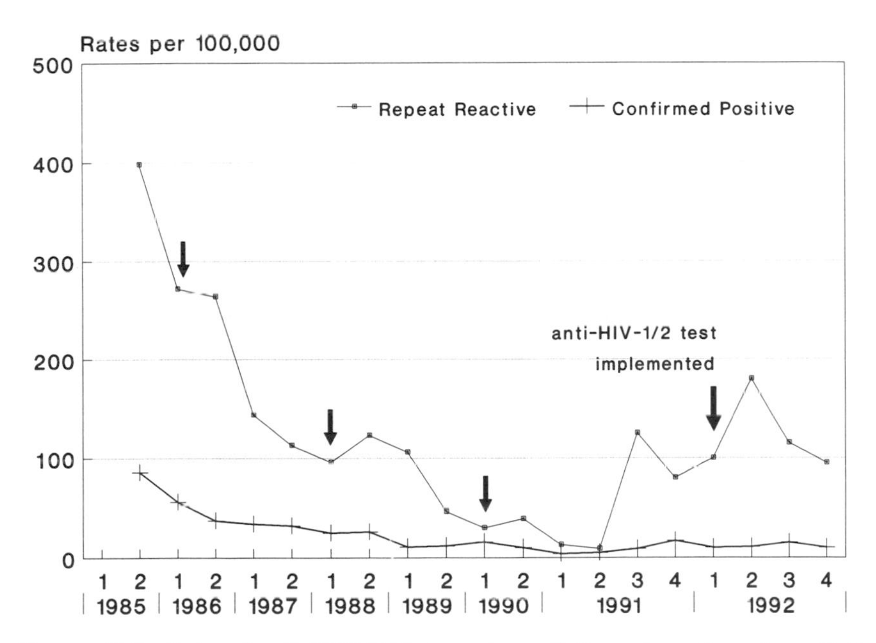

FIGURE 12.3. Rate of anti-HIV EIA repeat reactivity and confirmed positivity among volunteer blood donations to Irwin Memorial Blood Centers, San Francisco. Arrows indicate the dates on which new versions of anti-HIV EIAs were implemented.

The rate of reactivity of anti-HIV screening EIAs used in blood banks from 1985 through 1992 is illustrated in Figure 12.3. Note the marked decline in the rate of false-positive results from 1985 through 1990. This resulted from the exclusion of both true- and false-positive donors from the repeat donor pool, and the improving specificity of EIAs used for donor screening. The increased rate of false-positive results observed during the second half of 1991 was partially attributable to nonspecific reactivity observed in donors who received influenza vaccines, although test manufacturing problems also contributed. The further increase in false-positive results during mid-1992 coincided with implementation of anti-HIV-1/HIV-2 combitests. These recombinant DNA antigen-based tests appear to detect a new population of donors as falsely reactive; as these donors are culled from the donor base, reactive rates should continue to decline. This effect of increased false-reactive rates following introduction of a new test shows how changing tests in the donor setting may result in significant additional unit loss and donor deferral.

Supplemental Assays

The use of enhanced sensitivity screening assays for anti-HIV-1/HIV-2 has precipitated new concerns related to the sensitivity and specificity of supplemental assays.[10,11] Some of the newer "third generation" EIAs employed in blood banks detect seroconversion earlier than the currently licensed anti-HIV-1 supplemental tests, which makes it difficult to be certain that we are correctly classifying seroconverting subjects as infected and false-positive donors as noninfected.[11,12] It has been suggested that to resolve this problem p24 antigen and polymerase chain reaction (PCR) assays may have to be employed in parallel with supplemental antibody assays in order to accurately identify early seroconvertors.[15]

On the other hand, the increased sensitivity of third generation EIAs to antibodies directed at *env*-related antigens has resulted in detection of nonspecific *env*-reactive blood donors. Although most of these donors are classified as indeterminate, some of them may be falsely classified as infected based on current Western blot interpretive criteria.[16] Lastly, as noted above, donors identified as reactive on anti-HIV-1/HIV-2 combitests require supplemental testing to rule out both HIV-1 and HIV-2, which is expensive, complex, and results in an increased number of indeterminate classifications.[3,4]

Current Risk of HIV Transmission from Screened Blood Transfusions

Four approaches have been employed to estimate the risk of HIV-1 transmission via HIV-1-antibody screened blood (Table 12.2). The first

TABLE 12.2. Estimates of risk of HIV-1 infection from anti-HIV-1-screened blood transfusions.

Type of study, time period and region	Risk of HIV-1 infection per unit transfused	Ref.
Statistical models		
1985, USA	1/99,000	17
3/85–2/87, USA	1/38,000	18
3/85–2/87, USA	1/153,000	14
Lookback models		
3/85–12/86, Los Angeles	1/68,000	19
1/88–1/91, USA	1/225,000	20
Prospective recipient seroconversion		
4/85–12/89, Baltimore and Houston	1/60,000	21
Prospective donor cell culture/PCR analysis		
11/87–12/89, San Francisco	1/88,561	22

Source: Busch,[11] by permission.

involves estimating the risk based on data on the incidence and prevalence of HIV-1 infection among blood donors and assumptions about the sensitivity of anti-HIV-1 assays and the length of the seronegative window phase of infection (i.e., time from infection to screening EIA positivity). Unfortunately, uncertainty in these parameters has resulted in point estimates ranging from 1 in 36,282 to 1 in 153,000 per unit.[14,17,18]

The second approach uses data from lookback investigations to estimate the number of HIV-1 transmissions that would occur from repeat donors who donate during the "window period." By tracing recipients of earlier seronegative donations given by repeat donors who seroconverted to anti-HIV-1 positivity, one can determine the rate of recipient infection by these preseroconversion transfusions and then estimate the duration of the seronegative window. One can then calculate the risk of HIV-1 transmission from repeat donors and, with some assumptions, from first-time donors. Using this approach, Kleinman and Secord estimated the risk in Los Angeles in 1987 at one infection per 68,000 units transfused.[19] An analysis of national lookback data from the CDC/American Red Cross HIV Blood Donor Study yielded an estimate of risk of the blood supply in 1990 of one infection for 225,000 units transfused and an estimated infectious window period of 42 days.[20]

Two prospective studies were initiated in 1986 that were specifically designed to acquire data for defining and monitoring the risk of HIV-1 from screened transfusions. The first study, carried out in the Baltimore and Houston areas, involved testing patients for anti-HIV-1 before and after cardiac surgery.[21] Observed seroconversions were investigated to rule out other risk factors, and donors in those cases were traced so as to document their seroconversion. Two cases of HIV-1 infection were documented after the transfusion of 120,000 units of blood. In the other study,[22] conducted in San Francisco, virus culture and PCR analyses were used to detect HIV-1 in pooled cells derived from extra tubes of blood collected from anti-HIV-negative donors. Using this methodology on 103,000 donations given between 1988 and 1990, only a single infected donation was identified. In a continuation phase of this study, an enhanced sensitivity "macro-PCR" method was applied to an additional 97,000 donations in San Francisco, with no additional virus-positive, antibody-negative donations detected.

Based on the CDC Lookback Study's national estimate of one HIV-1 transmission for every 225,000 units transfused, the CDC has projected that approximately 70 recipients per year may be transfused with infected blood.[20] The CDC further projected that of these infected recipients only 12 to 15 would be expected to develop AIDS-related diseases prior to dying from other causes. The risk of HIV-2 transmission has been estimated at less than 1 in 10 million.[4] These risks are 2000-fold lower than what existed at the peak of the transfusion-related AIDS epidemic between 1982 and 1984.[23]

Need for Direct Virus Detection Assays

Evaluating the need for tests that could detect HIV-infected persons earlier than current antibody assays by targeting viral antigens or nucleic acids is a difficult challenge for blood banks.[11] Because most blood banks do not store serum (or cells) from all donations, specimens from preseroconversion donations (or units implicated in transmission by transfusions) are not available for use in such evaluations. This situation makes determination of the yield of a new assay in the donor setting difficult. One approach has been to implement a new test in parallel with existing antibody tests on hundreds of thousands of units. This approach has been used to evaluate the utility of p24 antigen assays for donor screening, with no p24-antigen-positive/antibody-negative donations detected in more than 1 million donations tested in the United States and Europe.[24,25] In contrast, antigen-positive/antibody-negative donations have been detected in Thailand, where the prevalence and incidence of HIV-1 among donors are 500-fold that in the United States[26]; p24 antigen screening is now routine in this unique setting.

An alternative approach for evaluating direct virus detection assays in the donor setting is to test donor repository specimens selected on the basis of donor demographics and date or geographic location of donations, or on the reactivity of donor sera on other tests. This approach has proved useful on several occasions.[11] For example, the Transfusion Safety Study was able to evaluate the utility of HIV p24 antigen assays for donor screening by testing fewer than 9000 specimens selected from a repository of 200,000 sera collected from donors in four high AIDS prevalence cities during late 1984.[27] Because the tested donations were selected as those given by donor subsets with high prevalence and incidence rates of HIV at the time (i.e., young men who lived in high HIV prevalence ZIP code areas), they were equivalent to more than 1 million contemporary donations with regard to frequency of early HIV infections. No antigen-positive/antibody-negative donors were identified.

The development of techniques for sensitive detection of HIV nucleic acids holds promise for more sensitive donor screening in the future, especially if assays can be formulated so several viral nucleic acids can be detected simultaneously.[28] Considerable research has been devoted to simplification of nucleic acid extraction methods and rapid detection of amplified RNA or DNA sequences. There has already been substantial progress in the development of HIV DNA PCR kits using 96-well format thermocyclers and microplate-based colorimetric readers rather than gel electrophoresis and radioactive probe detection systems. Several significant problems and questions remain to be resolved, however, before PCR or alternative nucleic acid amplification techniques are seriously considered for donor screening. Although the problem of frequent false-positive PCR results, primarily attributable to "carryover" of amplified product from positive to negative samples, has been essentially eliminated through implementation of stringent technical precautions and strategies for sterilization of amplified product, false positivity due to blood cross-contamination may present a problem when processing thousands of samples per day. The cost of adding nucleic acid detection assays to donor screening would clearly be substantial, and the incremental safety benefit that could be achieved is unclear. Studies suggest that both cell-associated HIV DNA and cell-free HIV RNA are detected no more than 1 to 2 weeks prior to the development of detectable anti-HIV.[11,15,29–32]

The fact that a test shortens the seroconversion window by days to weeks is not enough to warrant its implementation as a donor screening assay. This information must then be extrapolated to the blood donor setting using data on the current incidence of HIV in the donor setting (i.e., the rate of seroconversions); then the frequency of infected seronegative donations per year that would be detected by the "improved" test must be projected. For example, it is estimated that screening donors using a p24 antigen test would reduce the average infectious preseroconversion window by 5 to 8 days (from approximately 33 days to 25 days).[10,11] Based on this reduction, and current incidence data, antigen screening would detect and prevent transfusion of approximately one HIV-infectious donation per 1.6 million screened donations,[33] thereby reducing the risk of an infected transfusion by about one-sixth. The net yield would be two or three infected recipients per year in whom HIV-related disease might be prevented if p24 antigen screening were implemented in U.S. blood banks. On the other side of the equation, the direct cost of testing the 12 million annual U.S. blood donations for p24 antigen would be between $50 million and $100 million. The antigen tests have repeat reactive (false-positive) rates among donors of approximately 0.1%,[24] so about 12,000 safe donations would be discarded annually and thousands of healthy donors indefinitely deferred. Although donor screening using PCR could narrow the window by 25% to 50%, thereby preventing as many as 10 transfusion-related AIDS cases per year, its cost in dollars and false-positive results would be proportionately greater. Weighing the cost versus benefit of implementing such tests for donor screening will be an increasing challenge in the years ahead.[34]

References

1. McCullough J. The nation's changing blood supply [editorial]. JAMA 1993; 269:2239–2245
2. Miller WV. Blood banks should use good manufacturing practices and the pharmaceutical manufacturing approach: pro. Transfusion 1993;33:435–438
3. Center for Biologics Evaluation and Research, FDA. Revised recommendations for the prevention of human immunodeficiency virus (HIV) transmission by blood and blood products. April 1992
4. O'Brien TR, George JR, Holmberg SD. Human immunodeficiency virus type 2 infection in the United States: epidemiology, diagnosis, and public health implications. JAMA 1992;267(20):2775–2779
5. FDA. Proposed change to 21 CFR Parts 606, 610 (Docket No. 91N-0152): current good manufacturing practices for blood and blood components; notification of consignees receiving blood and blood components at increased risk for transmitting HIV infection. Fed Register 1993;58(124):34962–34970
6. Busch MP. Let's look at human immunodeficiency virus lookback before leaping into hepatitis C virus lookback! Transfusion 1991;31:655–661
7. Eble BE, Busch MP, Khayam-Bashi H, et al. Resolution of infection status of HIV-seroindeterminate and high-risk seronegative individuals using PCR and virus-culture: absence of persistent silent HIV-1 infection in a high-prevalence area. Transfusion 1992;32:503–508
8. Jackson JB. Human immunodeficiency virus (HIV)-indeterminate Western blots and latent HIV infection. Transfusion 1992;32:497–499
9. Donegan E, Stuart M, Niland JC, et al. Infection with human immunodeficiency virus type 1 (HIV-1) among recipients of antibody-positive blood donations. Ann Intern Med 1990;113:733–739
10. Gallarda JL, Henrard DR, Liu D, et al. Early detection of antibody to HIV-1 using an antigen conjugate immunoassay correlates with the presence of IgM antibody. J Clin Microbiol 1992;30:2379–2384
11. Busch MP. Retroviruses and blood transfusions: the lessons learned and the challenge yet ahead. In Nance SJ (ed): Blood Safety: Current Challenges

[transcription of the Emily Cooley Award/AABB 1992 annual seminar]. Bethesda: American Association of Blood Banks, 1992:1–44
12. Zaaijer HL, von Exel-Oehlers P, Kraaijeveld T, Altena E, Lelie PN. Early detection of antibodies to HIV-1 by third-generation assays. Lancet 1992; 340:770–772
13. Kleinman S. Donor selection and screening procedures. In Nance SJ (ed): Blood Safety: Current Challenges [transcription of the Emily Cooley Award/ AABB 1992 annual seminar]. Bethesda: American Association of Blood Banks, 1992:169–200
14. Cumming PD, Wallace EL, Schorr JB, Dodd RY. Exposure of patients to human immunodeficiency virus through the transfusion of blood components that test antibody-negative. N Engl J Med 1989;321:941–946
15. Henrard D, Phillips J, Windsor I, Fortenberry D, Korte C, Fang C, Williams A. Detection of HIV-1 p24 Antigen and plasma RNA: Relevance to Indeterminate Serologies Transfusion (in press)
16. Healey DS, Bolton WV. Apparent HIV-1 glycoprotein reactivity on Western blot in uninfected blood donors. AIDS 1993;7:655–658
17. Ward JW, Holmberg SD, Allen JR, et al. Transmission of human immunodeficiency virus (HIV) by blood transfusions screened as negative for HIV antibody. N Engl J Med 1988;318:473–478
18. Anonymous. Appendix C: risk of HIV transmission from blood transfusion. In: Confronting AIDS. New York: Institute of Medicine, National Academy of Sciences, 1986:309–313
19. Kleinman S, Secord K. Risk of human immunodeficiency virus (HIV) transmission by anti-HIV-negative blood: estimates using the lookback methodology. Transfusion 1988;28:499–501
20. Petersen LR, Satten G, Dodd RY, et al. Current estimates of the infectious window period and risk of HIV infections from seronegative blood donations. Presented at the 5th National Forum on AIDS, Hepatitis, and Other Blood-Borne Diseases, Atlanta, March 1992
21. Nelson KE, Donahue JG, Munoz A, et al. Transmission of retroviruses from seronegative donors by transfusion during cardiac surgery; a multicenter study of HIV-1 and HTLV-I/II infections. Ann Intern Med 1992;117:554–559
22. Busch MP, Eble BE, Khayam-Bashi H, et al. Evaluation of screened blood donations for human immunodeficiency virus type 1 infection by culture and DNA amplification of pooled cells. N Engl J Med 1991;325:1–5
23. Busch, M, Young M, Samson S, et al. Risk of human immunodeficiency virus (HIV) transmission by blood transfusions before implementation of HIV-1 antibody screening. Transfusion 1991;31:4–11
24. Alter HJ, Epstein JS, Swenson SG, et al. Prevalence of human immunodeficiency virus type 1 p24 antigen in U.S. blood donors—an assessment of the efficacy of testing in donor screening. N Engl J Med 1990;323:1312–1318
25. Baeker U, Weinauer F, Michel P, Weise W. HIV antigen screening in blood donors [abstract PO-C21-3135]. In: Proceedings of the IX International Conference on AIDS (Vol. II), Berlin, 1993:739
26. Chiewsilp P, Isarangkura P, Poonkasem A, et al. Risk of transmission of HIV by seronegative blood [correspondence]. Lancet 1991;338:1341
27. Busch MP, Taylor PE, Lenes BA, et al. Screening of selected male blood donors for p24 antigen of human immunodeficiency virus type 1. N Engl J Med 1990;323:1308–1312

28. Sunzeri FJ, Lee T-H, Brownlee RG, Busch MP. Rapid simultaneous detection of multiple retroviral DNA sequences using the polymerase chain reaction and capillary DNA chromatography. Blood 1991;77:879–886
29. Bruisten SM, Koppelman MHGM, Dekker JT, et al. Concordance of human immunodeficiency virus detection by polymerase chain reaction and by serologic assays in a Dutch cohort of seronegative homosexual men. J Infect Dis 1992;166:620–622
30. Farzadegan H, Vlahov D, Solomon L, et al. Detection of human immunodeficiency virus type 1 infection by polymerase chain reaction in a cohort of seronegative intravenous drug users. J Infect Dis 1993;168:327–331
31. Sheppard HW, Busch MP, Louie PH, Madej R, Rodgers GC. HIV-1 PCR and isolation in seroconverting and seronegative homosexual men: absence of long-term immunosilent infection. J AIDS 1993;7:381–388
32. Read S, Cassoll S, Coates R, et al. Detection of incident HIV infection by PCR compared to serology. J AIDS 1992;5:1075–1079
33. Petersen L, Busch M, Satten G, Dodd R, Henrard D. Narrowing the window period with a third generation anti-HIV-1/2 enzyme immunoassay makes p24 antigen screening of blood donors unnecessary [abstract PO-C17-3001]. In: Proceedings of the IX International Conference on AIDS (Vol. II), Berlin, 1993:717
34. Heymann SJ, Brewer TF, Fineberg HV, Wilson ME. How safe is safe enough? New infections and the U.S. blood supply. Ann Intern Med 1992; 117:612–614

13 HIV Testing for Organ and Tissue Transplantation

R.J. SIMONDS

Transmission of the human immunodeficiency virus (HIV) through transplantation of organs and tissues has been reported infrequently, in most cases from donors whose sera had not been tested for HIV antibody.[1] Screening of potential organ and tissue donors for HIV infection has greatly reduced the risk of HIV transmission through transplantation: Only 11 transplant recipients have been reported to have been HIV-infected through transplantation since the availability of donor serologic screening in 1985. Of these recipients, one was transplanted urgently with a liver before the result of the donor's positive antibody test was available[2]; one received a kidney from a living donor who seroconverted after having been screened 8 months before organ donation[3]; two received organs from a donor whose serum tested falsely negative after hemodilution from transfusion of a large volume of blood[4,5]; and seven received organs or tissues from a seronegative donor who was thought to have very early HIV infection.[6]

Prevention of HIV transmission to recipients of transplanted organs and tissues relies on screening potential donors for behavioral risk factors and laboratory markers for HIV infection and, for certain tissues, allograft processing that may inactivate HIV. This chapter discusses laboratory testing for HIV infection in organ and tissue transplantation, focusing on aspects of HIV testing unique for transplantation. Testing issues surrounding the donation of human milk, semen, and bone marrow closely resemble those having to do with blood donation and are not specifically addressed here.

Special Considerations for HIV Testing for Transplantation

The goal of screening organ and tissue donors for behavioral risk factors and laboratory markers of HIV infection is identical to that for blood donors: identification and exclusion from donation of persons who have HIV infection or who have a high risk of being HIV-infected. Several issues regarding HIV testing in organ and tissue transplantation are important for developing screening strategies to prevent HIV transmission through transplantation.

Types of Transplant

The risk of HIV transmission through transplantation differs for different types of allograft. The risk to recipients of transplanted organs and highly vascular tissues is high. Transmission has been reported only to recipients of organs, including kidney, liver, heart, and pancreas, as well as vascular tissues, including large marrow-containing bone pieces and skin. Other transplanted tissues, including other musculoskeletal tissue, corneas, and dura mater, appear to carry a much lower risk of HIV transmission, probably due to their relative avascularity or to the processing that many of these tissues undergo, such as treatment with radiation, ethanol, or other chemicals, that may inactivate HIV.[7] The risk of transmission through transplants that may be lifesaving (e.g., heart, liver) but in short supply also must be considered when developing a donor screening strategy to ensure a high level of safety while not unnecessarily limiting the supply of acceptable donors.

Donor Risk Screening

The presence of HIV risk factors in potential organ and tissue donors, many of whom are brain-dead or deceased, may be difficult to detect. In contrast to living donors, assessment of behavioral risk factors for HIV infection in brain-dead or cadaveric donors relies on reviewing existing medical and other records and interviewing family members and others. Such persons may not be able to provide an adequate history of HIV risk behaviors because they lack such knowledge and because they may be grieving the death or impending death of the donor. Because of the potential limitations of behavioral risk assessment, the screening of brain-dead or cadaveric donors relies heavily on laboratory testing.

The risk assessment for living donors is similar to that for blood donors. Also, because some tissues, such as musculoskeletal tissues, can be frozen and stored for many months before transplantation, the seronegative living donor can be retested for HIV antibody several months after the tissues are procured to exclude undetected early HIV infection at the time of procurement. Quarantine of tissue pending donor retesting 3 to 6 months after procurement has been recommended for living donors of semen and bone.[8,9]

Time Constraints for Donor Screening

Organ and tissue donor screening often must be done rapidly. Because the period of organ viability following cessation of the donor's cardiopulmonary function is limited, solid organs must be transplanted as soon as possible after the donor has been declared brain-dead. Therefore donor screening procedures that require more than 24 hours to complete may not be useful for screening cadaveric or brain-dead organ donors. Moreover, facilities for performing HIV testing must be available to transplant programs at night and on weekends.

Testing Cadaveric Donors

Several factors may affect the results of HIV screening tests in critically ill and cadaveric donors. First, HIV antibody tests may be falsely negative due to hemodilution. By the time of screening for possible organ and tissue donation, many brain-dead and cadaveric donors have already received numerous transfusions or other infusions to treat their terminal illness or injury. A large volume of infusate administered shortly before a blood specimen is obtained may result in hemodilution and falsely negative tests for HIV antibodies. One case of HIV transmission from a screened donor to two recipients of organs was attributed to an HIV antibody screening test result that was falsely negative because of such hemodilution.[4,5] In this case, a serum specimen taken after transfusion of 56 units of blood and blood components tested negative for HIV antibody by enzyme immunoassay (EIA) (the sample/control optical density ratio was 0.103:0.131), whereas a specimen that had been obtained before the transfusions and tested subsequently was positive by EIA (optical density ratio 0.556:0.126). Because of this concern it is recommended that specimens for donor HIV antibody screening be obtained before infusions or transfusions are given but as close to the time of organ or tissue procurement as possible.[5]

Second, because the licensed screening and confirmatory tests for HIV antibody and antigen were designed for use on specimens obtained from living persons, few data are available to address the sensitivity and specificity of these tests when they are used on postmortem blood specimens, which may be hemolyzed. Two studies have compared the sensitivity of several tests for detecting HIV antibody in blood from cadavers with an autopsy diagnosis of acquired immunodeficiency syndrome (AIDS). In one study of 35 cadavers, the sensitivity was 94% to 97% using three different EIA kits and 94% using the Western blot assay.[10] In another study of 30 cadavers, the sensitivity was 100% by EIA.[11] The specificity of HIV p24 antigen assays used on postmortem specimens may also be low, with high rates of nonspecific reactivity leading to falsely positive results.[10,12]

Although not currently recommended, the potential use of body fluids or tissues other than blood for brain-dead or cadaveric donors also has been considered. Specifically, the use of aqueous and vitreous humors of the eye for HIV antibody screening has been studied because proteins, such as immunoglobulins, are less likely to be affected by decomposition or hemolysis in these fluids. One study found a sensitivity of only 16% to 26% when testing aqueous humor for HIV antibody by EIA and 79% by Western blot assay in 19 cadavers.[13] Using vitreous humor, one study of 30 cadavers found a sensitivity of 83% using an EIA,[11] and another study of five cadavers found a sensitivity of 100% using the EIA.[14] Additional study is needed to determine the usefulness of assays on ocular fluids for donor screening, especially for corneal transplantation. Other tissues that may be accessible in cadaveric donors, such as lymph nodes, may also have a potential role for screening cadaveric donors.

Immunosuppression in Organ Recipients

Nearly all recipients of transplanted organs receive immunosuppressive chemotherapy (which primarily affects cellular immunity) to prevent rejection of the allograft. Because organ recipients may be pharmacologically immunosuppressed, detection of HIV infection, if it occurs, may be delayed if signs and symptoms of immunosuppression are attributed to antirejection therapy rather than HIV infection.[6] Concern also has been raised as to whether detection of HIV infection through HIV antibody testing may be impaired in organ recipients receiving antirejection immunosuppressive chemotherapy. In a review of published reports of 55 HIV-infected organ recipients with known dates of seroconversion; however only one infected donor tested HIV-seronegative more than 6 months after transplantation before seroconverting.[1]

HIV Testing of Organ and Tissue Donors

The U.S. Food and Drug Administration (FDA) has issued interim regulations that require screening of all tissue donors for HIV infection.[15] Other recommendations for screening organ and tissue donors also have been published in the United States,[5,8,9,16,17] Canada,[18] and Europe.[19] The revised U.S. Public Health Service (PHS) guidelines for preventing HIV transmission through transplantation recommend that all prospective donors of organs or tissues be screened for HIV infection by (1) a history and record review to assess the donor's risk for HIV infection; (2) a physical examination for signs of HIV infection or risk behaviors for HIV infection; and (3) testing the donor's blood for antibodies to HIV.[17]

After appropriate consent, all prospective donors should have a blood specimen taken for HIV antibody screening. The specimen should be obtained before administration of any transfusions but as close as possible to the time of organ or tissue retrieval. It should be tested for antibody to HIV-1 and HIV-2 using separate tests for HIV-1 and HIV-2 or a combination HIV-1/HIV-2 test. All testing should be performed as described in the package insert, including retesting of an initially reactive specimen. In some cases, time constraints due to the urgent need to perform life-saving organ transplantation may preclude following the conventional algorithm for screening. In these extreme cases, an initial sample should be set up in triplicate for EIA and the screening test considered positive if it is reactive in two or more of the three assays. A more rapid licensed test may also be used in triplicate when the severe time constraints make even testing by EIA impractical.

Assays for HIV antibody may be negative when a donor is recently infected with HIV.[6] To screen organ and tissue donors for earlier HIV infection than can be detected by antibody testing, other assays have been considered. The HIV p24 antigen assay has identified rare instances of HIV-infected blood donors who are antigen-positive/antibody-negative.[20] However, this assay is not licensed for screening; and because such antigen-positive/antibody-negative donors are exceedingly rare, screening blood donors for antigen has not been recommended in the United States.[21] Although its usefulness for organ and tissue donor screening is unknown the p24 antigen assay is part of the screening panel in some organ transplant centers.[22] When used, the possibly lower sensitivity and specificity of this assay with postmortem specimens should be considered.[10,12] In many donor populations, most initially positive screening tests for p24 antigen are likely to be falsely positive. Therefore it is critical to confirm the presence of p24 antigen using a neutralization assay.[23] Other promising assays that are still experimental, such as the polymerase chain reaction, should also be evaluated for their usefulness for donor screening when they become available.

Individuals should be excluded from organ or tissue donation in most cases if the screening assay for HIV-1 or HIV-2 antibody is repeatedly reactive or if risk factors for HIV infection are identified regardless of the screening test result.[17] (In severe cases, when the risk to the recipient of not receiving the transplant is thought to be greater than the risk of HIV transmission, donation may occur despite possible HIV risk factors. The recipient should provide specific informed consent in this situation.) Risk factors include a recent history of nonmedical drug injection, male–male sexual contact, exchange of sex for money or drugs, receipt of human-derived clotting factor concentrates, sexual contact with someone with any of the above risk factors, or a percutaneous, nonintact skin or mucous membrane exposure to HIV-infected blood. Anyone who cannot be tested for HIV antibody for any reason should also be excluded from organ or tissue donation.

A repeatedly reactive test using any screening assay should be confirmed with supplemental testing, including the Western blot assay or an immunofluorescence assay for HIV-1 antibody and a neutralization assay for HIV-1 antigen. No confirmatory tests for HIV-2 antibody have been licensed, although research assays are available in some centers.[24] All test results should be handled confidentially. If HIV infection is confirmed, prospective living donors or the spouses or known sexual partners of brain-dead or cadaveric donors should be notified in accordance with state law and counseled regarding HIV infection.

The HIV testing and exclusion protocol recommended by the Eurotransplant Foundation in Europe differs from that recommended in the United States.[19] With this protocol, organ and tissue donation is permitted (1) if the donor has no risk factors for HIV infection and the first screening antibody test result is negative or (2) if the first screening test is positive or indeterminate and two subsequent screening tests, done on different donor blood samples if available, are both negative. Donation of tissues, but not organs, is permitted if the screening antibody tests are positive but a confirmatory test is negative. This screening strategy is designed to maximize the number of organs and tissues cleared for transplantation.

HIV Testing of Organ and Tissue Recipients

It has been recommended that recipients of transplanted organs be routinely tested for HIV following transplantation. This recommendation is based on several benefits of such testing, including prompt identification of potentially infected donors, who may have donated other organs and tissues, and early identification of infected recipients so interventions, such as antiretroviral therapy and prophylaxis for opportunistic infections, may be offered in a timely manner.

The PHS guidelines recommend that organ recipients be tested for HIV antibodies immediately before and 3 months after transplantation.[17] Recipient testing should be with consent and not mandatory. If a recipient is found to have HIV infection, consistent with the particular state's law, the state health department, tissue bank, and in the case of organ recipients the United Network for Organ Sharing should be notified immediately.

References

1. Simonds RJ. HIV transmission by organ and tissue transplantation. AIDS 1993;7(suppl 2):S35–S38
2. Samuel D, Castaing D, Adam R, et al. Fatal acute HIV infection with aplastic anaemia, transmitted by liver graft [letter]. Lancet 1988;1:1221–1222

3. Quarto M, Germinario C, Fontana A, Barbuti S. HIV transmission through kidney transplantation from a living related donor [letter]. N Engl J Med 1989;320:1754
4. Ward JW, Schable C, Dickinson GM, et al. Acute human immunodeficiency virus infection—antigen detection and seroconversion in immunosuppressed patients. Transplantation 1989;47:722–724
5. Centers for Disease Control. Human immunodeficiency virus infection transmitted from an organ donor screened for HIV antibody—North Carolina. MMWR 1987;36:306–308
6. Simonds RJ, Holmberg SD, Hurwitz RL, et al. Transmission of human immunodeficiency virus type 1 from a seronegative organ and tissue donor. N Engl J Med 1992;326:726–732
7. Czitrom AA. Principles and techniques of tissue banking. Instruct Course Lect 1993;42:359–362
8. Centers for Disease Control. Semen banking, organ and tissue transplantation, and HIV antibody testing. MMWR 1988;37:57–58, 63
9. Centers for Disease Control. Transmission of HIV through bone transplantation: case report and public health recommendations. MMWR 1988;37:597–599
10. Pepose JS, Buerger DG, Paul DA, et al. New developments in serologic screening of corneal donors for HIV-1 and hepatitis B virus infections. Ophthalmology 1992;99:879–888
11. Klatt EL, Shibata D, Strigle SM. Postmortem enzyme immunoassay for human immunodeficiency virus. Arch Pathol Lab Med 1989;113:485–487
12. Novick SL, Schrager JA, Nelson JA, Baskin BL. A comparison of two HBsAg and two HIV-1 (p24) antigen EIA test kits with hemolyzed cadaveric blood specimens. Tissue Cell Rep 1993;1:2–3
13. Pepose JS, Pardo F, Kessler JA, et al. Screening cornea donors for antibodies against human immunodeficiency virus. Ophthalmology 1987;94:95–100
14. Grupenmacher F, Silva FM, Abib FC, et al. Determination of cadaveric antibody against HIV in vitreous humor of HIV-positive patients: potential use in corneal transplantation. Ophthalmologica 1991;203:12–16
15. Kessler DA, Shalala DE. Human tissue intended for transplantation. Fed Register 1993;238:65514–65521
16. Centers for Disease Control. Testing donors of organs, tissues, and semen for antibody to human T-lymphotropic virus type III/lymphadenopathy-associated virus. MMWR 1985;34:294
17. Centers for Disease Control and Prevention. USPHS guidelines for prevention of transmission of HIV through transplantation of human tissue and organs. MMWR (in press)
18. Anonymous. Guidelines for prevention of HIV infection in organ and tissue transplantation. Can Dis Wkly Rep 1989;15(suppl 4):1–17
19. Patijin GA, Strengers PFW, Harvey M, Persijn G. Prevention of transmission of HIV by organ and tissue transplantation. Transplant Int 1993;6:165–172
20. Irani MS, Dudley AW, Lucco LJ. Case of HIV-1 transmission by antigen-positive antibody-negative blood. N Engl J Med 1991;325:1174
21. Busch MP, Taylor PE, Lenes BA, et al. Screening of selected male blood donors for p24 antigen of human immunodeficiency virus type 1. N Engl J Med 1990;323:1308–1312

22. Ndimbie OK, Riddle PB. Serological assessment of the prospective organ donor. Lab Med 1993;24:103–106
23. Simonds RJ, Aguanno J, Hoxie NJ. HIV-1 from a seronegative transplant donor [letter]. N Engl J Med 1992;327:564–565
24. Centers for Disease Control and Prevention. Testing for antibodies to human immunodeficiency virus type 2 in the United States. MMWR 1992;41(RR-12):1–9

14
Programs for Routine, Voluntary HIV Counseling and Testing of Patients in Acute-Care Hospitals

ROBERT S. JANSSEN and ELIZABETH A. BOLYARD

Counseling hospital patients about and testing them for human immunodeficiency virus (HIV) can identify HIV-infected persons in a setting in which clinical evaluation and appropriate therapy are readily available. Infected persons may require treatment with antiretroviral agents, prophylaxis against *Pneumocystis carinii* pneumonia, tuberculin skin testing and prophylaxis, or other therapy to delay opportunistic infections associated with HIV infection.[1–4] In addition, counseling may help reduce the spread of HIV by helping some persons change high risk behaviors.[5]

Approximately 25 million individuals, or 10% of the U.S. population, were hospitalized in an acute-care hospital in 1990.[6] Of those patients, an estimated 225,000 (0.9%) were HIV-positive, of whom an estimated 163,000 had undiagnosed HIV infection.[7] Therefore offering counseling and testing in acute-care hospitals would benefit a large number of HIV-infected persons.

Strategies for Counseling and Testing

In a study to determine which patients should be routinely offered HIV counseling and testing, the patients' age, sex, race, and presenting medical condition were examined.[7] Age was found to be the most important determinant because the largest single group of HIV-positive patients were 15 to 54 years of age. Although targeted, routine, voluntary counseling and testing could be based on risk behaviors, accurate risk information may be difficult to obtain in health care settings,[8] and not all HIV-infected patients have discernible risk factors.[9]

Determining which hospitals should offer routine HIV counseling and testing can be based in part on the prevalence of HIV infection in the hospital. In an HIV serosurvey in 20 U.S. acute-care hospitals during,

1989–1991, the hospital-specific prevalence of HIV was approximately 10.4 times higher than the acquired immunodeficiency syndrome (AIDS) diagnosis rate [(annual number of individual AIDS patients diagnosed in a hospital and reported to the health department/annual number of discharges) × 1000].[7] In this study, the AIDS diagnosis rate was the only one of 13 hospital-specific characteristics associated with hospital-specific HIV seroprevalence. For 1990 the mean AIDS diagnosis rate in U.S. hospitals was 0.4 per 1000 discharges (range 0–78 per 1000 discharges). Of the 5558 U.S. acute-care hospitals, only 11% (593) had an AIDS diagnosis rate of 1 or more per 1000 discharges (estimated prevalence $\geq$1%).[7]

Because of the potential medical and public health benefits of recognizing early HIV infection and because of the large number of hospitalized patients with undiagnosed HIV infection, the Centers for Disease Control and Prevention (CDC) has updated recommendations for HIV counseling and testing in acute-care hospitals.[10,11] The CDC recommends that hospitals should encourage health care providers to inquire routinely about patients' HIV status and risks for HIV infection and then offer counseling and voluntary testing services for patients at risk. Hospitals with an HIV seroprevalence of at least 1% or an AIDS diagnosis rate of at least 1 per 1000 discharges should strongly consider offering HIV counseling and testing routinely to all patients ages 15 to 54 years.[11] If all 593 such hospitals adopted this recommendation, nearly 110,000 persons with undiagnosed HIV infection, or 11% of all HIV-infected persons in the United States, could be identified during the first year of testing.[7]

Factors for Consideration

Voluntary patient participation is critical to the success of a program that offers routine counseling and testing. High refusal rates could reduce the effectiveness of a counseling and testing program, particularly because persons at high risk may be more likely to refuse testing than those at low risk.[12,13] Programs that offer routine counseling and testing must determine which approaches work best in their own settings. Even within the same hospital, the proportion of persons who consent to testing may vary in different settings. In the only reported study of such a program in a hospital, the proportion of patients who agreed to testing varied from 11% on the psychiatric service to 91% on the cardiovascular surgery service.[12] Which personnel should offer counseling and testing and when should it be offered are important decisions. Primary health care providers incorporating HIV counseling and testing into routine medical care is one consideration.

In addition to some patients refusing to be tested, obstacles for programs that routinely offer HIV counseling and testing to hospital patients include the cost of the program, the efficiency of identifying undiagnosed HIV-infected patients after the first year, and the accuracy of testing. Little information is available on the cost efficiency of hospital-based testing programs, but one study suggests that such programs are cost-efficient when the prevalence of HIV in the population being tested is higher than 0.5%.[14] A hospital testing program may detect the highest rate of HIV-positive patients during the first year of the program. Because many patients go to hospitals repeatedly over time, those patients infected before the beginning of the program are likely to be detected during the first year of a program. Studies are needed to determine how much the efficiency of a program may decrease after the first year so policies can be modified if necessary.

Although the combination of enzyme-linked immunoassay and Western blot tests has high sensitivity and specificity, the accuracy of such tests decreases in inexperienced hands.[15,16] Of the 593 hospitals with an AIDS diagnosis rate of at least 1 per 1000 discharges, more than 80% had laboratories testing for HIV in 1990. Because most of these hospitals have considerable experience with HIV testing, they are not likely to produce poor quality testing.

Development of a Counseling and Testing Program

Some hospitals have already developed strategies for routine counseling and testing, but most hospitals do not have such policies. A survey of a random sample of U.S. hospitals determined that 83.4% of hospitals had adopted a written HIV-testing policy. Policies varied among institutions, with 9.6% requiring HIV testing of some or all patients at admission.[17]

An important consideration for deciding to offer routine HIV counseling and testing is the estimated prevalence of HIV infection in the hospital. Hospitals may use the AIDS diagnosis rate, which is based on the number of patients with AIDS who meet the 1987 CDC AIDS case definition.[18]* If a hospital does not track the number of AIDS cases diagnosed and reported within a year, it can obtain that information from the local or state health department. Hospitals that want more detailed information on the distribution of HIV infection among their patients may want to

*The association between the AIDS diagnosis rate using the 1993 CDC AIDS case definition[19] and hospital-specific HIV seroprevalence has not been evaluated. It is likely that the hospital prevalence is less than 10.4 times the AIDS diagnosis rate because using the 1993 AIDS case definition increases the AIDS diagnosis rate. Therefore the AIDS diagnosis rate should be used to estimate hospital prevalence before January 1, 1993. This is likely to be a reasonable estimate of prevalence for several years, as overall hospital prevalence is not likely to change significantly (CDC, unpublished data).

conduct an anonymous, unlinked serosurvey of HIV prevalence. Guidelines for conducting such surveys have been developed by the CDC.[20] Prevalence rates based on CDC guidelines for rapid assessment of HIV seroprevalence[20] overestimate the prevalence of undiagnosed HIV infection because the estimate includes patients with known infection and those with undiagnosed infection. Although information is limited, two studies have estimated that, in general, two-thirds of HIV-infected patients in hospitals have undiagnosed HIV infection.[7,21]

Policies for counseling and testing programs should be developed in accordance with local and state laws and regulations and should consider the provisions of the Americans With Disabilities Act of 1990.[22] The policies should define the various components of the program, including ordering of tests, type and documentation of consent, persons who will provide counseling, and the type of information to be included in both pretest and posttest counseling. The policies should specify how confidentiality will be maintained, including how the test results will be recorded in the laboratory and the medical record.

General Principles

Programs offering counseling and testing on a routine basis should be structured to facilitate confidential, voluntary patient participation and should provide pretest information on the testing policies of the institution or physician, basic information about the medical implications of the test, access to further information, and the documentation of informed consent. General guidelines to improve patient participation include the provision of testing as part of routine patient care and ensuring the confidentiality of test results and the public's confidence in that ability.

Once it has been decided that the hospital will offer routine, voluntary HIV counseling and testing to specified group(s) of patients, other factors should be evaluated to define the program components. Estimates of the expected number of persons to be tested can help determine what resources may be needed. Additional laboratory supplies and personnel may be needed if testing is done by the hospital laboratory. High quality laboratory services should be available, and hospital laboratories should participate in quality assurance programs such as the CDC's Model Performance Evaluation Program.[15,16]

Pretest and posttest counseling for patients is an integral part of the program. There should be sufficient numbers of adequately trained personnel to provide counseling to the patients. Knowledge that pretest counseling may take 10 to 45 minutes and posttest counseling 10 to 30 minutes for HIV-negative patients and 15 to 90 minutes for HIV-positive patients can help when estimating personnel needs. If additional personnel are necessary, the availability of quality training programs should be assessed. Local and state health departments can provide technical assistance and training for hospital staff responsible for counseling and testing services in acute-care settings.

The HIV counseling and testing should be offered in nonemergent situations, when patients can make an informed and voluntary decision regarding HIV testing. Offering testing to patients who are too ill to understand the pretest information or give informed consent should be delayed.

All patients should be informed of their test results, whether negative or positive. Test results should be provided to patients in a confidential manner and forwarded to state health departments in accordance with local law. Plans should be developed for providing posttest counseling for patients whose test results are not available until after the patient is discharged. Posttest counseling for infected patients and those at increased risk should be performed in accordance with CDC recommendations.[10,23]

The hospital should ensure that HIV-positive persons can obtain rapid, appropriate medical care and that they are not subjected to discrimination. The hospital should determine the availability of medical care providers for HIV-positive individuals. If adequate resources are not available at the hospital, private physicians, other hospitals and clinics, or other community resources may need to be used. HIV-positive patients may also require social services, including additional counseling and support services, legal services, financial assistance, substance abuse treatment programs, and housing assistance. Coordination among the resource groups should be ongoing. State and local health departments can provide partner notification services as well as additional prevention services for uninfected patients who are at high risk for HIV infection.

Patients who decline testing must not be denied needed medical care, not be provided suboptimal care, nor be tested surreptitiously. However, such persons should be given information about anonymous testing sites if such places are available.

Use of Test Results

In hospitals, HIV test results are sent to the patient's physician and to the patient's medical record. These confidential results are communicated to the patient during posttest counseling and are used to direct the patient's care. In addition, aggregated, not individual, HIV test results can used by infection control personnel or the hospital AIDS coordinator to estimate resources needed to care for HIV-infected persons. Knowledge of the prevalence of HIV infection in a hospital may also enable the hospital to obtain greater resources for HIV services. However, knowledge of these results should not replace universal precautions to protect health care workers from becoming infected by their patients.[24,25] The purpose of a routine counseling and testing program is to identify persons with HIV infection for early treatment and to communicate prevention messages.

The purpose is not for decreasing transmission to health care workers. Studies have demonstrated that knowledge that a patient is HIV-positive does not alter health care worker techniques consistently, which might decrease the potential for exposure and transmission of HIV.[26,27]

Conclusion

Because a large number of uninfected and HIV-infected persons could benefit, the CDC has recommended that hospitals with an HIV seroprevalence of at least 1% or an AIDS diagnosis rate of at least 1 per 1000 discharges should strongly consider routinely offering HIV counseling and testing to all patients ages 15 to 54 years.[11] Hospitals offering such services must provide high quality pretest and posttest counseling, access to quality laboratory support, protection of confidentiality of HIV test results, and, for HIV-infected persons, access to rapid and appropriate medical care.

References

1. Rhame FS, Maki DG. The case for wider use of testing for HIV infection. N Engl J Med 1989;320:1248–1254
2. Volberding PA, Lagakos SW, Koch MA, et al. Zidovudine in asymptomatic human immunodeficiency virus infection: a controlled trial in persons with fewer than 500 CD4-positive cells per cubic millimeter. N Engl J Med 1990; 322:941–949
3. Centers for Disease Control. Recommendations for prophylaxis against Pneumocystis carinii pneumonia for adults and adolescents infected with human immunodeficiency virus. MMWR 1992;41(No. RR-4):1–11
4. Centers for Disease Control. Screening for tuberculosis and tuberculous infection in high-risk populations and the use of preventive therapy for tuberculous infection in the United States: recommendations of the Advisory Committee for Elimination of Tuberculosis. MMWR 1990;39:RR-8
5. Higgins DL, Galavotti C, O'Reilly KR, et al. Evidence for the effects of HIV antibody counseling and testing on risk behaviors. JAMA 1991;266:2419–2429
6. American Hospital Association. 1990 AHA Annual Survey of Hospitals [electronic data tape]. Chicago: American Hospital Association, 1991
7. Janssen RS, St. Louis ME, Satten G, et al. HIV infection among patients in U.S. acute-care hospitals: strategies for the counseling and testing of hospital patients. N Engl J Med 1992;327:445–452
8. Castro KG, Lifson AR, White CR, et al. Investigations of AIDS patients with no previously identified risk factors. JAMA 1988;259:1338–1342
9. Kelen GD, Fritz S, Qaqish B, et al. Unrecognized human immunodeficiency virus infection in emergency department patients. N Engl J Med 1988;318: 1645–1650

10. Centers for Disease Control. Public Health Service guidelines for counseling and antibody testing to prevent HIV infection and AIDS. MMWR 1987; 36:509–514
11. Centers for Disease Control and Prevention. Recommendations for HIV testing services for inpatients and outpatients in acute-care hospital settings. MMWR 1993;42(No. RR-2):1–6
12. Harris RL, Boisaubin EV, Salyer PD, Semands DF. Evaluation of a hospital admission HIV antibody voluntary screening program. Infect Control Hosp Epidemiol 1990;11:628–634
13. Centers for Disease Control. Pilot study of a household survey to determine HIV seroprevalence. MMWR 1991;40:1–5
14. McCarthy BD, Wong JB, Munoz A, Sonnenberg R. Who should be screened for HIV infection? Arch Intern Med 1993;153:1107–1116
15. Taylor RN, Przybyszewski VA. Summary of the Centers for Disease Control human immunodeficiency virus (HIV) performance evaluation surveys for 1985 and 1986. Am J Clin Pathol 1988;89:1–13
16. Centers for Disease Control. Update: serologic testing for HIV-1 antibody—United States, 1988 and 1989. MMWR 1990;39:380–383
17. Lewis CE, Montgomery K. The HIV-testing policies of US hospitals. JAMA 1990;264:2764–2767
18. Centers for Disease Control. Revision of the CDC surveillance case definition for acquired immunodeficiency syndrome. MMWR 1987;36(suppl 1S):1S–15S
19. Centers for Disease Control and Prevention. 1993 Revised classification system for HIV infection and expanded surveillance case definition for AIDS among adolescents and adults. MMWR 1992;41(RR-17):1–19
20. Schwartlander B, Janssen RS, Satten GA, et al. Guidelines for designing rapid assessment surveys of HIV seroprevalence among hospitalized patients. Public Health Rep 1994;109:53–59
21. Gordin RM, Gibert C, Hawley HP, Willoughby A. Prevalence of human immunodeficiency virus and hepatitis B virus in unselected hospital admissions: implications for mandatory testing and universal precautions. J Infect Dis 1990;161:14–17
22. Americans with Disabilities Act, 1990. Public Law 101–336, 104 Stat. 327, 42 U.S.C. 12101 et seq. 101st Congress
23. Centers for Disease Control and Prevention. Technical guidance on HIV counseling. MMWR 1993;42(no. RR-2):11–17
24. Centers for Disease Control. Recommendations for the prevention of HIV transmission in health-care settings. MMWR 1987;36(suppl 2S)
25. Centers for Disease Control. Update: universal precautions for prevention of transmission of human immunodeficiency virus, hepatitis B virus and other bloodborne pathogens in health-care settings. MMWR 1988;37:377–388
26. Tokars J, Bell D, Culver D, et al. Percutaneous injuries during surgical procedures. JAMA 1992;267:2899–2904
27. Gerberding JL, Littell C, Tarkington A, et al. Risk of exposure of surgical personnel to patients' blood during surgery at San Francisco General Hospital. N Engl J Med 1990;322:1788–1793

15
HIV Testing for Life Insurance

Nancy J. Haley and Barry S. Reed

Acquired immunodeficiency syndrome (AIDS) presents a great danger to public health in the United States and is a major concern for the health care and insurance industries. In 1992, U.S. Life and other health insurance companies paid a record $1.4 billion in AIDS-related claims, a 7% increase from the prior year.[1] Of the 1992 figure, $768 million were paid on life insurance claims and $645 million on accident and health-related policies.[1] Total claims have risen each year since 1986 when the first figures were compiled and the total payment amount was $292 million.[2] Employers have also been adversely affected, as more than 85% of the insurance written in the United States is through employer-financed plans. In 1987, it was estimated that AIDS could cost employers more than $14 billion in benefits payments over the following 5 years.[2] Today the financial burden imposed by insurance benefit packages is perceived to be the major problem faced by employers in metropolitan areas.[2]

Testing for human immunodeficiency (HIV) status is not conducted on the millions of persons who receive benefits through employee plans. Large group plans or those who are self-insured attempt to cover the costs of insuring employees through adjustments in the pricing of benefit packages, with the cost of insurance reflecting the overall health experience of the group.[3] Insurance tests for HIV status affect a relatively small proportion of the insured population and, for the most part, are limited to those persons applying for individual life, health, and disability income coverages or small group plans (usually two to five lives).

Typically, individual life insurance is insurance purchased by a person to provide security and protection for survivors or their estates at the death of the insured. In its early stages, the debate over testing for HIV antibodies in persons applying for insurance raised the sensitive issues of testing methods, confidentiality of results, and availability of coverage. However, safeguards on appropriate testing, release of results, and counseling are carefully monitored by insurance commissioners in all states.[4] As with any other disease of epidemic proportions, testing for HIV is seen to be essential to preserve the integrity of the system.

Basis of Insurability

Life insurance responds to the needs of persons for security and is based on the principles of legal contract in which the insurer agrees to pay a certain amount of money at the death of the insured. The insured pays an amount of money referred to as a premium, and so long as the policy remains in force the insurer makes the payment upon the death of the insured.[5] The basic insurance contract has changed little since the earliest days of the industry, although the types of policies and price packages offered by life insurance companies have changed to reflect the needs of the times. Changes in life styles, incomes, educational levels, and the economic climate alter the shape of the industry and insurance policies. The insurance industry's ability to respond to changing conditions while preserving its primary objective of financial security is the most important factor in the industry's growth and protection for the millions of insured lives.[5]

To allow for this protection, insurance medicine analyzes the insurability of applicants and their expected longevity. Most important to the medical selection of risks is the expected rate of mortality of the applicants who apply for insurance. The risk selection and mortality must be predictable so the entire pool is protected against catastrophe. This point is the basis for underwriting the risks and protecting against identifiable premature mortality.

To determine how much money will be needed to pay future claims, an insurer must be able to estimate with accuracy the number and timing of deaths. The first serious attempt to establish population mortality rates was made by John Graunt, who analyzed records of christenings and burials in the City of London in 1662. He published "The Natural and Political Observations made upon the Bills of Mortality" and compiled a table of survivors.[6] Some years later, the Astronomer Royal, Edmund Halley, made a study of records of births and deaths kept for more than a century and produced the first table that predicted life expectations. Halley's work along with that of Newton and De Moive, laid the foundations of actuarial science.[6]

Because life expectancy with various disease states has changed over the years with new treatments, early diagnosis, and preventive measures, actuarial science has had to respond to changes in the predictability of mortality from various causes. From the time of discovery of antibiotic therapy until the 1980s, infectious diseases were not prominent in the medical selection of risks for life insurance. AIDS is now a worldwide epidemic that can only increase in severity for the forseeable future. At this time, the zenith of the epidemic cannnot be determined because of the long time lag between infection and clinical illness, the continued spread of the virus, and the lack of an effective vaccine.[7]

Testing for HIV Status

Since the first cases of AIDS were described in 1981, the epidemic has become a modern-day version of the plague of the Middle Ages. The number of HIV-positive individuals in the United States has continued to increase despite the fact that the modes of transmission are, for the most part, documented, and preventive measures have been published in all forms of lay literature and scientific journals. The demographic databases compiled by insurers have been helpful for following changes in this epidemic.

The decision by an insurer to test for HIV in a given population is based in part on the risk that the insurer is prepared to take in that class of insurance applicants.[3] Financial considerations on the cost of testing, potential impact on business, and anticipated claims are components of the actuarial model used by an insurer when deciding who should be tested. The use of mortality tables is essential for determining if an applicant should be tested for HIV or have a cardiogram for heart disease.

The Society of Actuaries of the United States has projected the potential effects of HIV claims on the insurance industry and has recommended that, as for any other identifiable disease state, prudent underwriting requires accurate testing for the disease state.[8] Therefore antibody testing is utilized to determine HIV status in proposed insureds. The methods for laboratory analysis of HIV status are standard U.S. Food and Drug Administration/Centers for Disease Control and Prevention (FDA/CDC)-approved laboratory practice. However, the requirements for confidentiality of results, pretest and posttest counseling, and individual state regulations ensure that HIV antibody testing by the insurance industry is a closely monitored activity.

The HIV test is a highly sensitive, specific test for antibodies against HIV. The significance of a positive test result has a profound impact on life plans of the tested person.[9] Testing for HIV status is conducted by the insurance industry in the same way it is in any clinical or public health laboratory. There are safeguards of informed consent for the person to be tested. There is confirmatory testing for individuals who screen positive by enzyme immunoassay and confidentiality of results with disclosure as mandated by law.[9]

Medical Underwriting for Life Insurance

Because the insurer is promising that all claims of insured persons will be paid, applicants to the pool of insureds are required to satisfy certain underwriting criteria prior to being accepted for an insurance policy. These criteria are determined by individual companies depending on actuarial models and the amount of risk the insurer is prepared to accept.[10] Small policies on young, healthy persons do not have the extensive underwriting requirements as a $1 million policy on an older person with a significant medical history.

Some medical underwriting is done on most applicants. It might entail a medical history, an attending physician's statement, and the collection of blood and urine for chemical analysis.[11] Each insurer may have variations in their requirements, but the guidelines are applied to each person in the pool in a nondiscriminatory manner.[10] Underwriting attempts to identify unacceptable risks and substandard risks. However, these choices cannot be based on ethnic background or gender preference.[5] HIV testing therefore is carried out in a nondiscriminatory manner according to the actuarial model, accounting for costs of testing, pricing of the policy, and the risk the insurer can take without catastrophic loss.[9]

The guidelines for testing as carried out in the United States differ from the HIV testing procedure used in the United Kindom. In the latter, insurers can ask questions of proposed insureds about behaviors that might put them at high risk for HIV and ask directly if they are intravenous drug users or hemophiliacs or if their sexual partners are homosexual or bisexual. Answers to these questions can then be used to decide who is to be tested for HIV.[12] Antidiscrimination laws do not allow these questions in the United States.

HIV Testing: Collection of Specimens

According to underwriting criteria, blood or urine (or both) are collected for analysis. The results of the analyses are then used as part of the risk assessment process. Medical histories of diseases such as diabetes or coronary artery disease are validated by blood tests such as the glucose or cholesterol level. Smoking status is verified by analysis of a metabolite of nicotine. The specific profile of tests varies by the insurance company and the face amount of the policy the applicant has requested.

Samples are collected by physicians, nurses, or technicians who usually are not employees of the insurance companies. These examiners are employed by independent companies that specialize in the acquisition of medical information and biologic specimens. The examiner may ask the applicant about health status, medications, and life style factors, such as smoking. They can measure height, weight, and blood pressure; and they can collect the samples needed by the company to complete its underwriting process.

Early insurance laboratory testing was confined primarily to urine samples. However, the utility of analytes from a basic serum screen (SMAC) such as glucose, cholesterol, and liver enzymes, and the need to determine HIV status has increased the collection of blood-based products for risk assessment. They include whole blood, serum, and recently dried blood spots. Alternate body fluids that can be tested for the presence of HIV antibodies include urine and saliva. However, these methods have not yet been approved by the FDA and are not used in MetLife Laboratories.

HIV Testing: Notice and Consent

The examiner conducting the insurance examination is supplied with prepackaged kits of collection materials and information relative to the type of examination being conducted. The examiner is directed to ensure that the applicant receives, reads, and understands the notice and consent form. This consent form must be signed before sample collection. The examiner is also instructed to provide a booklet on AIDS and HIV testing. This booklet was written by the CDC and explains behaviors that place a person at risk for infection, documents the tests that will be performed, and provides a source for further information through the AIDS hotline.

A notice and consent for testing form explains how results will be handled by the laboratory and insurance company. All test results are confidential. Results are reported by a commercial insurance testing laboratory to the appropriate insurance company. There are four major commercial insurance testing laboratories that perform testing for U.S. insurance companies. However, as one of the largest insurance companies, MetLife maintains its own laboratory, which reports results directly to physician underwriters. Data collection for underwriting usually stays within the individual company, but occasionally information is shared when companies are co-writing or sharing the risk on large-amount cases. Information on applicants can be maintained by the Medical Information Bureau, which guarantees the confidentiality of all client information.

Disclosure of an abnormal HIV antibody test is restricted to persons with a need to know. These individuals vary with state requirements but generally are restricted to include the applicant's physician or health care provider who can give the appropriate counseling and treatment. Some states require that HIV-positive results be reported in a confidential manner to state authorities to ensure that appropriate follow-up is completed. Insurance companies comply as required by state law.

The notice and consent form completed by the examiner and signed by the applicant contains bar-coded labels that are affixed to all blood and urine samples of the applicant. The techniques developed for proper chain of custody by forensic and drug testing laboratories are employed when labeling and sealing all samples. Chain of custody procedure documents the location and handling of the specimen from collection to disposal.[11] The kit, with signed consent forms and samples sealed with tamper-evident tape, is sent to the testing laboratory.

HIV Testing: Shipment and Receipt of Samples

Kits containing whole blood or serum are shipped to the laboratory via overnight courier service or mail. Extensive studies have been conducted to ensure the stability of these samples for HIV antibody testing. Extreme changes in temperature and delays of up to 1 week have been evaluated on HIV-positive and HIV-negative samples to guarantee that the shipment procedures are appropriate. Upon receipt at the laboratory, all samples are processed under strict quality controlled procedures and according to accepted methods of chain of custody.[11]

When the dried blood spot (DBS) is used for analysis, the sample is taken by capillary puncture and expressed onto six spots on a filter paper card (Fig. 15.1). The sample must be sufficient to completely fill and saturate each spot on the filter paper. This card is signed by the applicant attesting to the fact that the blood spots are his or hers; the card is allowed to dry for 10 minutes, placed in an envelope, and a tamper-evident label affixed to the seal. This tamper-evident tape is initialed by the applicant and the examiner, and the card is shipped to the laboratory with an accompanying urine sample, which is used for testing of other analytes needed for underwriting.

Sample Accession at MetLife Laboratory

Laboratory procedures vary with the size of the laboratory and the sample mix that is accessed on a daily basis. Most laboratories utilize procedures that are similar; those detailed below represent the procedures in place at MetLife. Samples received at the laboratory are assigned a unique accession number, and the personal identification of the applicant is removed. The kit number assigned to the applicant by the examiner remains on the sample as a cross-reference to the accession number. The notice and consent forms are processed to guarantee that the proper consent has been provided and the sample collected with the proper chain of custody. The policy information is entered into the

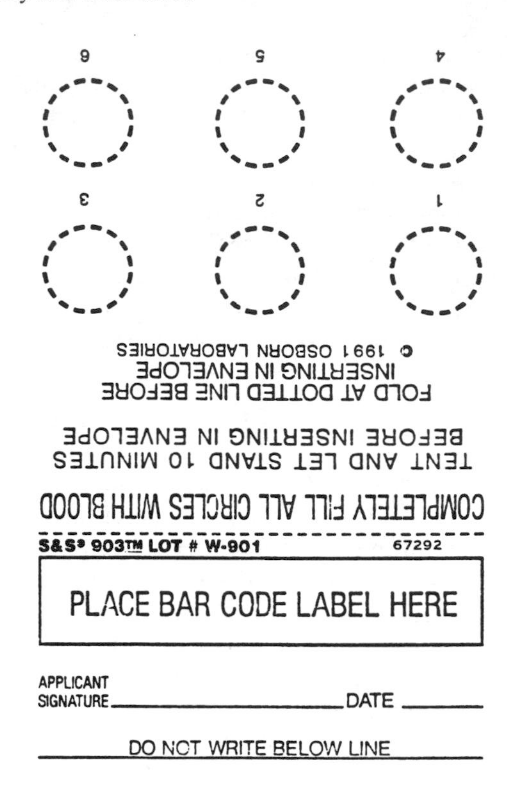

FIGURE 15.1. Card for collection of dried blood spot sample. All six circles are saturated with blood from the applicant.

MetLife underwriting data base to match the original information taken by the sales agent. The accession number becomes a part of this file and is used for any further tracking of the specimens.

All serum and urine samples are bar-coded with the accession number and racked for robotic aliquoting from the original collection tube. Enzyme immunoassay (EIA) methods are employed for HIV antibody testing, and appropriate aliquots of serum are placed in 96-well plates by automated methods. The primary tube is retained and stored for any further testing that might be indicated.

The DBS cards are prepared by scanning the bar code and punching out the first spot on the card into the appropriate well of a 96 well plate. Currently, automated equipment that can read the bar code and punch the spot into the plate is being evaluated. It should increase the efficiency of the setup part of this analysis. A map of the plate with accession numbers is placed under the plate as the samples are punched. The accession numbers are then appropriately matched to the computer-verified work lists.

Dried blood samples do not require refrigeration[13,14] and when protected from moisture can be stored for many months.[15] Once the initial spot has been removed from the card and the sample deemed to be nonreactive, all cards are stored in plastic bags according to laboratory accession number.

Immunologic Testing

Insurance testing laboratories have historically used the 96-well format for a variety of EIA analyses involved in risk assessment, and most prefer to keep the same standard protocol for as many tests as possible. Although bead technology for HIV testing has been evaluated and exhibits a high degree of sensitivity and specificity, the 96-well plate format performs equally as well and is easier to implement in our laboratory. When HIV testing was begun at MetLife Laboratory, an FDA-approved test for HIV-1 supplied by Genetic Systems was chosen for the screening test. In September 1992 MetLife implemented a change in serum testing to the combination test for the detection of HIV-1/HIV-2; this test kit is supplied by the same manufacturer. Combination tests have been used successfully and have identified HIV-2-positive sera not identified using HIV-1 screening tests. There is no loss of sensitivity in detecting antibodies to either virus, and in some cases an increased sensitivity to HIV-1 has been demonstrated.[16] Prior to use of the FDA-approved combination test, two samples that were EIA-reactive for HIV-1 were found to be indeterminate on atypical Western blot assays. Research methods identified these samples to be HIV-2-positive. Similar results for HIV-2 have been reported by commercial insurance testing laboratories.

Tests on DBS are carried out with reagents supplied by Genetic Systems for the detection of antibodies to HIV-1. At this point, the combination test for HIV-1/HIV-2 has not received FDA approval for use with DBS. The advantages of the DBS format outweigh this limitation of testing for HIV-1 alone, as collection of capillary blood is less of an inconvenience to the proposed insured, the dried samples are less of a biologic hazard to laboratory personnel, and there are bar-coded numbers identifying the applicant on the filter paper card.

The algorithms for testing samples with the combination test or the HIV-1 test are identical with recommendations of the Association of State and Territorial Public Health Laboratory Directors (ASTPHLD). A single determination is made on a sample, and if nonreactive the results are released as nonreactive. A reactive sample is repeated in duplicate with serum withdrawn from the primary tube of blood collected from the applicant. If these tests are again reactive, confirmation testing is conducted.

Confirmation procedures include the use of a Western blot analysis with reagents supplied by Bio-Rad. The Western blot assay uses separated HIV proteins stabilized in nitrocellulose to determine the antibodies present in a sample found to be reactive in the EIA. MetLife has also utilized the FDA-approved immunofluorescent antibody (IFA) procedure, but we have conducted this test in parallel with the Western blot analyses. Reports to physician underwriters and physicians treating proposed insureds have suggested that currently there is a better understanding of the Western blot results than those provided by the IFA.[17]

The findings of reactive samples allows us to evaluate other tests that might be introduced into future testing algorithms. Results of these research efforts are not reported to the physician underwriters but permit us to determine the sensitivity and specificity of tests that might be introduced to the testing market. Testing supplemental to the Western blot assay and the IFA would be desirable especially during attempts to resolve indeterminate results of the Western blot test. Furthermore, there is a need for confirmatory testing for HIV-2 infection. We concur with the recommendations of the Panel on Testing at the 1993 ASTPHLD meeting, which requested the establishment of acceptable alternatives to the Western blot assay and the IFA. Our research work with peptide-based assays has suggested a high degree of specificity with this method and no need for the subjective readings by the laboratorian. Further work in this area is desirable.

Quality Control Procedures

Modern technology for bar coding, robotic sampling, and computer downloading of data have greatly improved the quality assurance of HIV antibody testing. Unique identifiers for the tubes in a blood or DBS kit matched with the notice and consent form greatly assists in the receipt of proper samples from the proper applicants.

Quality control for serologic testing employs positive and negative controls purchased from manufacturers as well as those prepared in house. Biologic positives are prepared by the repeated analysis of reactive samples by at least two EIA procedures supplied by different manufacturers and confirmed by the Western blot assay. False-positive controls or samples providing atypical Western blot results are also prepared. A test panel for the training of laboratory technologists utilizes these varied control materials as well as samples prepared from diluted or adulterated serum. In-service training is an essential part of the quality assurance program.

Quality controls for the Western blot assay include manufacturers' controls and biologic controls prepared from samples found to be positive or with varying band patterns and reported as indeterminate. Laboratorians are tested regularly on the interpretation of the Western blot results. After the technologist and quality assurance team leader have signed off on the results, the laboratory director or associate director reviews all results, including the EIA absorbances, Western blots, and supplemental tests, as well as the chain of custody before the results are sent to the physician underwriter.

Confidentiality of Results

Reporting HIV-positive results does not follow the usual release of laboratory results from the risk assessment profile. All results on blood abnormalities are blocked from general access by computer, and any cases with abnormal HIV findings are routed to the medical department. The sample is identified by accession number, and an electronic mail message is sent to the appropriate physician assigned to the case. This message informs the physician that nonspecific abnormal test results are being sent to his or her attention. It is only a physician who can put the file back together, establish the client's identity, and match the results of the tests with the individual. If the test results are positive for HIV infection, the applicant is informed in a confidential manner.

The current practice in the U.S. insurance industry is to send a registered letter to the applicant indicating that his application has been declined because of an abnormal laboratory result and then asking for permission to mail the results to the applicant's physician. After signed authorization is returned, the results are forwarded to the personal physician, who provides notification and counseling. Some states allow positive results to be sent directly to the proposed insured. In these instances the notification letter should include a referral to a local AIDS counseling organization and appropriate literature regarding the significance of a positive test.[7] With either form of notification, the proposed insured and his or her physician is strongly encouraged to have the HIV test repeated.

If there is no response to the letter, a second letter is sent to the applicant. Depending on the state of residence, if there is no response to the second notice no further correspondence might be necessary. Where state law requires disclosure to the Department of Health in the absence of a physician being authorized to receive the results, the appropriate letter is sent by a physician underwriter to the Department of Health.

If an indeterminate Western blot result is sent to the physician underwriter, he or she must take a different action depending on the prevailing state regulations. In certain states, further consideration of the case may be postponed for up to 6 months or another test may be ordered to see if there is a change in the Western blot test result.

Indeterminate Western blot results with a p17 or p24 band alone occur in as many as 15% of the Western blots performed.[15] These results can be released as negative by the physician underwriter. Repeating these tests could cause undue anxiety in the proposed insureds, and experience has shown that these patterns generally do not change and do not indicate infection with the HIV virus.[15] Persons with such persistent Western blot patterns usually do not manifest laboratory signs of immune deficiency, nor do they develop disease states typical of HIV infection.

Results of HIV Testing in the Insurance Industry

The growth of this pandemic can only mean that the industry will continue its testing for HIV status among proposed insureds. The data compiled by the American Council on Life Insurance show a steady increase in claims each year. Including 1992 claims, the insurance industry has paid $6.3 billion in claims related to AIDS over a 7-year span.[1] Most of these claims were from group policies or individual life policies put into force before HIV testing or with policy amounts below the testing thresholds of most companies.

Although the amount paid in claims increased over this time period, of greater concern is the change in the ratio of AIDS claims to all ordinary life claims. In 1986, fewer than 1% (0.95%) of claims were AIDS-related; this percentage has more than doubled to 2.24% in 1991. The MetLife experience has seen a growth of AIDS-related claims to 1.6% of total claims in 1992 from 0.40% in 1986.[18]

Figure 15.2 presents the experience of MetLife with regard to AIDS claims from 1985 to 1989. For group life insurance in 1985, the company processed 60 death claims as a result to AIDS. For 1989, the number had

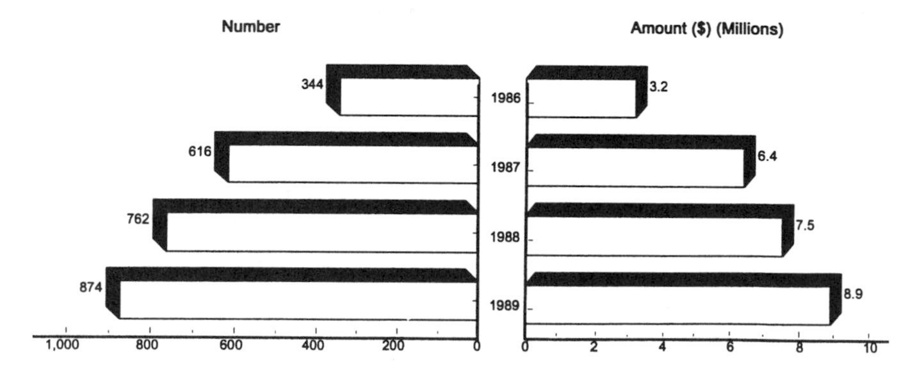

FIGURE 15.2. Number of AIDS claims and amount (in dollars) of MetLife Personal Life Insurance, 1986–1989. (Courtesy of Metropolitan Life Insurance Company, Actuarial Division, AIDS Surveillance Report.)

risen to 689, a 10-fold increase. In contrast to the group insurance area, individual life insurance policies, which have more extensive underwriting and now require testing for HIV status, have not seen the staggering increase in death claims from AIDS-related diseases.[18] As a result of underwriting and testing for HIV, the cost of life insurance to the consumer has not seen a substantial increase as a result of the AIDS epidemic.[19]

Individual life insurance AIDS-related death claims paid by MetLife in 1992 on policies issued after testing began has shown the effectiveness of HIV testing. All claims have been on policies issued without HIV testing. An important point to remember here, however, is that once a life insurance policy is issued it remains in force so long as the premium is paid regardless of change in the medical condition of the insured. Persons seroconverting after the insurance contract is issued are protected by the contract.

The prevalence of HIV-positive findings in the U.S. insurance industry is lower than is seen in the general population. Numbers taken from the reports of insurance testing laboratories show a positive rate of about 0.08%, which is approximately one-fourth the rate estimated by the CDC for HIV infection in the United States. There are high prevalence states that closely parallel those identified by the CDC for reported AIDS cases.[20] These high prevalence areas (which include 11 states and the District of Columbia) have rates in excess of 0.5 per 1000. Therefore, as mentioned early in the chapter, where the insurance is being issued is important when determining when to test for HIV status.

The amount of the insurance policy (face amount) has been inversely proportional to positive HIV findings. Additionally, the age of the applicants demonstrates different rates. Men between the ages of 20 and 39 have the highest rate (0.11), with women of the same age presenting with a rate of 0.03. A larger proportion of individual life insurance is written on men than on women. There has been an increase in the percentage of female HIV positive applicants. In 1989 about 10% of MetLife's AIDS-related death claims were for women.[12] Insurance testing laboratory statistics for the same year showed that 6% of HIV-positive applicants were women. In 1992 the percentage of female HIV-positive applicants had risen to 15%, a figure that compares closely with the rise reported by the CDC during this period.

Conclusion

Testing for HIV status by the U.S. insurance industry is conducted with safeguards of confidentiality, informed consent, and quality laboratory practices. With the availability of sensitive, accurate assays to detect the presence of antibodies to HIV, the use of these tests by the insurance industry can help identify HIV-positive individuals in generally low risk populations, hopefully slowing the spread of disease.

References

1. Wall Street Journal. Dow Jones News and Retrieval, Dow Jones & Company, New York, August 10, 1993
2. Survey of AIDS Related Claims. Washington, DC: American Council of Life Insurance/Health Insurance Association of America, 1986
3. Pricing of Life Insurance. In Long GL, Mortin GA (eds): Principles of Life and Health Insurance (2nd ed). Atlanta: Life Office Management Association, 1988:19–37
4. National Association of Insurance Commissioners Advisory Committee on AIDS. Medical/Lifestyle Questions on Applications Underwriting Guidelines Affecting AIDS and ARC. 1986:para II(B)
5. Insurance and the Insurance Industry. In Long GL, Mortin GA (eds): Principles of Life and Health Insurance (2nd ed). Atlanta: Life Office Management Association, 1988:1–18
6. Brackenridge RDC, Brown AE. A historical survey of the development of life insurance. In Brackenridge PDC, Elder WJ (eds): Medical Selection of Life Risks (3rd ed). Basingstoke, UK: Macmillan, 1992
7. Gleeson RK. HIV and AIDS. In Brackenridge RDC, Elder WJ (eds): Medical Selection of Life Risks (3rd ed). Basingstoke, UK: Macmillan, 1992
8. Cowell JJ, Haskens W. AIDS, HIV Mortality and Life Insurance. Itasca, IL: Society of Actuaries, 1987
9. Bergkamp L. Life insurance and HIV testing: insurance theory, discrimination and solutions. Med Law 1989;8:567–580
10. The policy is insured. In Long DL, Mortin GA (eds): Principles of Life and Health Insurance (2nd ed). Atlanta: Life Management Institute Association, 1988:99–127
11. Brackenridge RDC, Croxson RS. The Medical Examination. In Brackenridge RDC, Elder WJ (eds): Medical Selection of Life Risks (3rd ed). Basingstoke, UK: Macmillan, 1992
12. Large RE. How informed is patient's consent to release of medical information to insurance companies? BMJ 1989;298:1495–1496
13. Bekets F, Kashamuka M, Pappacoanou M, et al. Stability of HIV type 1 antibodies in whole blood dried on filter paper under various tropical conditions in Kinshasa, Zaire. J Clin Microbiol 1992;30:1179–1182
14. Van der Akker R, Kooy H, van der Meyden H, Lumey BH. Recovery of HIV antibodies in eluates from plasma and erythrocytes dried on filter paper and stored under various conditions. AIDS 1990;4:90
15. George JR, Schochetman G. Serologic tests to detect HIV infection. In Schochetman G, George JR (eds): AIDS Testing: Methodology and Management Issues. New York: Springer-Verlag, 1992
16. Constantine NT. Serologic tests for retroviruses: approaching a decade of evolution. AIDS 1993;7:1–13
17. Gellert GH, Moore DF, Weismuller PC, Greenwood R, Maxwell RM. Office-based test systems for HIV antibody. N Engl J Med 1993;328:211–216

18. Pickett NA, Drewry SJ, Comer EL. AIDS: MetLife's experience 1986–1989. Stat Bull 1990;71:2–9
19. Kay BR, Kita MW. The use and interpretation of laboratory derived data. In Brackenridge RDC, Elders WJ (eds): Medical Selection of Life Risks (3rd ed). Basingstoke, UK: Macmillan, 1992
20. Centers for Disease Control and Prevention. HIV/AIDS surveillance report third quarter edition. MMWR 1993;5:3–4

16
HIV Infection in Children

MARTHA F. ROGERS and GERALD SCHOCHETMAN

Children represent an increasingly important part of the human immunodeficiency virus (HIV) epidemic. As of December 1992, there were 4249 HIV-infected children under the age of 13 years who had been reported to the Centers for Disease Control and Prevention (CDC), representing 1.7% of all reported acquired immunodeficiency syndrome (AIDS) cases. During 1992 cases due to perinatal (mother-to-child) transmission and those due to heterosexual contact had the largest increases in number of reported AIDS cases compared with all other transmission groups.[1]

Prepubertal children acquire HIV primarily through two routes: (1) perinatally, through exposure to their infected mothers; and (2) parenterally, through transfusions of blood or blood products. Transmission through sexual abuse[2] or through percutaneous exposure to contaminated needles[3] is possible but rarely reported. Of the 4249 children reported, 3665 (86%) acquired HIV perinatally, 306 (7%) acquired HIV through transfusion of blood or blood products, 188 (4%) had coagulation disorders, and 90 (2%) had undetermined modes of transmission (usually due to inadequate information).

The laboratory and clinical diagnosis of HIV infection in children presents some unique problems. Because of the passive transfer of maternal antibody in utero, HIV infection in infants born to HIV-infected mothers cannot be confirmed with the anti-HIV antibody assays commonly used for diagnosis in adults. Certain clinical features, such as disease manifestations and the incubation/latency period to the development of AIDS, are different in children than in adults. Understanding these differences is important for diagnosing HIV infection in children. This chapter briefly reviews the modes of transmission and clinical features of HIV infection and discusses the laboratory diagnosis in the pediatric population.

Modes of Transmission in Children

Perinatal (Vertical) Transmission

Because donor deferral and screening of the blood supply has virtually eliminated transmission via blood or blood product transfusion in most developed countries, perinatal (vertical) transmission accounts for nearly all HIV infections in prepubertal children. HIV can be transmitted from mothers to infants (1) during pregnancy through transplacental passage of the virus, (2) during labor and delivery through exposure to infected maternal blood and vaginal secretions or possibly from maternal-fetal transfusions, and (3) during the postpartum period through breast-feeding. In utero transmission has been documented by identification of HIV in fetal tissue as early as 8 weeks' gestation,[4] although the exact timing of transmission during gestation is unknown. Approximately 30% to 50% of HIV-infected infants have detectable virus at the time of birth, suggesting in utero transmission.[5,6]

Intrapartum transmission has been suspected in infants who test virus-negative at birth but test virus-positive later. Additional evidence for intrapartum transmission comes from a study of twins born to HIV-infected mothers. This study concluded that a substantial proportion of mother–infant transmission occurred during the intrapartum period based on analyses showing that among twins discordant for HIV infection status the first-born twin (twin A) had a higher likelihood of becoming HIV-infected than the second-born twin (twin B).[7] The authors speculated that twin A was at greater risk for infection because of the greater exposure (both volume and duration in the birth canal) to vaginal fluids than twin B. Evidence for intrapartum transmission also comes from studies showing a somewhat lower transmission rate among infants delivered by cesarean section than among infants delivered vaginally.[8] However, other studies have not shown a difference in transmission rates between these two groups,[9] and cesarean section is not recommended as a method for reducing maternal–infant transmission. Additional studies may clarify its role, if any, in preventing maternal–infant transmission.

Several cases of HIV transmission to infants have been documented in mothers who seroconverted to HIV while breast-feeding their infants.[10–12] HIV has been isolated from cell-free breast milk.[13] In addition, studies of mothers who were infected during or before pregnancy have shown an increase in the maternal–infant transmission rate among those who breast-feed compared with those who do not. A meta-analysis of six studies indicated an additional risk of about 14%.[14] Breast-feeding is not recommended for infants of HIV-infected mothers in the United States, where adequate alternative nutrition is readily available.[15] This recommendation does not apply in situations where alternative nutrition and clean water to reconstitute infant formula are not available.

The rate of transmission from mothers to infants varies from study to study, with the larger, multicenter prospective studies reporting rates of 12% to 39%.[9,16–18] Several of these studies found that mothers with lower CD4 counts and more advanced stage of disease were more likely to transmit HIV. Other studies indicate that other factors may play a role in that chorioamnionitis was associated with an increased risk of transmission.[19,20] Three studies found lower transmission rates in infants born to mothers with antibody to certain epitopes of the V3 loop of gp120 compared with infants born to mothers without these antibodies,[21–23] but later studies failed to confirm these results.[24]

The epidemiologic characteristics of children with perinatally acquired HIV infection reflect characteristics of women with the infection. HIV infection in women is largely associated with intravenous drug use or sexual contact with men who use drugs intravenously.[1] Rates of HIV infection among intravenous drug users (IDUs) have been highest in the northeastern states, particularly New York and New Jersey.[25] A disproportionate number of IDUs in these areas are African-American or Hispanic. Thus most children with perinatally acquired infection reside in these areas and are of the same two ethnic groups. However, reports emphasize the increasing importance of heterosexual contact in the spread of HIV among women and the increased number of women with HIV in the southeastern part of the United States and in rural areas.[1]

Transfusion and Hemophilia-Associated Transmission

Transfusion of HIV-infected blood or blood products is the most efficient means of HIV transmission. Studies have shown that about 90% of individuals receiving an HIV-seropositive transfusion have become infected.[26] One study also found shorter incubation/latency periods to the development of AIDS in individuals who received blood or blood products from donors who developed AIDS soon after donation than in those who received transfusions from donors in earlier stages of the disease.[27]

The epidemiologic characteristics of children with transfusion-associated AIDS reflect the population of children being transfused. Most cases have occurred in children who received transfusions during the newborn period for perinatal conditions.[28] Cases have also occurred in children with congenital anemia (sickle cell, thalassemia) and other conditions that require multiple transfusions. Although most of these children are white (52%), a disproportionate number of children have been African-American (22%) or Hispanic (24%), probably due to the higher rate of low birth weight and other perinatal problems that place these children at greater risk of receiving blood or blood product transfusions. The geographic distribution of pediatric transfusion-associated AIDS cases within the United States reflects the seroprevalence in donor populations prior to donor screening. More than half (61%) of the cases have been reported in California, New York, New Jersey, Florida, or Texas.[28]

Children with coagulation disorders who received clotting factor between 1978 and 1985 were at particularly high risk of HIV infection. Up to 80% of the children who received large-pool, non-heat-treated factor products from plasma donated before HIV screening have developed HIV infection.[29] More severe hemophilia and greater factor usage have been associated with a greater risk of HIV infection.[29] Although some investigators have speculated that some hemophiliacs may have become "immunized" through receipt of virus inactivated during the factor separation process, studies have shown that virus can be isolated from virtually all HIV-seropositive hemophiliacs.[30]

The epidemiologic characteristics of children with hemophilia-associated HIV infection are more reflective of the general U.S. population: 69% are white, 13% are African-American, 15% are Hispanic, and 3% are of other race or ethnicity. Geographically, these cases are dispersed throughout the United States (determined by the CDC's national AIDS surveillance).

Measures to control transmission of HIV through blood or blood products have virtually eliminated this form of transmission. Although blood donations from newly infected persons donating during the "window" period between initial infection and seroconversion would not be eliminated by antibody screening, studies estimate that the risk of acquiring HIV from screened blood or blood products is around 1 per 225,000 units.[31] AIDS cases, however, are continuing to occur in individuals who received blood or blood products prior to donor screening because of the long incubation period for the disease.

Clinical Features of HIV Disease in Children

Although the pathophysiology of HIV infection in children is similar to that in adults, there are important differences in the disease manifestations for the two age groups. *Pneumocystis carinii* pneumonia (PCP) is the most common opportunistic infection in both children and adults; another pulmonary condition, however, lymphoid interstitial pneumonitis, occurs in many pediatric cases but is rare in adults. Moreover, recurrent bacterial infections have been reported more commonly in children.[32] Neurologic disease that results in dementia in adults is likely to cause developmental delay or loss of developmental milestones in young children. Weight loss and wasting in adults may appear as failure to thrive in children. Immunologic parameters also differ. CD4 counts are markedly higher in infants and young children compared to those in adults.[33] Children do not attain standard normal adult ranges for CD4 counts until 6 years of age. Thus CD4 values used for initiating PCP prophylaxis[34] and antiretroviral therapy[35] are much higher for young children than for adults.

In general, the incubation/latency period for AIDS is shorter in children than in adults.[28] Children who acquire HIV from their mothers or from blood transfusions given during the first year of life have the shortest median incubation period, often developing AIDS within the first 3 years of life. Median incubation periods in adults have generally ranged from 7 to 10 years, and adults rarely develop signs or symptoms of the disease during the first 3 years after infection.[36]

Laboratory Diagnosis of HIV Infection in Children

Antibody to HIV

Enzyme Immunoassay and Western Blot Assay

As in adults, tests for antibody to HIV are the primary means of laboratory diagnosis of HIV infection in children who acquire the virus through nonperinatal routes (i.e., transfusion/hemophilia-associated infection). Children with perinatally acquired infection, however, present unique problems. Because maternal IgG antibody to HIV is passed transplacentally to virtually all infants born to HIV-seropositive women, the presence of antibody in the infant does not necessarily indicate virus infection. Studies have shown that about 20% to 30% of infants born to infected women acquire HIV infection, whereas close to 100% test positive by enzyme immunoassay (EIA) or Western blot assay during the neonatal period (0–28 days of life), and maternal antibody can persist for up to 15 to 18 months of age (median age at seroreversion is about 10 months). The standard EIAs and Western blot assays most efficiently detect IgG and cannot differentiate between mother- and infant-derived antibody.

Several approaches have been suggested to overcome this problem. One of the simplest approaches is to look for new bands on serial Western blot assays on samples obtained during the first year of life. New bands indicate antibody production by the infant and thus infection, but they are not observed in most infected infants and may not appear until several months of age.[37]

Another approach that has been used for other congenital infections is to monitor antibody titer. A rising titer indicates antibody production by the infant and thus infection, and a falling titer indicates noninfection. Although theoretically applicable to HIV, studies have shown that antibody titer as measured by the optical density reading of the EIA is not always correlated with the infection status of the infant.[37]

HIV-Specific IgA and IgM Assays

A more promising approach is to test for HIV-specific antibodies that do not cross the placenta, such as IgM or IgA. Studies have shown that assays for HIV-specific IgA may be useful for diagnosing perinatally infected infants. Using a panel of seroconversion samples tested for HIV-specific IgG, IgA, and IgM, Weiblen et al. found that IgA appeared as early as IgG but, compared with IgM, persisted longer, reacted with more protein bands, and showed more intense staining.[38] Several studies indicate that HIV-specific IgA assays can detect most HIV-infected infants by 6 months of age but are generally negative in infected infants under 3 months of age, presumably because HIV-specific IgA antibody is not produced in sufficient quantity to be detected during the first months of life in many infants.[39,40]

Assays for IgM have been less useful. One study found that results of HIV-specific IgM antibody tests did not correlate with cord blood cultures or the presence of signs and symptoms of AIDS in infants during the first year of life.[41] In another study, only four of eight HIV-infected infants had detectable anti-HIV IgM antibody.[37] These problems with the IgM assay could be due to the interference from abundant IgG, which can produce spurious results. Removal of IgG using protein G prior to performing the IgM or IgA assay, as was done by Weiblen et al.[38] in the studies cited above, appears to be important for improving the sensitivity and specificity of these assays.

Importance of Clinical Follow-up

Although the standard EIA and Western blot assay alone cannot identify infected infants, the use of these tests serially over the first 2 years of life coupled with careful medical follow-up is the most readily available and accurate method for definitive diagnosis. Infants who develop CDC-defined AIDS, those with milder signs and symptoms of HIV disease who have the classic immune abnormalities, and those with persistent antibody beyond 15 to 18 months of age regardless of symptoms are considered infected.[42] These definitions are outlined in more detail in the CDC Classification System for HIV in children under 13 years of age.[42]

Infants who remain well and lose maternal antibody are considered uninfected, although a few cases of virus-positive/antibody-negative children have been reported.[43,44] In some cases, children who became antibody-negative later seroconverted to antibody-positive.[37,45] Prospective studies of infants born to HIV-infected women indicate that the frequency of occurrence of antibody-negative/virus-positive children is rare.[46,47]

p24 Antigen

The standard p24 antigen assay has been a helpful but limited method for diagnosing pediatric HIV infection. Studies of infants born to HIV-infected mothers have found few infants to be antigen-positive early in the course of infection due to the apparently low levels of antigen during the first month of life and the presence of excess maternal antibody, which complexes to any free p24 antigen present.[48] The standard p24 antigen assay does not detect p24 antigen complexed with antibody, and therefore the test may not be positive unless there are sufficient levels of circulating free p24 antigen.

Despite its limitations for early diagnosis, the standard antigen test can be helpful in settings where virus is likely to be present but antibody is at low levels or absent. Hypogammaglobulinemic infants with HIV infection can have positive antigen tests.[49] Patients (other than infants born to HIV-infected mothers) with recently acquired infection can have detectable antigen levels during the seroconversion period between initial infection and prior to the production of antibody.[50]

The p24 antigen can be detected in serum from patients in the end stage of their disease process when anti-HIV p24 antibodies may disappear and levels of p24 antigen are high owing to increased virus replication. p24 antigen assays have been used to monitor the effect of antiviral therapy in patients with advanced disease[51] and as a prognostic marker for progression to AIDS.[52] The antigen assay is also used as the standard method to detect HIV growth in co-cultures.[53]

Studies have shown that modification of the standard p24 antigen assay by acidification of the sample to dissociate the immune complexes can increase the sensitivity of the assay to detect HIV infection in infants.[54–56] The immune complex dissociation p24 antigen assay has been shown to be highly sensitive and specific for diagnosis of HIV infection in infants, although samples obtained during the first week of life have been somewhat problematic in that false-positive and false-negative tests have been observed. Licensed assay kits using this methodology are not yet available in the United States, but ongoing clinical trials in infants of different ages and at different disease stages should establish the advantages and limitations of this assay for pediatric diagnosis of HIV infection in the near future.

HIV Culture

Virus culture is one of the most sensitive techniques for detecting HIV infection in infants and has been used extensively in research settings. Microcultures in 12-, 24-, and 96-well plates are equally sensitive and specific compared with standard flask methods, and they require far less blood.[57] Virus cultures may be difficult to perform, as they require considerable experience and special facilities to deal with biosafety precautions to prevent exposure of laboratory personnel. HIV culture is not useful as a rapid diagnostic test, as cultures typically take 7 to 28 days or more to complete.

Polymerase Chain Reaction

The polymerase chain reaction (PCR) is one of the most sensitive diagnostic techniques for detecting HIV infection. HIV is an RNA retrovirus that transcribes its RNA into DNA using the viral enzyme reverse transcriptase after entry into the human host cell. These DNA sequences, termed proviral DNA, can then integrate into the human cellular DNA. Because the amount of proviral DNA is small compared with the amount of human DNA and because the number of infected cells in the blood is small, direct detection of proviral DNA with standard molecular biologic techniques is not feasible. PCR is a method for amplifying proviral DNA as much as one million times or more to increase the probability of detection.

The PCR can be used for diagnosing HIV infection in infants by detecting proviral DNA in peripheral blood mononuclear cells (PBMCs). Studies evaluating the use of PCR for early diagnosis of HIV infection in infants have shown that approximately 30% to 50% of HIV-infected infants test positive around the time of birth.[58] This percentage increases to nearly 100% by 1 to 3 months of age. These sensitivities are comparable to those of virus culture.[8] In situ PCR techniques that detect the presence of HIV is specific cells or tissues[59] and quantitative PCR techniques to measure virus load have also been developed.

The PCR can also be used to measure active virus replication by detecting viral RNA in plasma.[58] Amplification of RNA virus sequences of HIV, as would be found in viral particles in plasma or from messenger RNA in infected cells, can be done by first making a DNA copy using a commercially available reverse transcriptase enzyme. RNA PCR is most useful for monitoring circulating virus and should not be used in place of DNA PCR for routine diagnostic purposes, as plasma viremia is not always detectable in HIV-infected persons. One study used RNA PCR to examine newborn plasma samples and found that 7 of 10 HIV-infected infants had detectable HIV-1 RNA at 4 to 9 weeks, whereas only one of these infants had detectable RNA at birth.[58]

In addition to diagnosing infected infants, both PCR and virus culture can be used to evaluate infants who lose maternal antibody. Several studies indicate that infants who lose maternal antibody and remain healthy are uninfected based on negative virus cultures and PCR.[46,47] Some laboratories have reported rare seroreverting children who occasionally test positive by PCR on a single specimen but negative on subsequent specimens.[46,47] The reasons these infants tested positive on one occasion are unclear. A mix-up in specimens or other laboratory error must be ruled out. Alternatively, transient PCR positivity may represent (1) true infection that has cleared, (2) a provirus blood level that is below the level detectable by the test, or (3) maternal blood cell contamination. Continued evaluation of these infants is important.

Because PCR detects the presence of HIV proviral DNA rather than antibody to the virus, it avoids the problem of persistent maternal antibody. PCR requires less than 1 ml of blood, an amount obtainable from a newborn. Unlike virus culture, which can take up to 4 weeks or longer to complete, PCR testing can be done in 1 to 2 days.

In Vitro Antibody Production Assays

Another technique for infant diagnosis is the in vitro antibody production assay (IVAP). This assay detects the presence of HIV-specific antibody-producing B lymphocytes in the infant, which indicates that the infant's immune system has been stimulated by HIV infection. IVAP assays require less time to complete than culture. The standard IVAP assays require 7 to 10 days to complete, and the Elispot (described below) can be done in 1 to 2 days.

Two methods for this technique have been described. With one method, PBMCs are separated from whole blood, carefully washed to remove plasma, placed in medium, and stimulated to produce antibody with either pokeweed mitogen[60] or Epstein-Barr virus.[61] HIV-sensitized B cells produce antibody that is released into the culture supernatant. HIV-specific antibody in culture supernatants can be detected using standard methods.

With an alternative method, called Elispot,[62] washed PBMCs are placed in wells with nitrocellulose membrane bottoms coated with HIV antigen. The cells are incubated, washed, and then treated sequentially with biotinylated anti-human IgG, horseradish-conjugated avidin, and enzyme substrate. The appearance of "spots" on the nitrocellulose membrane indicates the presence of HIV-specific antibody-producing lymphocytes in the patient.

Both types of assay for detecting antibody-producing cells have been successful in identifying HIV-infected infants, but there are some important limitations. False-positive tests in uninfected infants have been reported during the first 2 months of life. These spurious results might come from the detection of maternal B lymphocytes that are producing antibody but may not harbor the virus, although this possibility has not been carefully studied. Additionally, in the presence of abundant maternal IgG in the infant's serum, false-positive tests may result from maternal anti-HIV antibody that has adhered to infant's B cells in the culture. The Elispot method may be able to differentiate between this carryover of maternal antibodies from true HIV antibody-producing cells.[63] The adsorbed antibody yields diffuse pink background staining that can be readily distinguished from the intense red spots that result from true antibody-producing cells.

False-negative IVAP tests have been observed in HIV-infected persons with advanced immunodeficiency who cannot produce antibody or in those with latent infection who do not have active virus replication. Active virus replication is the stimulus for the replication of antibody-producing cells. Suppression of virus replication by immune mechanisms or by antiviral therapy could suppress the production of antibody-producing cells below the level of detection. In one study, 22 (63%) of 34 asymptomatic HIV-seropositive persons, 12 (75%) of 16 symptomatic patients, and 1 (8%) of 13 AZT-treated symptomatic patients were Elispot-positive.[62] Another study found that whereas some AZT-treated patients were IVAP-negative at 3 to 5 months after starting therapy most were positive when tested after more than 6 months of treatment.[61]

Factors Affecting the Performance of Diagnostic Assays

Several factors must be considered when choosing assays to diagnose perinatal HIV infection. Probably the most important factor is the age of the infant at the time of testing. Although the precise kinetics of virus replication and antibody production remain to be defined, several studies indicate that the virus load is probably lowest at the time of birth in most infants and increases during the first few months of life.[6,58] IgA antibody production appears also to be low during the first few months of life.[39,40] IVAP assays may yield false-positive results during the first 1 to 2 months of life (see previous discussion in IVAP section). Thus during the neonatal period most studies have shown that the highest percentage of infected infants are detected using sensitive virologic assays such as PCR or virus culture, and several assays can detect infection by 3 to 6 months of age.[64] Even the most sensitive assays can detect only about 50% or fewer infected infants around the time of birth. Several reasons might account for the low detection rate at birth: (1) infection has only recently been transmitted in the case of intrapartum transmission; (2) the virus load is low; or (3) virus is suppressed or sequestered in other tissues. Figure 16.1 summarizes the sensitivity of the various assays to detect HIV infection, by the age of the infant at the time of testing.

The stage of disease also affects the likelihood that a given assay can detect HIV infection in perinatally infected infants and in older children and adults. As the disease progresses, the virus load in terms of the plasma level and the percentage of infected cells increases,[65,66] and it has been reported that the phenotypic properties of the virus change from slow-growing to rapid-growing, syncytia-forming.[67] Assays that detect the virus, its proteins, or the viral genome are more likely to be positive

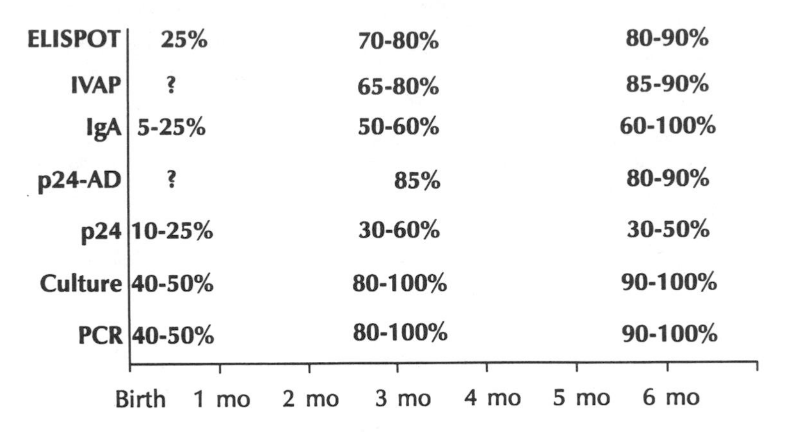

FIGURE 16.1. Sensitivity of various diagnostic tests by the age of the infant at the time of testing. Percentages represent the sensitivity of the diagnostic test to identify HIV-infected infants at a given age shown on the x-axis. In general, sensitivity is lowest at birth and increases thereafter. IVAP = in vitro antibody production assay; IgA = HIV-specific IgA assay; p24-AD = acid dissociation p24 antigen assay; PCR = polymerase chain reaction.

during later stages of the disease than during the earlier, asymptomatic stages in infants. Some severely immunosuppressed adults and children with end-stage disease may lose antibody, and some perinatally infected infants may be hypogammaglobulinemic and so do not produce antibody.

Some antiretroviral therapies may affect the results of diagnostic tests by lowering the virus load. Patients who tested p24 antigen-positive prior to therapy may become antigen-negative after therapy. Reducing the virus load decreases the stimulus for antibody production, and IVAP/Elispot assays may also become negative.[61,62]

Finally the timing of transmission may also affect the likelihood of early detection of infection in perinatally infected infants. Theoretically, infants infected in utero should be positive at birth, whereas those infected late in the pregnancy or during the intrapartum or postpartum periods may not become positive until some time after birth.

Practical Aspects of Diagnostic Assays

Several practical aspects of the various diagnostic assays must be considered, including costs, the technical difficulty of assay and sample preparation, the time required to complete the assay, commercial availability, and FDA licensure.[64] The routine use of certain assays varies depending on the sophistication of the laboratory and the available technology and resources. Many developing countries have additional considerations, such as availability of equipment, consistent electrical power, purified water, and trained laboratory personnel.

Table 16.1 summarizes several of these practical aspects. Unfortunately, the most sensitive and specific assays for the earliest diagnosis (e.g., virus culture) are also the most technically difficult and costly. Although the current availability of PCR is limited, test kits are under development. These kits have greatly simplified sample preparation and require only 0.1 to 0.5 ml of blood, which is critical for testing infants for whom obtaining blood samples is often problematic. The cost of these kits, however, is prohibitively expensive for most developing countries. Other, less technically difficult assays, such as the serologic tests, may prove to be more practical for these settings.

The type of sample and sample preparation must also be taken into consideration. Virus culture and the currently used methods for PCR, IVAP, and Elispot require that mononuclear cells be separated from whole blood, which is more difficult and time-consuming than separating serum or plasma for serologic tests, such as p24 antigen or antibody EIA. PCR kits under development offer the advantage of using whole-blood samples. Although venipuncture samples are often difficult to obtain and their volume is limited in newborns, cord blood samples are not generally recommended for use in diagnosis unless careful attention is paid during collection to prevent maternal blood contamination[68]—often an impossibility in hospital delivery room settings.

TABLE 16.1. Practical aspects to be considered when choosing diagnostic tests for infants born to HIV-infected mothers.

Assay	Overall sensitivity/ specificity[a]	Technical difficulty	Time to result (days)	Specimen required	Cost[b]
Culture	++++/++++	High	15–35	PBMCs	$250
PCR	++++/++++	High[c]	1–2	PBMCs[c]	$175
p24 Ag[d]	++/+++[e]	Low	1–2	Serum/plasma	$25
IgA	+++/+++[e]	Low	1–2	Serum/plasma	$10–50
IVAP	+++/++[e]	Moderate	/ 10	PBMCs	$50–75
Elispot	+++/+++	Moderate	1–2	PBMCs	$20–30

Source: Report of a Consensus Workshop,[64] by permission.

PBMCs = peripheral blood mononuclear cells; PCR = polymerase chain reaction; IVAP = in vitro antibody production assay; Ag = antigen assay.

[a] Sensitivity of the assay for detecting HIV-infected infants, specificity of assay in giving negative results for uninfected infants, both taking into account the reliability of the test at different ages. ++++ = excellent; ++ = good; ++ = moderate; + – poor.

[b] Costs reflect commercial costs and are approximations.

[c] Current methods for PCR are technically difficult and require PBMCs. Methods under development are of moderate technical difficulty and can be done using whole blood.

[d] Refers to standard p24 antigen assay that does not use acid hydrolysis.

[e] Less specific in infants under 1 to 2 months of age.

A diagnostic algorithm to be used as a guide for choosing diagnostic tests is shown in Figure 16.2. This algorithm takes into account the patient's age at the time of testing, whether the child is symptomatic or asymptomatic, and the predictive value of the assay at a given age. The algorithm assumes that the infants are not being breast-fed (no ongoing exposure after birth) and that the health care provider would choose the most cost-efficient assay whenever possible. Tests with greater sensitivity (but also greater cost), such as PCR or virus culture, would be used to clarify negative or inconclusive results derived from less expensive assays.

For children 0 to 6 months of age, PCR and virus culture are recommended because of the greater sensitivity of these tests at this age. HIV IgA or p24 antigen assays are another option, particularly in children 3 to 6 months of age, but these tests are less likely to be positive in infected infants under 3 to 6 months of age as discussed earlier. The IVAP assays should not be used for diagnosis in children under 2 months of age unless technical modifications under study are found to eliminate the problem of maternal antibody adherence.

For children 6 to 18 months of age an initial standard antibody test should be done to determine whether seroreversion (loss of maternal

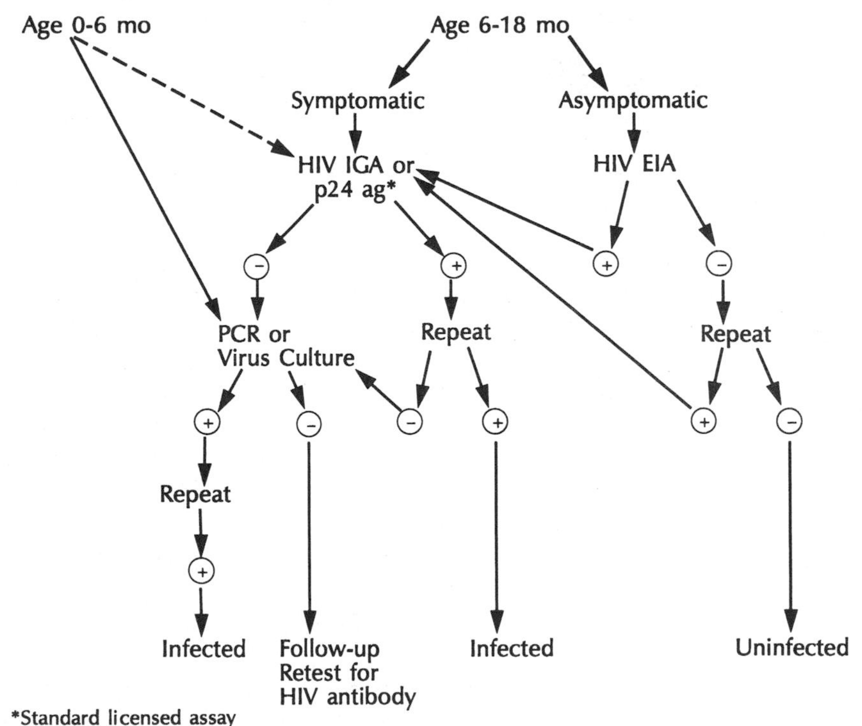

FIGURE 16.2. Diagnostic testing algorithm for infants born to HIV-infected mothers.

antibody) has already occurred. If the antibody test is negative on initial and repeat testing, this child should be considered uninfected unless additional follow-up reveals other signs or symptoms of HIV disease. Children 6 to 18 months with positive antibody tests should be tested with other assays as indicated in the algorithm. Until the reliability of these assays for early diagnosis is firmly established, the tests should be repeated; and when possible another test should be used to confirm the diagnosis.

References

1. CDC. Update: acquired immunodeficiency syndrome—United States, 1992. MMWR 1993;42:547–557
2. Gellert GA, Durfee MJ, Berkowitz CD, Higgens KV, Tubiolo VC. Situational and sociodemographic characteristics of children infected with human immunodeficiency virus from pediatric sexual abuse. Pediatrics 1993;91:39–44
3. CDC. HIV infection in two brothers receiving intravenous therapy for hemophilia. MMWR 1992;41:228–231
4. Lewis SH, Reynolds-Kohler C, Fox HE, Nelson JA. HIV-1 in trophoblastic and villous Hofbauer cells, and haematological precursors in eight-week fetuses. Lancet 1990;335:565–568
5. Rogers MF, Ou CY, Rayfield M, et al. Use of the polymerase chain reaction for early detection of the proviral sequences of human immunodeficiency virus in infants born to seropositive mothers. N Engl J Med 1989;320:1649–1654
6. Burgard M, Mayaux M-J, Blanche S, et al. The use of viral culture and p24 antigen testing to diagnose human immunodeficiency virus infection in neonates. N Engl J Med 1992;327:1192–1197
7. Goedert JJ, Duliege AM, Amos CI, et al. High risk of HIV-1 infection for first-born twins. Lancet 1991;338:1471–1475
8. Villari P, Spino C, Chalmers TC, Lau J, Sacks HS. Cesarean section to reduce perinatal transmission of human immunodeficiency virus. J Curr Clin Trials [serial online] 1993;2:doc. no. 74
9. Blanche S, Rouzioux C, Moscato MG, et al. A prospective study of infants born to women seropositive for human immunodeficiency virus type 1. N Engl J Med 1989;320:1643–1648
10. Oxtoby MO. Human immunodeficiency virus and other viruses in human milk: placing the issues in broader perspective. Pediatr Infect Dis J 1988; 7:825–835
11. Hira SK, Mangrola UG, Mwale C, et al. Apparent vertical transmission of human immunodeficiency virus type 1 by breast-feeding in Zambia. J Pediatr 1990;117:421–424
12. Van de Perre P, Simonon A, Msellati P, et al. Postnatal transmission of human immunodeficiency virus type 1 from mother to infant. N Engl J Med 1991;325:593–598
13. Thiry L, Sprecher-Goldberger S, Jonckheer T, et al. Isolation of AIDS virus from cell-free breast milk of three healthy virus carriers. Lancet 1985;2: 891–892

14. Dunn DT, Newell ML, Ades AE, Peckham CS. Risk of human immunodeficiency virus type 1 transmission through breastfeeding. Lancet 1992; 340:585–588
15. CDC. Recommendations for assisting in the prevention of the perinatal transmission of human T-lymphotropic virus type III/lymphadenopathy-associated virus and acquired immunodeficiency syndrome. MMWR 1985;34: 721–726, 731–732
16. Ryder RW, Nsa W, Hassig SE, et al. Perinatal transmission of the human immunodeficiency virus type 1 to infants of seropositive women in Zaire. N Engl J Med 1989;320:1637–1642
17. European Collaborative Study. Risk factors for mother-to-child transmission of HIV-1. Lancet 1992;339:1007–1012
18. Gabiano C, Tovo P-A, de Martino M, et al. Mother-to-child transmission of human immunodeficiency virus type 1: risk of infection and correlates of transmission. Pediatrics 1992;90:369–374
19. St. Louis ME, Kamenga M, Brown C, et al. Risk for perinatal HIV-1 transmission according to maternal immunologic, virologic, and placental factors. JAMA 1993;269:2853–2859
20. Nair P, Alger L, Hines S, et al. Maternal and neonatal characteristics associated with HIV infection in infants of seropositive women. J Acquir Immune Defic Syndr 1993;6:298–302
21. Goedert JJ, Mendez H, Drummond JE, et al. Mother-to-infant transmission of human immunodeficiency virus type 1: association with prematurity or low anti-gp120. Lancet 1989;2:1352–1354
22. Rossi P, Moschese V, Broliden PA, et al. Presence of maternal antibodies to human immunodeficiency virus 1 envelope glycoprotein gp120 epitopes correlates with the uninfected status of children born to seropositive mothers. Proc Natl Acad Sci USA 1989;86:8055–8058
23. Devash Y, Calvelli TA, Wood DG, Reagan KJ, Rubinstein A. Vertical transmission of human immunodeficiency virus is correlated with the absence of high affinity/avidity maternal antibodies to the gp120 principal neutralizing domain. Proc Natl Acad Sci USA 1990;87:3445–3449
24. Parekh BS, Shaffer N, Pau C-P, et al. Lack of correlation between maternal antibodies to V3 loop peptides of gp120 and perinatal HIV-1 transmission. AIDS 1991;5:1179–1184
25. Hahn RA, Onorato IM, Jones TS, Dougherty J. Prevalence of HIV infection among intravenous drug users in the United States. JAMA 1989;261:2677–2684
26. Donegan E, Stuart M, Niland JC, et al. Infection with human immunodeficiency virus type 1 (HIV-1) among recipients of antibody-positive blood donations. Ann Intern Med 1990;113:733–739
27. Ward JW, Bush TJ, Perkins HA, et al. The natural history of transfusion-associated infection with human immunodeficiency virus. N Engl J Med 1989;321:947–952
28. Jones DJ, Byers R, Bush T, Rogers MF. The epidemiology of transfusion-associated AIDS in children in the United States, 1981–1988. Pediatrics 1992;89:123–127
29. Jason J, McDougal S, Holman RC, et al. Human T-lymphotropic retrovirus type III/lymphadenopathy-associated virus antibody. JAMA 1985;253:3409–3415

30. Jackson JB, Sannerud KJ, Hopsicker JS, et al. Hemophiliacs with HIV antibody are actively infected. JAMA 1988;260:2236–2239
31. Petersen LR, Simonds RJ, Koistenen J. HIV transmission through blood, tissues, and organs. AIDS 1993;7(suppl 1):S99–S107
32. Bernstein LJ, Krieger BZ, Novick B, Sicklick MJ, Rubinstein A. Bacterial infection in the acquired immune deficiency syndrome of children. Pediatr Infect Dis J 1985;4:472–475
33. European Collaborative Study. Age-related standards for T-lymphocyte subsets based on uninfected children born to human immunodeficiency virus 1-infected women. Pediatr Infect Dis J 1992;11:1018–1026
34. Centers for Disease Control. Guidelines for prophylaxis against Pneumocystis carinii pneumonia for children infected with human immunodeficiency virus. MMWR 1991;RR-2;40:1–13
35. Working Group on Antiretroviral Therapy, National Pediatric HIV Resource Center. Antiretroviral therapy and medical management of the human immunodeficiency virus-infected child. Pediatr Infect Dis J 1993;12:513–522
36. Hessol NA, Lifson AR, O'Malley PM, et al. Prevalence, incidence, and progression of human immunodeficiency virus infection in homosexual and bisexual men in hepatitis B vaccine trials; 1978–1988. Am J Epidemiol 1989;130:1167–1175
37. Johnson JP, Nair P, Hines SE, et al. Natural history and serologic diagnosis of infants born to human immunodeficiency virus-infected women. Am J Dis Child 1989;143:1147–1153
38. Weiblen BJ, Schumacher RT, Hoff R. Detection of IgM and IgA HIV antibodies after removal of IgG with recombinant protein G. J Immunol Methods 1990;126:199–204
39. Quinn TC, Kline RL, Halsey N, et al. Early diagnosis of perinatal HIV infection by detection of viral-specific IgA antibodies. JAMA 1991;266: 3439–3442
40. Landesman S, Weiblen B, Mendez H, et al. Clinical utility of HIV-IgA immunoblot assay in the early diagnosis of perinatal HIV infection. JAMA 1991;266:3443–3446
41. Ryder RW, Hassig SE. The epidemiology of perinatal transmission of HIV. AIDS 1988;2:583–589
42. Centers for Disease Control. Classification system for human immunodeficiency virus (HIV) infection in children under 13 years of age. MMWR 1987;36:225–236
43. Italian Multicentre Study. Epidemiology, clinical features, and prognostic factors of pediatric HIV infection. Lancet 1988;2:1043–1045
44. European Collaborative Study. Mother-to-child transmission of HIV infection. Lancet 1988;2:1039–1042
45. Aiuti F, Luzi G, Mezzaroma I, Scano G, Papetti C. Delayed appearance of HIV infection in children. Lancet 1987;2:858
46. Jones DS, Abrams E, Ou C-Y, et al. The lack of detectable human immunodeficiency virus (HIV) infection in antibody-negative children born to HIV-infected mothers. Pediatr Infect Dis J 1993;12:222–227
47. Wiznia A, Conroy J, Liu HK, Nozyce M. Virus isolation, PCR, and neurodevelopmental delay in children who are HIV seroreverters (P-3) [abstract ThC 1578]. Presented at the VIII International Conference on AIDS, Amsterdam, July 1992

48. Borkowsky W, Krasinski K, Paul D, et al. Human immunodeficiency virus type 1 antigenemia in children. J Pediatr 1989;114:940–945
49. Borkowsky W, Krasinski K, Paul D, et al. Human-immunodeficiency-virus infections in infants negative for anti-HIV by enzyme-linked immunoassay. Lancet 1987;1:1168–1171
50. Ward JW, Holmberg SD, Allen JR, et al. Transmission of human immunodeficiency virus (HIV) by blood transfusions screened as negative for HIV antibody. N Engl J Med 1988;318:473–478
51. Chaisson RE, Allain J, Volberding PA. Significant changes in HIV antigen level in the serum of patients treated with azidothymidine. N Engl J Med 1986;315:1610–1611
52. Fahey JL, Taylor JMG, Detels R, et al. The prognostic value of cellular and serologic markers in infection with human immunodeficiency virus type 1. N Engl J Med 1990;322:166–172
53. Feorino PM, Forrester B, Schable C, Warfield D, Schochetman G. Comparison of antigen assay and reverse transcriptase assay for detecting human immunodeficiency virus in culture. J Clin Microbiol 1987;25:2344–2346
54. Palomba E, Gay V, de Martino M, et al. Early diagnosis of human immunodeficiency virus infection in infants by detection of free and complexed p24 antigen. J Infect Dis 1992;165:394–395
55. Lee F, Nesheim S, Sawyer M, Slade B, Nahmias A. Early diagnosis of HIV infection in infants by detection of p24 antigen in plasma specimens after acid hydrolysis to dissociate immune complexes [abstract PoB 3702]. Presented at the VIII International Conference on AIDS, Amsterdam, July 1992
56. Miles SA, Balden E, Magpantay L, et al. Rapid serologic testing with immune-complex-dissociated HIV p24 antigen for early detection of HIV infection in neonates. N Engl J Med 1993;328:297–302
57. Alimenti A, Luzuriago K, Stechenberg B, et al. Quantitation of HIV-1 in the blood of vertically infected infants and children. J Pediatr 1991;266:3443–3446
58. Krivine A, Firtion G, Cao L, et al. HIV replication during the first weeks of life. Lancet 1992;339:1187–1189
59. Bagasra O, Hauptman SP, Lischner HW, Sachs M, Pomerantz RJ. Detection of human immunodeficiency virus type 1 provirus in mononuclear cells by in situ polymerase chain reaction. N Engl J Med 1992;326:1385–1391
60. Amadori A, De Rossi A, Giaquinto C, et al. In-vitro production of HIV-specific antibody in children at risk of AIDS. Lancet 1988;1:852–854
61. Pahwa S, Chirmule N, Leombruno C, et al. In vitro synthesis of human immunodeficiency virus-specific antibodies in peripheral blood lymphocytes of infants. Proc Natl Acad Sci USA 1989;86:7532–7536
62. Lee FK, Nahmias AJ, Lowery S, et al. Elispot: a new approach to studying the dynamics of virus-immune system interaction for diagnosis and monitoring of HIV infection. AIDS Res Hum Retroviruses 1989;5:517–523
63. Nesheim S, Lee F, Sawyer M, et al. Diagnosis of human immunodeficiency virus infection by enzyme-linked immunospot assays in a prospectively followed cohort of infants of human immunodeficiency virus-infected women. Pediatr Infect Dis J 1992;11:635–639
64. Early diagnosis of HIV infection in infants: report of a consensus workshop, Siena, Italy, January 17–18, 1992. J Acquir Immune Defic Syndr 1992;5: 1169–1178

65. Ho DD, Moudgil T, Alam M. Quantitation of human immunodeficiency virus type 1 in the blood of infected persons. N Engl J Med 1989;321:1621–1625
66. Alimenti A, O'Neill, Sullivan JL, Luzuriaga K. Diagnosis of vertical human immunodeficiency virus type 1 infection by whole blood culture. J Infect Dis 1992;166:1146–1148
67. Tersmette M, Gruters RA, de Wolf F, et al. Evidence for a role of virulent human immunodeficiency virus (HIV) variants in the pathogenesis of acquired immunodeficiency syndrome: studies on sequential HIV isolates. J Virol 1989;63:2118–2125
68. Bryson YJ, Luzuriaga K, Wara D. Proposed definitions for in utero versus intrapartum transmission of HIV-1. N Engl J Med 1992;327:1246–1247

17
Molecular Epidemiology of AIDS

Barbara H. Bowman and Thomas J. White

Determining the possible modes of transmission of the human immunodeficiency virus (HIV) is a major public health concern. In cases of putative transmission by a novel means, it is essential to determine whether such transmission really did occur or whether the infection arose through a different, perhaps more commonly known source. Understanding the overall history and structure of the epidemic may be of value for predicting its course and developing public health policy.

Molecular sequence data can be used to complement epidemiologic information in transmission studies. This is possible because HIV, like any genetic entity, evolves over time. Variants of the virus in two individuals with related infections (e.g., the donor and recipient of an HIV-infected unit of blood) have much more similar DNA or RNA sequences than do variants from individuals whose infections are not related.

The ability to determine with confidence the relatedness—or nonrelatedness—of two HIV variants can also help to answer questions important to understanding HIV infection in general. For example, if two very different variants are found within a single individual, can we be sure they represent a case of superinfection, that is, that a second infection has occurred from a different source than from that of the first infection? So far, such a case has not been documented for HIV-1. More subtly, when two variants from one individual look similar to two variants in another person with a related infection, is it a case in which two lineages of HIV have successfully infected the second person, or has the same sequence simply arisen from a common ancestor by chance in both individuals?

The methods of molecular epidemiology are based on our understanding of the evolution of the acquired immunodeficiency syndrome (AIDS) virus at the nucleic acid sequence level. We propose a model for how that evolution may proceed and then describe how knowledge of HIV evolution can be used to investigate HIV transmission and to determine other information about the nature of the epidemic. The details of molecular evolutionary analysis are beyond the scope of this chapter and have been ably covered in recent reviews.[1–3]

Evolution of HIV at the Molecular Level

As is true for any genomic nucleotide sequence, HIV evolves over time by accruing nucleotide substitutions as well as insertions and deletions of bases. HIV acquires these mutations at a faster rate than do most nucleic acid sequences, but otherwise the process is similar. Generally, the greater the amount of time that has passed, the more mutations will have accrued.

A viral genome can be thought of as having "offspring" during the course of an infection. These offspring are new viral genomes created when the "parent" virus is copied. Each offspring of the parent virus can then become the parent for a new generation of offspring, which in turn become parents themselves, and so on. The parent and offspring viruses are linked in an ancestral relationship much like a "family tree." Figure 17.1 is a schematic illustration of this descent from one viral generation to another, beginning with a single viral genotype infecting an individual.

Because errors are often introduced when the viral genome is copied, each new genome produced may be identical to its parent or may contain mutations relative to its parent. Furthermore, viruses that share the same parent may differ from each other somewhat, as any mutations acquired by one offspring are independent of mutations acquired by the others. Still, viruses that share the same parent may well be identical or nearly so. Thus in Figure 17.1, HIV variants A^1 and A^2 will differ little.

The second generation of offspring usually differ from each other more than does the first; they have begun with parents that are already slightly different from each other and give rise to progeny each of which has acquired its own, independent set of mutations. Thus over time and over many viral generations, HIV variants in one individual may come to differ considerably from each other, even though they shared a common ancestor at the beginning of an infection. Thus in Figure 17.1, variants B^1 and C, which are descended from A^1 and A^2, respectively, may differ by as much as a few percent, if the time passed has been several years. At the same time, there are variants within the individual that are recently descended from the same parent (such as B^1 and B^2), and these variants will be similar to each other.

In some regions of the genome, notable among them the "V3 loop,"* selection may cause the nucleic acid sequences of distantly related variants to become similar to each other.[4] This pattern is an exception to the normal one of random divergence in separate lineages. Such convergent substitutions can also be expected to occur at the same position by chance on occasion in variants infecting different people; but because their occurrence is random, they should not create a consistent pattern that

*The term V3 loop is used here to describe the approximately 36 amino acid coding region which makes up roughly the center third of the V3 variable region.

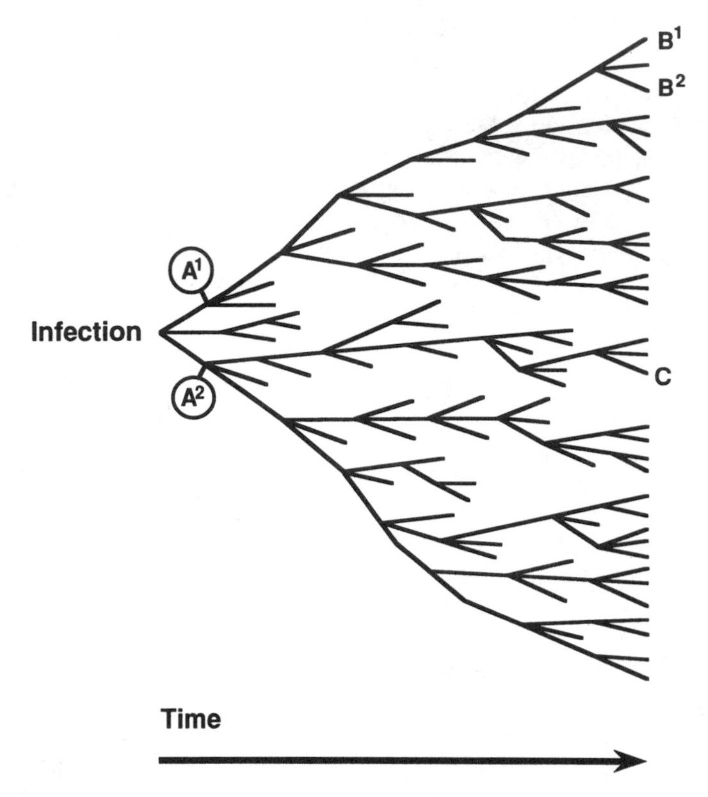

FIGURE 17.1. Evolutionary tree illustrating a model for the descent of a single variant of HIV infecting one individual. The progression of time is shown from left to right. A^1 and A^2 represent two descendants of the infecting virus. B^1, B^2, and C represent later descendants of those lineages. Branches that terminate represent lineages that did not produce offspring.

would cause any two particular lineages to appear related when they are not. Random convergence is more likely in a genomic region that is accruing mutations at a high rate. For this reason, rapidly evolving regions may exhibit more "noise" during data analysis when comparing epidemiologically unlinked lineages, such as those from individuals typically used as "controls."

Evolution of Genomic Regions

The rate at which base substitutions are incorporated by the reverse transcriptase that copies the RNA strand into DNA is roughly 1 every 1700 to 4000 bases,[5] or an average of perhaps two to three substitutions every time the approximately 9 kb genome is reverse-transcribed.

Some of these substitutions will prove deleterious or lethal: They will make the viral variant defective or unable to produce viable offspring. For example, substitutions that cause amino acid replacements that impair binding of the virus to CD4 cells or interfere with cDNA synthesis by the reverse transcriptase create variants that are unable to support replication and will not be passed on to future virus generations. Note that in Figure 17.1 a random selection of viruses has been shown as becoming extinct (failing to produce offspring) during the course of infection. Such extinction could be for cause, such as acquiring a lethal base substitution, insertion, or deletion, or by selective neutralization by the immune system; or it could be simply the result of random extinction of lineages.

On the other hand, some substitutions may prove advantageous, such as those in regions of the *env* gene that allow evasion of the immune response. Viruses that have by chance acquired such mutations are more likely to produce successful offspring. Many mutations have no significant effect on the vitality of the virus and are therefore considered selectively "neutral."

The observed result of this panoply of deleterious, advantageous, and neutral mutations is that separate regions of the viral genome evolve at different rates. When HIV variants are sampled from different individuals, some regions of the genome may be very similar or identical, as substitutions in these regions may produce defective viruses. Yet regions in which mutations are neutral or even advantageous may be so different as to be unrecognizable. This heterogeneity in rates of evolution is also evident when viral variants are sampled from within a single patient, though the overall differences are smaller than those between randomly chosen individuals.

Evolutionary Basis of Molecular Epidemiology

The same processes of evolution are at work on viral variants transmitted from one individual to another. In Figure 17.2, transmission from the first individual to the second has occurred at time 2. This schematic drawing shows the descent of a single viral variant from person 1 infecting person 2; that infecting variant is a descendent of variant "D" in person 1. (In this scenario, we assume that variant D has had multiple offspring.)

Shortly after this variant infects person 2 (immediately after time 2), its descendants will be very similar to each other. Importantly, all of the descendants of D in person 2 will still be very similar to other descendants of D in person 1. Variants in person 2 are also moderately close relatives of the other variants (such as E) in person 1. Even after some time has elapsed (time 3) the viruses in the two individuals will still be closer than either is to a nonlinked infection. It is for this reason that it is possible to determine whether one individual (or another source of infection, such as a unit of blood) is the source of infection in a second individual.

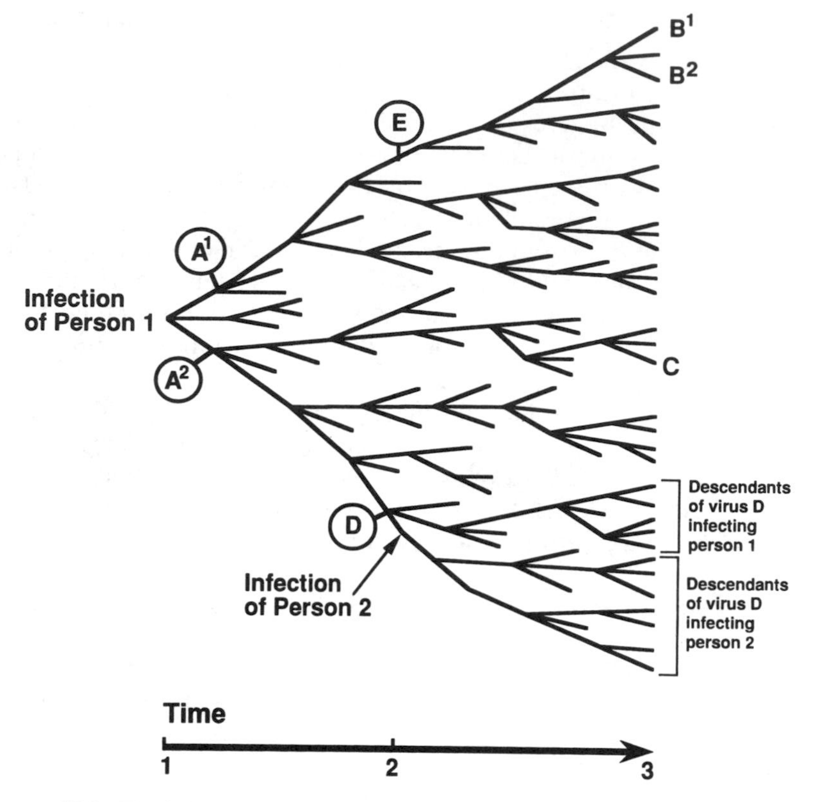

FIGURE 17.2. Evolutionary tree illustrating a model for transmission of one HIV variant from one individual to a second. At time 2, an immediate descendant of variant D infects a second individual, whereas other descendants of D remain in the first individual. Immediately after infection (right after time 2) the HIV variants in the second person all share a very recent common ancestor with the descendants in the first person. They share a recent but earlier ancestor (at time 1) with variant E. Thus choosing variant E from the first person for comparison with variants in person 2 would significantly overestimate the length of time since infection. At time 3, despite the increasing number of substitutions, insertions, and deletions in the viruses of each of the two individuals, the evolutionary relation showing their epidemiologic linkage should still be discernible.

Methods

Sequences of HIV variants recovered from individuals contain information about the evolutionary history of the virus in those individuals; collectively, HIV sequences contain information about the evolutionary history of the HIV epidemic, from person-to-person transmission to a global overview. This information can be retrieved and interpreted; the outline of this procedure and the results obtainable from it constitute the remainder of this chapter.

In order to obtain enough HIV DNA for sequencing (it is also possible to perform this protocol starting with RNA), it is necessary to amplify the proviral DNA present in the patient sample using the polymerase chain reaction (PCR). The product of the amplification reaction can be sequenced directly to produce a consensus sequence (a compilation of the majority base at each position) reflective of the most abundant variant present, or it can be cloned and individual clones sequenced. Direct sequencing requires that a significant fraction (25–33%) of the amplified products contain the variant sequence.[6] The choice between these two alternatives depends on the question being asked and is discussed below. The resulting sequence data are evaluated in several ways.

The earliest means of estimating the relatedness or nonrelatedness of two infections was simply to compare the percent sequence difference in a selected genetic region among the variants under study, and between them and sequences that were expected to be epidemiologically unlinked (e.g., those from the databases). It was shown by 1990 in the United States that sequences from epidemiologically linked infections ranged from 3.5% to 5.0% different in the *env* region, whereas sequences from unlinked infections ranged from 8% to 20% different (G. Schochetman, personal communication). These numbers are approximations, however; and it would be difficult to exclude the relatedness of two sequences based simply on a high percent divergence between them: Over time even HIV sequences from related infections continue to diverge, and there is currently no practical way to set an upper limit on how different they can become. In one instance of known perinatal transmission from mother to child, HIV sequences sampled from the two individuals 11 years after transmission were as different, by this criterion, as each was from selected database sequences, though evolutionary tree analysis confirmed their relatedness.[7]

It should be possible, however, to determine whether transmission has taken place within a recent, defined period. For example, an estimate for substitution rate within the *env* gene V1/V2 region is 0.4% to 1.6% per year.[8] If transmission was alleged to have taken place within the year (e.g., if the putative recipient had only recently seroconverted), a high degree of divergence in the V1/V2 region (e.g., 10%) would rule out a proposed recent source for the infection. Similarly, knowing the amount of divergence between variants from two individuals within a certain defined genomic region of the virus, it is possible to set an approximate upper limit on how long ago transmission could have occurred.

Analysis of Clones

To place a limit on the time since transmission, it is necessary to know the *minimum* difference between viral variants in the two individuals. To be certain of finding the most similar sequences, HIV sequences from at least the putative donor (or from both patient samples if the direction of transmission is not known) must be cloned and individually sequenced, rather than sequenced directly from amplified DNA. The reason is clear when examining Figure 17.2. If it happens that descendants of variant D predominate in patient 1, the direct, consensus sequence inferred from that patient sample, when compared with any sequence from patient 2, will accurately reflect the most recent common ancestry of the variants in the two patients—and thus the transmission time—at time 2. However, should another variant be more abundant in patient 1, such as B^2 or C, the common ancestry of the viral sequences obtained by direct sequencing from the two individuals would be near time 1 and would overestimate the difference between viruses from the two patients, therefore overestimating the time since the transmission occurred. By sequencing enough clones from the donor one can increase the likelihood of finding the sequence most closely related to that of the recipient and therefore of being able to obtain a minimum estimate of the time since transmission. It does not matter which variant is chosen from the recipient, as any variant chosen should give the same divergence date as does any other when compared to any chosen sequence from the donor.

Molecular Phylogeny in Transmission Cases

The most theoretically sound method of determining relatedness of viral variants in cases where transmission may have occurred is to use the techniques of molecular phylogenetic reconstruction to see if the sequences in question form a *monophyletic* group, that is, if they share a common ancestor to the exclusion of all unrelated sequences. Truly related infections do have this property: The sequences from the individuals must be each other's closest relatives. The only sequences that would be expected to have descended from within such a monophyletic group would be those from other related infections. These statements reflect the actual evolutionary ancestry of the sequences and are true. *Proving* the existence of such a monophyletic group with a high degree of confidence, on the other hand, has sometimes been difficult. We discuss below some considerations that may be of value in such analyses and what problems and solutions may exist.

Convergence

Convergence is the term used for the occurrence of the same base substitution independently on unrelated lineages. Such a mutation, taken by itself, would (wrongly) suggest a close relationship between the two lineages on which it occurred. Ordinarily in molecular phylogenetic studies, convergence is assumed to have a minor effect on the analysis because substitutions are assumed to be fixed at random, and independently, in separate lineages; the result is that it is unlikely that many convergent substitutions would occur, all of which suggested the same, incorrect relation among sequences.

However, cases are known where convergence at the molecular level has occurred because of selection pressure favoring a particular sequence at several positions,[9] and it may also be the case in certain areas of the HIV genome. The V3 loop has been proposed to be a region undergoing convergent selection.[4] For example, positively charged amino acids are predominant in certain positions of variants that induce syncytium formation.[10] In such regions of the virus, convergent substitutions might be present in sufficient quantity to give the illusion that two variants are related when they are not. This situation could occur if the portions under selection accounted for a major proportion of the substitutions among the sequences being analyzed. It is more likely that convergence, either the random, intermittent kind or the sort due to selection, will simply make parts of the data set contradict the true phylogeny, thereby eroding the strength of support for any phylogenetic conclusion that might have been drawn from the data. Such "noise" is more common in regions of the genome that are accruing mutations rapidly than in regions that are less divergent. An additional difficulty arises when certain lineages are acquiring substitutions at a much greater rate than others. Convergence is much more likely on such lineages.[11]

A conflict arises here: For transmission studies, one must use a rapidly evolving region that provides sufficient variability that distant relatives are clearly different, whereas close relatives are recognizably similar. On the other hand, a rapidly evolving region may be so noisy that it is difficult to arrive at a well supported phylogenetic conclusion. If this proves to be the case, it may be necessary to examine more than one genomic region in order to pick the one most appropriate for the study. A slowly evolving region, such as the p17 region of the *gag* gene, may aid in determining the deeper evolutionary branching orders among HIV sequences. That is, the relationships among more distant relatives can be described. At the other end of the spectrum, the V1/V2 variable region of the *env* gene is an extremely rapidly evolving area not thought to undergo convergent selection, and it may be useful for discriminating among recently related variants. It is also important when sequencing for phylogenetic purposes to acquire sequences of sufficient length that there is substantial phylogenetic information available in case an unrecognized convergent area is part of the data set.

Methods of Molecular Phylogenetic Analysis: Cautionary Tales

Phylogenetic analysis of nucleic acid or protein sequence data is performed using one of two basic analytic methods (for review, see Hillis et al.[1] and Swofford and Olsen[3]). The first of these methods is genetic distance, in which each viral sequence in the data set is compared with each other sequence, and a single number representing their "differentness" is determined; this is called the "pairwise distance." Tree-drawing algorithms attempt to optimize the branching order and branch lengths on the tree according to the criteria of the particular program. For example, some programs attempt to minimize the total length of the tree, that is, to make as small as possible the total of all substitutions assigned to all branches of the tree to explain the current differences among the sequences. A variety of algorithms have been developed, most of which are highly reliable in simulation studies where the data set consists of sequences evolving at similar rates and having a relatively balanced base composition. Algorithms that assume a molecular clock, such as the unweighted pair-group method of analysis (UPGMA), are often subject to error and should be avoided.

All tree-drawing algorithms make assumptions, however, and the degree to which these assumptions are violated may influence the reliability of the answers they provide. For example, a strongly biased tendency for one particular base substitution to occur in preference to others violates the assumptions of most algorithms, unless specific corrections are applied.

A second common tree-drawing method is known as "parsimony" (for review, see Stewart[2]). Rather than combining all the differences among pairs of sequences into single numbers, parsimony makes use of the *pattern* of bases that is found at each sequence position throughout the region under study. Considering each sequence position independently, it is assumed that all variants having the same base at that position are closest relatives, whereas those sharing a different base are another set of close relatives. Parsimony algorithms act by evaluating all sequence positions that are informative (i.e., where at least two sequences share one base and at least two others share a different base) and creating the branching order that most often groups the sequences having like bases. In practice, it is done by choosing the branching order that requires the fewest number of substitutions (i.e., the most parsimonious) to explain the actual sequence data observed in the HIV variants. Most of the informative sequence positions must be consistent with this branching order.

Testing for Strength of Support

When performing and critically reviewing genetic distance and parsimony analyses, it is necessary to keep in mind that the "best" tree chosen by the optimality criterion of the method may be poorly supported by the data and may have no strong claim to being correct. It is necessary to evaluate the strength of support for particular branching orders, notably the grouping of purportedly related HIV variants to the exclusion of others, rather than taking the branching order at face value. Any program will give *a* tree; whether that tree has any claim to validity is determined by subsequent analysis.

The problem is that in many cases the sequence data are so divergent that any phylogenetic signal that might have been present has been obscured by noise. A second difficulty lies at the other end of the spectrum: when so few sequence positions are variable there is no usable

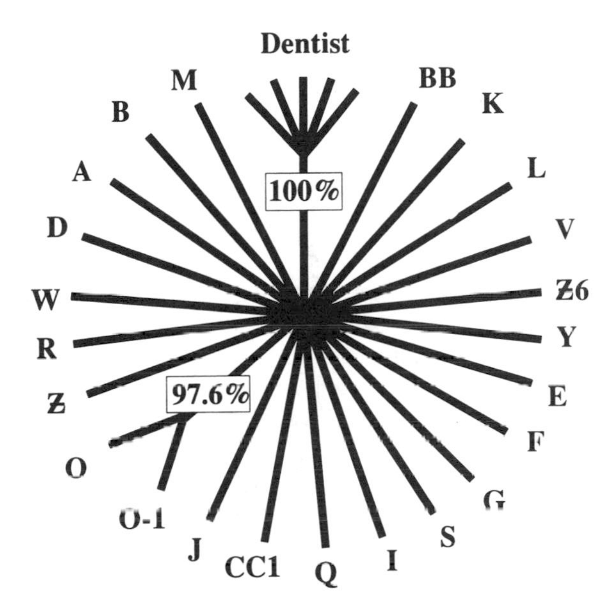

FIGURE 17.3. Evolutionary tree with unsupported branches collapsed to a multifurcation. This tree shows the relationships among HIV sequences from a dentist and 20 of his patients, plus 2 other individuals with possible epidemiologic linkage and a sequence from the databases (Z6). The tree was generated using the neighborjoining algorithm (a genetic distance method). The percentage of bootstrap replications in which two sets of branches were grouped is shown. No other grouping appeared in 70% or more of the replications. Thus the only sequences shown to be related are O and O-1, who were sexual partners, and five sequences from different HIV-1 variants infecting the dentist. There was no evidence of relatedness between the dentist's and any patient sequences or between sequences from any patients. Nonsupported branching orders were collapsed to a multifurcation to emphasize that no relations other than the two shown could be inferred from the data.

information present. In both cases one or more branching orders are, by the optimality or other criterion of the tree-drawing algorithm used, the "best"; but this best is meaningless.

This is true of both entire trees and parts of trees. It is possible, for example, that the data strongly support one particular grouping of sequences to the exclusion of others, but that no other branching order in the tree is well defined. For example, HIV sequences from a donor and several recipients of a contaminated unit of blood might be strongly grouped together to the exclusion of control sequences in the analysis, but the particular branching order among the sequences within that group, or among the controls, may not be determinable. So long as no conclusions are drawn from unsupported branching orders, their existence is not a problem. It is recommended that trees be drawn in which branches with low support are collapsed to multifurcations (Fig. 17.3), rather than present a bifurcating tree that suggests branching orders that are not supported.

Determining Strength of Support

The most common method used to determine strength of support for particular branches within the tree is "bootstrapping."[11] Bootstrapping algorithms take the original data set and resample from it multiple times, choosing sequence positions at random to create new replicate data sets the same size as the original. In each new data set, a sequence position from the original may appear once, several times, or not at all. If support for a particular branch in the tree is pervasive and consistent throughout the data set, most or all of the resampled data sets will return that branch when used to construct a tree. However, if a particular branching order was only the best of a bad compromise based on inadequate data, many or most of the data sets will produce different compromises. Bootstrap programs report the percentage of replicate trees in which a particular branch was returned. Those branches with strong, pervasive, and consistent support are returned in many trees and receive strong bootstrap support.

A predicted branch that is returned in 95% or more of the replicate trees is considered strongly supported. This number does not represent a true confidence interval, however, and therefore is not strictly a representation of "statistical support." Hillis and Bull[12] have shown in computer simulations of parsimony bootstrapping that with data sets meeting certain criteria for orderliness and good behavior a bootstrap result of 95% represents an almost 100% chance that the branch reflects a true evolutionary relationship, and that bootstrap values as low as 70% may reflect a confidence of 95% in the branch.

Maximum likelihood methods can also supply information about the support for a branching order. For example, a proposed tree can be compared to a tree representing a null hypothesis to determine if the data are significantly more likely given the proposed tree. Maximum likelihood can also be used to compare alternative branching orders to see if one indicates a significantly greater likelihood of the data than does the other. If two hypotheses are being compared (e.g., that HIV infecting one individual came from a particular unit of blood versus that it was the result of sexual transmission), maximum likelihood comparison of the branching orders may be used to see if one hypothesis significantly maximizes the likelihood of the sequence data above the other hypothesis.

Large Data Sets

One difficulty encountered by researchers performing molecular phylogenetic studies of HIV is that the number of sequences involved is usually large compared to the number that sequence analysis programs can typically handle. The number of different bifurcating trees that can be drawn using only 10 sequences is about 2 million; for 53 sequences, it is 10^{80}, which is approximately the number of elementary particles in the universe. There are exact algorithms for determining the most parsimonious tree (e.g., simply drawing all possible trees and seeing which one requires the fewest mutational events). However, the time required to perform exact parsimony algorithms becomes prohibitive when more than about 20 to 25 sequences are being analyzed. Distance programs are faster and so can usually handle more sequences, but they still are limited. Note that bootstrapping multiplies the program's execution time by the number of replications performed.

The necessary response to this problem is to use faster methods that use heuristic approaches, that is, algorithms that approximate, but may not always succeed in finding, the optimal tree. Typically, a heuristic algorithm adds sequences in the order it encounters them in the data set, discovering the best place for each on the tree-in-progress but not altering the positions of previous branches. Branch rearrangements are performed, exchanging either near branches or distant ones, to minimize the possibility that the "best" tree found will in fact be a "local optimum" from which the true "best" tree could not be reached during execution of the algorithm. The larger the data set, the more likely it is there is a "best" tree that will not be found in a single run. Multiple runs of the program, using a different input order of sequences for each run, are necessary when using heuristic algorithms with a large data set.

Uses of Molecular Phylogeny in Epidemiology

Case of Transmission

An example of phylogenetic analysis in determining transmission from one individual to others is the case of a dentist whose HIV sequence was also found in several of his HIV-positive patients who had no other risk factors for the disease.[13] The evolutionary tree relating the sequences grouped them on a single branch. Bootstrapping revealed moderate strength of support for that grouping, which excluded HIV sequences from other HIV-positive individuals in the surrounding geographic area as well as patients of the dentist who had other known risk factors for the disease. This showed that there was reason to accept the idea of HIV transmission from the dentist to his patients.

The strength of the original analysis depended on the percentage of times that sequences from *all* the patients in question grouped with those of the dentist. Hillis and Huelsenbeck[14] reanalyzed the original data, asking if there is evidence of transmission between dentist and patient for each patient individually, rather than for all the patients as a group. When this analysis is done, much higher bootstrap numbers are returned, showing strong support for the dental transmission hypothesis.

Structure of the Epidemic

Molecular epidemiology is ultimately the study of transmission from one individual to another and the summation of those transmissions. The Los Alamos Database[15] has suggested in unpublished analyses that there are five major lineages of HIV-1 sequences, which they call subtypes, some of which are geographically distributed. Phylogenetic analysis suggests that these subtypes represent the descendants of lineages that separated early in the epidemic and have been evolving independently ever since.

A particularly striking phenomenon has been observed in Thailand, where the epidemic arose rather late, striking injecting drug users in 1988, female prostitutes in 1989, their heterosexual clients in late 1989 to 1990, and subsequently female partners of those men.[13] A reasonable assumption from this pattern would be that individuals infected during the first wave in 1988 subsequently infected the female prostitute population, and so on. However, phylogenetic analysis shows that the strain affecting injecting drug users clusters with the U.S./European B subtype, whereas the strain that is predominantly sexually transmitted clusters with the subtype E viruses.[13,15] The simple percentage differences within and between the groups are consistent with the hypothesis of two introductions of HIV into Thailand—and only two. Within each group the *env* C2/V3 sequences differ little, suggesting that each of the groups diverged from its own common ancestor very recently; sequences compared between the two groups showed average percent differences more than six times as high as the within-group differences.[13] The recent introduction of each subtype into Thailand creates a potential difficulty when analyzing transmission cases using the C2/V3 region. All the infections within one subgroup share a common ancestor so recently they may appear related, or it may be difficult to isolate truly related infection pairs from those related only a bit more distantly. In such a case, where HIV strains can be expected to share a recent common ancestor despite being epidemiologically unlinked, it may be necessary to use a region of extreme variability such as the *env* V1/V2 region to detect cases of direct transmission.

Population Structure Analysis

A number of epidemiologic and virologic questions require an assessment of the relative populations of different variants of HIV present in an individual over time, in different tissues of an individual, or in the same tissues in different individuals. For example, it is of interest to know if (and if so, how) (1) the distribution of HIV variants within an individual changes over the course of an infection; (2) HIV sequences selectively converge on a particular motif during the course of infection; (3) particular sequences preferentially infect the brain (for example), whereas others infect other organs; (4) particular strains are preferentially T-cell or macrophage trophic; or (5) a type of variant (e.g., syncytium-inducing or non-syncytium-inducing) appears or disappears during the course of infection.

Such studies have typically been attempted by amplifying a segment of viral RNA or DNA, then cloning the amplified products and sequencing a small number of clones, typically 2 to 20. Unfortunately, inferences about the distribution of variants in the provirus population cannot reliably be drawn from such a small number of clones. Most observed distributions of clones (e.g., 15 of one type and 5 of another) could have been drawn from a wide range of possible virus populations within the individual. Thus such a small sample cannot aid in determining the distribution of variants in the population from which the sample was drawn. To obtain a reliable estimate of the original virus population would require sequencing a much larger number of clones—and this from each of the samples necessary to the study—which is highly impractical. A related problem is the inference of the *absence* of a particular variant sequence from a patient sample. For example, a sequence present in an early sample from an individual is sometimes inferred to have disappeared based on a failure to detect it in a small number of clones from a later sample from the same patient. Such an inference is risky. The chance of missing a variant present as a respectable, though not predominant, fraction of the population is shown in Table 17.1.

A potentially simpler, more accurate way to determine the distribution of variants in a particular sample, or the presence/absence of a particular

TABLE 17.1. Probability (p) of missing a variant when sequencing n clones.[a]

Actual fraction (p) of clones with sequence X	p (probability of missing sequence X), by no. of clones analyzed		
	10 Clones	20 Clones	30 Clones
0.50	0.00098	0.000001	9.3×10^{-10}
0.30	0.028	0.0008	0.000022
0.10	0.35	0.12	0.04
0.05	0.60	0.36	0.21

[a] $p = (1 - \mathrm{p})^{\mathrm{n}}$.

sequence, is the phage or colony blot method. Amplified DNA is cloned into a phage or plasmid vector, and 500 or more plaques or colonies are plated. After several clones are sequenced to determine unique signatures within the amplified region, the plates are blotted and sequentially probed with a series of oligonucleotide probes that are complementary to the individual signature sequences. The percentage of matching clones among the approximately 500 could provide a more statistically accurate picture of their representation within the original sample than does sequencing a small number of clones (e.g., 20). Similarly, plates can be probed to determine the presence or absence of a particular sequence. The assertion of the absence of a sequence, in particular, would be much stronger because of the ability to detect sequences present as a low fraction of the population.

Leong and White (personal communication) have shown, in preliminary mixing experiments using the colony blot method with varying ratios of bacterial DNAs, that the percentage of clones from each input DNA is a good approximation of its representation in the original population of DNAs. If similar model experiments with HIV indicate that preferential amplification of certain variants does not occur,[16] the phage or colony blot approach may provide a more reliable estimate of the frequency of each variant in samples taken at different times, from different tissues, or from related infections.

Conclusion

The combination of molecular sequence analysis with epidemiologic information provides us with powerful tools for understanding the mode and tempo of HIV transmission and evolution during the course of infection. Molecular phylogenetic methods that were developed for other organisms can also be used to determine the relatedness of HIV variants. Because of the rapid sequence evolution of HIV and the complexities introduced by analyzing a population of variants, or viral quasispecies, there is a need for further improvements in methods of tree-building, such as accounting for compositional bias. However, current methods already provide the means for identifying particular sequence motifs that may be subject to selection or convergence and may therefore be of importance in vaccine research or for understanding cell tropism.

Appendix: Synopsis of Methods in Molecular Epidemiology

Similarity Comparisons

For similarity comparisons, aligned portions of sequences of interest are compared, and the percentage of sequence positions containing substitutions between two sequences is determined. Such percentages are comparable between two studies only if exactly the same aligned sequence positions are being compared. A low divergence in a highly variable region of the virus can be used to limit the time elapsed since variants from two individuals diverged (i.e., since the time of infection). A high degree of divergence does not rule out relatedness of infections but does rule out recent transmission. This approach is rapid but provides limited information.

Phylogenetic Analysis

1. *Genetic distance.* Pairwise distances—that is, the number of substitutions inferred to have taken place since two variants shared a common ancestor—are determined for each pair of sequences under study. Genetic distance programs use heuristic algorithms (e.g., neighbor-joining) or optimality criteria (e.g., Fitch least-squares analysis) to determine the tree with the shortest overall branch lengths. These methods are fast.

2. *Parsimony.* The pattern of bases at each alignable sequence position is inspected, and overall homoplasy (the occurrence of the same substitution independently on separate lineages) is minimized. The branching order that requires the fewest substitutions is considered the optimal tree. Exact algorithms are guaranteed to find the "best" tree but are time-consuming and are prohibitively slow for large data sets. Heuristic algorithms are fast, but they depend on the input order of sequences and are not guaranteed to find the "best" tree; multiple runs must be performed, reordering the sequences in the data file each time.

3. *Maximum likelihood.* For any branching order, the maximum likelihood approach determines the likelihood of the observed data. It is extremely slow as a means to determine a "best" tree but is often used to assess whether one of two branching orders is significantly superior to the

other. It requires varying parameters, such as transition/transversion ratios, in order to optimize likelihood.

4. *Bootstrapping—testing for strength of support*. Every phylogenetic program returns a tree. In studies of highly variable regions of HIV, most branches are so poorly supported they should be considered artifactual. Bootstrapping is a simple way to determine the strength, in consistency and pervasiveness throughout the data set, of support for a particular branch on a tree. Bootstrapping may be used with any tree-drawing algorithm. The original data set is sampled, with replacement, to create a new data set that is the same size as the original. This process is repeated many times (at least 100, but commonly up to 1000). From each resampled data set a tree is constructed, and the program reports in what percentage of these trees a particular grouping occurred.

References

1. Hillis DM, Allard MW, Miyamoto MM. Analysis of DNA sequence data: phylogenetic inference. Methods Enzymol 1993;224:456–487
2. Stewart CB. The powers and pitfalls of parsimony. Nature 1993;361:603–607
3. Swofford DL, Olsen GJ. Phylogeny reconstruction. In Hillis DM, Moritz C (eds): Molecular Systematics. Sunderland, MA: Sinauer, 1990:411–501
4. Korber BTM, Farber RM, Wolpert DH, Lapedes AS. Covariation of mutations in the V3 loop of human immunodeficiency virus type 1 envelope protein: an information theoretic analysis. Proc Natl Acad Sci USA 1993; 90:7176–7180
5. Vaishnav YN, Wong-Staal F. The biochemistry of AIDS. Annu Rev Biochem 1991;60:577–630
6. White TJ. Amplification product detection methods. In Persing DH, Smith TF, Tenover FC, White TJ (eds): Diagnostic Molecular Microbiology. Washington, DC: American Society for Microbiology, 1993:138–148
7. Burger H, Weiser B, Flaherty K, et al. Evolution of human immunodeficiency virus type 1 nucleotide sequence diversity among close contacts. Proc Natl Acad Sci USA 1991;88:11236–11240
8. Chang SYP, Bowman B, Weiss JB, Garcia RE, White TW. The origin of HIV-1 isolate HTLV-IIIB. Nature 1993;363:466–469
9. Stewart CB, Schilling JW, Wilson AC. Adaptive evolution in the stomach lysozymes of foregut fermenters. Nature 1987;330:401–404
10. Groenink M, Fouchier RAM, Broersen S, et al. Relation of phenotype evolution of HIV-1 to envelope V2 configuration. Science 1993;260:1513–1516
11. Felsenstein J. Confidence limits on phylogenies: an approach using the bootstrap. Evolution 1985;39:783–791
12. Hillis DM, Bull JJ. An empirical test of bootstrapping as a method for assessing confidence in phylogenetic analysis. Syst Biol 1993;42:182–192
13. Ou CY, Takebe Y, Weniger BG, et al. Independent introduction of two major HIV-1 genotypes into distinct high-risk populations in Thailand. Lancet 1993;341:1171–1174

14. Hillis DM, Huelsenbeck JP. Support for dental HIV transmission. Nature 1994;369:24–25
15. Myers G, Rabson AB, Berzofsky JA, Smith TF, Wong-Staal F. Human Retroviruses and AIDS. Los Alamos, NM: Los Alamos National Laboratory, 1993
16. Larder BA, Kohli A, Kellam P, et al. Quantitative detection of HIV-1 drug resistance mutations by automated DNA sequencing. Nature 1993;365:671–674

18 HIV Counseling and Testing: What Is It and How Well Does It Work?

LYNDA S. DOLL and MEAGHAN B. KENNEDY

In 1985 the U.S. government, in collaboration with state and local health departments, established a network of publicly funded human immunodeficiency virus (HIV) counseling and testing programs. By 1991 there were 65 programs in 50 states, 6 cities, 7 territories, the District of Columbia, and Puerto Rico that offered counseling and testing services through facilities such as freestanding counseling and testing centers; sexually transmitted disease (STD), family planning, prenatal, and tuberculosis clinics; drug treatment centers; and prisons. The number of HIV antibody tests provided through these programs has grown steadily from approximately 79,000 in 1985 to 2,090,635 in 1991; most of the tests have been performed at freestanding sites (38%) or STD clinics (27%).[1]

The public health goals for counseling and testing programs have changed with the acquired immunodeficiency syndrome (AIDS) epidemic.[2] Following the original licensure of the HIV antibody test in 1985, some individuals sought testing at blood donation centers where donor screening programs were in place. In response, a nationwide alternate test site (ATS) program was initiated to discourage the use of blood centers as testing sites. Over time, the counseling provided through ATS programs was increasingly emphasized. By 1987 ATS sites were renamed HIV counseling and testing sites to stress the importance of the counseling component of the process. That same year, the U.S. Public Health Service (PHS) published counseling guidelines[3] that emphasized the prevention goal: helping individuals initiate and maintain behavioral change. Voluntary notification and counseling and testing of partners were an important element of this prevention goal. Finally, with the increased availability of medical interventions for HIV-infected persons, a fourth goal became more salient, namely, referral for medical treatment and psychosocial support.

Many important questions have been generated concerning the most effective ways to implement counseling and testing programs in the United States.[4–6] A key concern, however, is the behavioral and psychological impact of the testing and counseling process and the extent to which the process, as it is now implemented, achieves the dual goals of facilitating the reduction of risk behaviors and encouraging health care-seeking behavior among HIV-seropositive persons. In this chapter we examine research evaluating the impact of HIV counseling and testing. We first describe (1) components of the counseling and testing process; (2) models of client–counselor interaction and behavioral change that have been proposed for use by counselors; and (3) available data showing the procedures and strategies actually followed by counselors. We then describe literature on the behavioral and psychological impact and briefly discuss efforts to compare the impact of different types of counseling and testing programs.

Counseling and Testing Process

The PHS guidelines recommend that the following minimal components be included in counseling and testing programs: pretest counseling, venipuncture, notification of results, and posttest counseling. In developing countries with few resources, there has been an interest in examining the effectiveness, as well as the ethics, of using counseling only or testing and notification only.[7] However, with the exception of mandatory screening programs (e.g., the armed forces, Job Corps, blood donation centers), persons seeking HIV testing in the United States also should receive both pretest and posttest counseling.

The emphasis on counseling underscores the expectations placed on the counseling and testing process to help prevent HIV transmission. During the pretest counseling session the advantages and disadvantages of having an HIV test and then learning the results are explored. PHS guidelines[8] also recommend that the pretest counseling session include a client-specific risk assessment and the development of a personalized plan for behavioral change. Enhancing personal risk perception may be particularly important for those persons who do not specifically go to a clinic or physician for testing but are offered and accept HIV testing as part of routine clinical care. Valdiserri et al.[9] suggested that if counselors are successful at enhancing clients' risk perception during pretest counseling the numbers of persons who fail to return for test results may be reduced. Approximately 37% of those tested at publicly funded clinics in 1990 failed to return for their test results.[1]

Components of the posttest counseling session differ depending on the outcome of the test. For seropositive individuals, the counseling involves elements of emotional support, prevention (encouraging personal behavior change and disclosure of serostatus to partners), and referral for medical care. For seronegative individuals, the emphasis continues to be on behavioral risk reduction and the need for the client to develop strategies and social support for behavioral change. For seronegative individuals with apparent low levels of risk but high anxiety (the "worried well"), referral for further counseling to help reduce unnecessarily high anxiety may be appropriate.[10]

Finally, there is a small group of individuals who are notified that their test results are inconclusive.[11,12] The Western blot confirmatory test occasionally yields indeterminate results in individuals whose test results are repeatedly reactive on the enzyme immunoassay (EIA). Indeterminate results can occur early in HIV infection when there is an incomplete antibody response or for those not truly infected with HIV.[13] Counseling for these individuals must address strategies for resolving the ambiguity of the test result, which includes seeking additional HIV testing, and where necessary helping the individual find support for coping with an inconclusive diagnosis.

Models of Client–Counselor Interactions

Several books and articles have been published outlining strategies for communicating with clients about HIV test results. These materials vary from concrete advice on what information to provide and when to address this information during a counseling session,[14–17] to discussions of the client–counselor interaction and behavioral or counseling theory.[8,18–20] In general, there is agreement on several points. First, the counseling session, whether pretest or posttest, should include both information dissemination and opportunities for the client to personalize the information to his or her life. Thus the client–counselor contact should be interactive and client-centered, with advice given by the counselor kept to a minimum. The counselor's goal should be to facilitate *client problem-solving* related to risk reduction or coping with HIV infection.

Inherent in the client-centered approach is the need to tailor each counseling session to the cultural, developmental, and emotional needs of individual clients. For discussions of HIV risk reduction, counselors also have been encouraged to assess the client's stage of behavioral change in order to assist in the development of a concrete action plan.[21] This approach recognizes that individuals vary in their perception of personal risk, intentions to change, and ability to adopt or maintain new behaviors. Thus counselors may be ineffective in encouraging the use of condoms if the individual does not perceive a personal risk for HIV.

A crisis intervention counseling approach may be an appropriate model for posttest counseling of individuals with positive or indeterminate test results.[18,22,23] Characteristics of crisis counseling include (1) emotional intensity; (2) a focus on a relatively limited but often complex set of issues; (3) information sharing by the counselor; (4) the need for a practical plan for coping with a crisis-laden event or circumstance; and (5) limited client–counselor contact after the initial session.[23]

Applying this model to posttest counseling, the counselor's goal is first to provide support for the client's emotional response to his or her serostatus and, second, to help with specific plans for coping. Coping strategies may include seeking appropriate medical care, disclosing one's serostatus to others, and adopting safer sexual practices. Several writers have questioned the value of lengthy discussions of coping strategies during posttest counseling sessions if the client's emotional reaction is one of fear, anxiety, or depression.[22,24] However, there are individual differences in these initial emotional reactions. Feelings of anxiety and helplessness during a crisis also may motivate some individuals to find solutions.[18] Thus for some seropositive persons, learning their serostatus may actually reduce ambiguity and serve as an impetus for behavioral change.[25]

Practice of HIV Counseling

As counseling and testing programs expand, it is important to assess how closely the practice of HIV counseling may approximate these recommendations for conducting effective counseling. In 1991 the Centers for Disease Control and Prevention (CDC) funded a case study and inventory of services provided by 43 publicly funded service delivery sites in five high AIDS prevalence areas of the United States.[26] Although it was not the goal of this study to evaluate the quality of services provided, the investigators observed counseling sessions in order to document approaches used for counseling.

Investigators found that the counseling sessions they observed often included factually correct and complete information provided through an interactive format. However, the extent and quality of individual risk assessments varied considerably. Furthermore, most counselors did not assist clients in the development of a personalized risk reduction plan.

In most clinics, seropositive individuals underwent one posttest counseling session of approximately 20 minutes' duration. Unlike counseling for seronegative individuals, this counseling session was often provided by someone with specialized training in HIV counseling. Many sites had extensive referral systems available for medical and social support of seropositive persons. The investigators noted that some counselors were uncomfortable discussing the need for the client to notify partners of their HIV status.

Posttest counseling sessions for seronegative individuals typically lasted approximately 10 minutes and often consisted of admonitions to reduce risk and discussions of the need to be retested in the future. Investigators noted the difficulty counselors faced in introducing discussions of behavioral change when the individual was experiencing overwhelming relief. Counselors also frequently commented on the lack of support services for seronegative persons.

It would be inappropriate to generalize from the type and quality of counseling identified through this study to all counseling programs in the United States. The results do, however, highlight areas in which publicly funded counseling programs may have difficulty complying with recommendations and need further technical or financial assistance. Furthermore, as programs evaluate the impact of their counseling on individual behavioral change, it may be important to consider the discordance between proposed guidelines for effective counseling and actual practice.

Although data are available on the number of HIV tests provided in publicly funded sites, there has been no systematic attempt to gather information on the number of tests conducted in settings outside this system. Data from the CDC's National Health Interview Survey (NHIS), a nationally representative, cross-sectional survey, showed that by 1990 approximately one-fourth of respondents had been tested at least once. Of these individuals, 15.9% were tested through blood donation and another 5.6% for other required purposes. Nearly 5% had voluntarily sought testing and more than two-thirds of these tests were obtained in private-sector settings, including 41% through physicians or health maintenance organizations (HMOs) and 23% through hospitals.[27] These results are consistent with results from a 1990 survey of more than 80,000 adults in 44 states which showed that when participants were asked where they could go to be tested for HIV 42% indicated the office of a private physician or HMO, 23% a hospital or emergency room, 11% a health department, and only 2% an AIDS testing site.[28] Similar results were found in a 1988 survey; 52% of those planning to be tested during the next year indicated they would go to a physician or HMO and 15% to a hospital, emergency room, or outpatient clinic.[29] These data suggest that the number of persons undergoing HIV testing in private settings may be high.

Whereas there are at least preliminary, qualitative data on the implementation of HIV counseling in publicly funded sites, there is limited information on the extent and quality of counseling provided in other settings. Data from the 1990 NHIS showed that individuals who obtained testing in private settings reported receiving pretest and posttest counseling less frequently than those who underwent testing in public settings (pretest counseling: public 58%, private 39%; posttest counseling: public 43%, private 25%).[27] Concerns have also been raised about the adequacy of counseling provided by health care providers in private settings.[30] Of particular concern are the noninteractive counseling methods used by some health care providers[20] and the reported discomfort on the part of some staff in private settings with homosexual behavior, drug use, and individuals with AIDS.[30,31] Training opportunities for learning HIV counseling techniques, such as those provided to primary care physicians through Project ADEPT,[32] may improve the quality of HIV counseling at these settings. However, the extent to which such opportunities are available to health care providers is unknown.

Methodologic Issues

Since licensure of the HIV antibody test in 1985, numerous studies have attempted to assess the behavioral and psychological impact of HIV counseling and testing. In this section we briefly review these studies, first for persons who learn they are seropositive and then for seronegative individuals. Most studies have evaluated sexual behavior change among homosexual men. We were able to find only one study that evaluated the specific impact of the counseling and testing process on subsequent health care-seeking behavior.[33] Given the increasing emphasis on this outcome of the counseling and testing process, we anticipate that additional evaluation studies will be forthcoming in this area. There is also a lack of research evaluating the impact of an individual learning that he or she has an indeterminate test result. Anecdotal information suggests that receiving such results may negatively affect those who receive them.

The outcome evaluation of HIV counseling and testing is methodologically complex and difficult to assess in field settings. Although the studies we review are useful, it is important to understand the limitations of this existing literature.[7,34] Many of the early attempts to evaluate counseling and testing were conducted using samples of homosexual men from preexisting cohorts. Participants were usually highly educated and motivated, white, and urban. This demographic profile is not representative of other homosexual male or intravenous drug user (IDU) populations, women, persons of color, those less educated, and those who live in rural settings.[35,36] Additionally, data from homosexual men who were offered access to antibody testing during participation in an ongoing cohort study cannot be equated with those choosing to undergo counseling and testing in other settings.[36]

In most studies we reviewed, participants were not randomly assigned to intervention groups. Moreover, control groups were often not available, thereby compromising the ability of researchers to make causal inferences. Most studies used self-reported measures of sexual activity or other risk behaviors as outcome measures.[35] These measures are subject to biases because of inaccurate recall of behaviors or socially desirable responses. Studies that examine the psychological impact of notification also may be biased by the fact that HIV infection can influence mental health functioning.[37]

Finally, measures of sexual behavior and psychological functioning, as well as the period of follow-up, varied considerably, making cross-study comparisons difficult. Short follow-up periods, such as 2 weeks, also limit the ability of researchers to evaluate sustained behavioral change.

In addition to these methodologic limitations, it is important to understand the context in which the AIDS epidemic has occurred and how it has affected the counseling and testing process. As the AIDS epidemic has emerged over time, the social and medical contexts also have evolved. The early portrayal of AIDS as a plague contrasts with its newly emerging portrayal as a chronic disease. The number and quality of interventions available to infected and uninfected persons have increased in many locales. Studies undertaken to assess the effects of counseling and testing on risk behaviors must be evaluated against a documented overall decline in risk behaviors, particularly among groups of homosexual men who have participated extensively in these interventions.[35,38,39] Similarly, when the HIV antibody test first became available, a positive result signaled limited treatment options and severe social isolation and stigma for many.[40,41] Now, with infected individuals living longer,[42] a positive test has come to signify chronic illness and not necessarily imminent death.[43] These changes may have significant effects on the psychological impact of notification of test results.

Impact of HIV Counseling and Testing

Notification of Seropositivity

Behavioral Impact

Before 1985 stored sera from homosexual men participating in previously assembled cohort studies of HIV were used, with consent, for the development of the antibody test. When the test was licensed in 1985, many of these men were invited to learn their test results, and several studies have examined the impact of test disclosure among them. Representative of this first group of studies is a study conducted by Fox and colleagues[35] that evaluated the impact of test result notification on a cohort of homosexual men living in the Baltimore/Washington, DC area. Beginning in 1984, at 6-month intervals these men had been asked to provide information about health and medical status, drug use, sexual practices, and depressive symptoms for the previous 6 months. Before the third follow-up visit, participants were mailed information about the availability of test results. Of 1001 eligible men, 67% chose to learn their results. The authors' striking observation at the fourth follow-up visit was the overall reduction in unsafe sexual practices, regardless of knowledge of serostatus or actual serostatus. Much of the decline occurred before test results were offered and thus would have occurred without individual knowledge of serostatus. Six months after notification the group of men who were informed that they were seropositive showed the largest decline in number of unprotected anal insertive and receptive partners, but this group did continue to engage in some unprotected anal intercourse. The authors conclude that because notified seropositives reported the largest decline in the number of anal sex partners, informing those who are seropositive of their serostatus may be a useful behavioral intervention.

Similar studies in San Francisco, Pittsburgh, Boston, New York City, Washington, DC, and Chicago also have shown declines in unsafe sexual behavior among homosexual men enrolled in cohort studies.[24,37,39,43–46] Although some early studies showed differential rates of decline by knowledge of serostatus,[35,43] several later studies did not find a relation between knowledge of serostatus and rates of decline in unsafe behavior among homosexual men.[24,37,39,45]

Studies among other exposure groups also have provided inconsistent results. In a cohort of IDUs recruited from methadone maintenance clinics in 1984 and notified of test results in 1986, most members reported reducing both sexual and drug use risk behavior before the start of the study.[47] The authors of that study reported fewer HIV risk behaviors (unprotected sex and intravenous drug use) among the seropositive than among the seronegative individuals 10 weeks after notification as well.

Two studies evaluated the behavioral impact in samples heterogeneous for gender, race, and sexual preference. The results are again inconclusive. Blood donors testing positive for HIV between June 1, 1986 and February 28, 1988 reported engaging in fewer unsafe sexual behaviors during the 2 weeks following notification than before notification.[48] However, in a group of 235 individuals seeking testing at anonymous testing sites in North Carolina in 1987 and 1988, no significant behavioral changes were seen at follow-up 1 year after notification.[49] Unfortunately, the use of different follow-up periods in these studies precludes comparing results.

Most HIV testing research has been conducted among groups of homosexual men, IDUs in methadone maintenance programs, and STD clinic patients. Research among other populations is often available only in conference abstracts. In one of the few studies among other groups, Futterman et al.[50] studied HIV testing experiences among seropositive individuals 15 to 21 years of age. Subjects were recruited from an urban clinic for high risk youths and were tested before enrollment. Among this population, the authors found that reductions in unsafe sex occurred more frequently with cessation of drug use than with notification of serostatus. Consistent safer sex practices were seen only in ongoing relationships.

Among heterosexual HIV discordant couples recruited in New Jersey, a significant increase in safer sexual activity has been noted following knowledge that one partner is HIV infected.[51] In that study, where couples were enrolled a median of 10 months after knowledge of one partner's seropositivity, 8% of couples who maintained sexual activity throughout the study period reported consistently using condoms before learning of their discordance. In contrast, 61% of these couples reported consistently using condoms after learning of one partner's seropositivity. However, among couples seen for 6-month follow-up interviews, a 25% decrease was seen in condom use among sexually active couples.

Most studies have used self-reported behavioral change as the outcome measure for assessing the impact of counseling and testing on sexual behavior, but some studies have used reinfection with an STD as a more objective measure of risky behavior. Otten et al.[52] compared the occurrence of an STD during the 6 months before and after HIV testing among a heterogeneous group of STD patients. Among patients testing positive and receiving their test results, there was a 29% decrease in gonorrhea (from 6.3% to 4.5%) and 12% decrease in any STD (from 7.9% to 7.0%) between the two time periods. In a study among STD patients, where the largest HIV risk behavior group was intravenous drug use, Zenilman et al.[53] matched seropositives to seronegatives according to age, sex, and month of HIV test. Ten percent of seropositive STD patients and a similar percent of seronegative patients returned to the clinic with a definitively diagnosable STD (syphilis, gonorrhea, or trichomoniasis) after undergoing HIV testing and posttest counseling. However, a significantly lower percent of seropositive than seronegative patients returned to the clinic with a probable STD, such as nongonococcal urethritis or pelvic inflammatory disease (3.9% versus 10.2%) or a partner infected with an STD (1.5% versus 3.3%).

Other studies have examined the impact of HIV counseling and testing on women's pregnancy decisions and found conflicting results. Sunderland et al.[54] compared 32 seropositive and 34 seronegative women who were notified of their antibody status early enough in pregnancy to have the option of abortion. The study population, recruited from several obstetric clinics and through a referral system, had heterogeneous HIV risk behaviors. Seroprevalence in the prenatal clinic populations varied from 40% in a clinic located in a drug treatment program to 6% in a clinic for Haitian women. Significantly more seropositive than seronegative women (18.8% versus 2.9%) chose to abort. Other studies have not found a difference in the proportion of women choosing abortion based on serostatus.[55,56] Selwyn et al.[56] studied pregnant IDUs informed of their HIV antibody status before 24 weeks' gestation. There was no significant difference in the percent of seropositive and seronegative women choosing to terminate their pregnancies (50% versus 44%).

In summary, no consistent behavioral impact of being notified that one is HIV-seropositive was noted across studies. Studies showed decreases in risk behaviors among homosexual men and IDUs; however, most of the reductions in unsafe behaviors occurred independent of serostatus notification. Some studies did indicate an effect on sexual risk behavior or pregnancy decisions after notification of seropositivity. Because of the varying results, it is difficult to draw conclusions about the behavioral impact of notification among those who are seropositive. The extent to which this ambiguity is a result of methodologic limitations of the studies, study populations, or the differing impact of counseling and testing over time is unclear.

Psychological Impact

Studies of the psychological impact of notification of seropositivity also have reported varying results. A study of homosexual men in Pittsburgh measured depression and general anxiety 1 week after testing (before results were available), 2 weeks after testing, and again 6 months to 1 year later.[24] Notified seropositive individuals were significantly more depressed than were seronegative individuals immediately after notification. Depression among seropositive individuals decreased at 6 months but was still above the normal range on the standardized scale.

Anxiety scores for all men were above average 1 week after testing but before results were available. Anxiety rose in the positive group and fell in the negative group immediately after learning test results. At 6 months, anxiety scores of those who received negative results fell below the group norm, whereas those receiving positive results had anxiety scores above the group norm at 6 months, though slightly below initial levels.

Similarly, in a cohort of homosexual men in Chicago, notification of seropositivity was significantly associated with higher levels of depression, anxiety, obsessive-compulsive behavior, and total distress 6 months after notification.[37] However, in an abstract report of long-term functioning in this cohort, it was found that the seropositive and seronegative groups did not differ on general measures of psychological distress at their 13th semiannual assessment.[57]

Other authors have not shown consistent psychological differences between seropositive and seronegative individuals during shorter follow-up periods. In a cohort study of homosexual men in San Francisco, notified seropositive individuals showed significantly more distress than seropositive individuals choosing not to learn their serostatus at 1 year after notification but not at any of the previous follow-up periods.[36] At 2 weeks, 3 weeks, and 6 months after notification, levels of distress were similar between notified and nonnotified seropositives. The authors suggest that at the time of the 1-year follow-up numerous accounts of the poor prognosis for seropositive individuals had begun to appear in the press, which may have contributed to the increased levels of distress reported at that time.

It is interesting that seropositive individuals, who had not received an AIDS diagnosis or experienced significant illness related to their HIV positivity, reported significantly lower levels of distress and hopelessness than men with an AIDS diagnosis or those who had become symptomatic, suggesting that distress was related more closely to symptomatology than notification of seropositivity.

In this cohort study, 73% of seropositive individuals correctly anticipated their seropositivity prior to receiving their test results. Because most seropositives correctly anticipated their serostatus, the lack of distress differences between notified seropositive and nonnotified seropositive individuals may be related to their expectations. The authors suggest that the negative effects of notification of seropositivity reported in other studies may have been related to incongruity between actual and expected test results among study participants. That is, notification of seropositivity may not induce significant distress among homosexual men whose expectations for the test result are confirmed.

Few studies have examined the psychological effects of notification in groups not composed entirely of homosexual or bisexual men. Perry et al.[58] studied 218 asymptomatic adults in a confidential clinic setting in New York City in 1987 and 1988. Subjects were evaluated 2 weeks before notification, immediately before and after notification, and 2 and 10 weeks after notification. Each participant reported at least one HIV risk behavior (homosexual intercourse, heterosexual intercourse with a possibly infected partner, intravenous drug use). At visits to the clinic, subjects completed scales measuring depression and anxiety, as well as four visual analogue scales measuring current feelings of depression, anxiety, fear of AIDS, and fear of having infected others. Immediately after notification, seropositive individuals did not report significantly increased visual analogue scale scores for depression. Additionally, at 10 weeks after notification, they reported significantly less distress, depression, anxiety, and psychiatric symptoms than before notification. Because the study examined only immediate and intermediate effects of serostatus notification, the authors caution against pooling their results with those from studies with longer periods of follow-up.

Results were similar for a group of IDUs assembled in New York City in 1984 and notified of test results in 1986.[47] Seropositive individuals did not report lasting increases in either depression or anxiety at the 12-week follow-up.

Most studies examining the psychological impact of notification have been among groups of participants who volunteered for testing. Little is known about the impact of notification for people who are tested as part of routine screening. Cleary et al.[33] studied blood donors who tested positive between June 1986 and February 1988 in New York City. Donors completed a questionnaire at notification that asked about the period just before notification and another questionnaire approximately 2 weeks later. At the 2-week follow-up, donors were asked about their use of health and social services since notification. Women had significantly higher depression scores than men at both visits, but no significant changes were seen in scores reported immediately after notification and at the 2-week follow-up for either men or women. More than one-fourth of the men and one-third of the women reported seeking psychological or psychiatric services following notification. Depression scores reported at notification and *not* being homosexual or bisexual were significantly related to seeking professional care after notification. The authors concluded that the results do not indicate any increase in symptoms associated with notification of seropositivity.

In summary, whereas some studies report a negative psychological impact of notification of seropositivity, many do not support these findings. It has been suggested that if expected HIV test results coincide with actual results, individuals suffer less distress. Additionally, if those being tested have a support network with other persons with similar AIDS-related concerns (e.g., homosexual men, IDUs involved in a methadone program), they may experience less distress. Cleary et al.[33] found that men who were homosexual or bisexual were less likely to seek professional support after notification of a positive result. Similarly, in the previously mentioned study where seropositive IDUs in a methadone maintenance program did not report lasting increases in anxiety or depression, the authors reported that seropositive individuals formed a support group after learning of their results.[47]

Notification of Seronegativity

Behavioral Impact

Some studies suggest that HIV counseling and testing may result in reductions in sexual risk behaviors, but examination of data on seronegative individuals has shown little effect of testing on subsequent behavior.[24,33,44] For example, despite an overall decline in unsafe sexual practices among all men, Fox et al.[35] found that seronegative individuals showed the least decline. Men notified of seronegative results showed significantly less decline in number of sexual partners and number of insertive and receptive anal sex partners than did seropositive men. The authors speculate that disclosure of a negative test result may have implied to the participant that he was protected because, despite previous sexual practices, he was not infected. Moreover, in some cases an actual increase in risk behaviors has been shown among seronegative individuals. Otten et al.[52] found that among 666 seronegative STD clinic patients, there was a 106% increase in newly diagnosed cases of gonorrhea (from 2.4% during the 6 months before HIV testing to 5% during the 6 months after HIV testing) and a 103% increase in any STD (from 5.0% to 10.1%) after HIV counseling and testing. In this clinic population, seronegative individuals were more likely than seropositive individuals to have a new STD during the 6 months after testing.

Other researchers have not observed this effect. In a cohort of homosexual men in San Francisco risk behaviors decreased between 1983 and 1987 for all men; and, after testing, the risk behaviors for seronegative men were similar to those of seropositive men.[39] The authors concluded from their data that learning that one is seronegative does not necessarily lead to an increase in high risk behavior. Similar results were seen in a study of HIV testing in women. The proportion of women who reported engaging in unprotected intercourse during the previous month was similar at baseline, immediately after pretest counseling, and at the 3-month follow-up after having learned of their seronegativity.[59] Consequently, no firm conclusions regarding the impact of notification of seronegativity can be drawn. Although some studies show increased risky behaviors after knowledge of seronegativity, others indicate no effect on sexual behavior.

Psychological Impact

Numerous studies have shown that there is a reduction in emotional distress following notification of a seronegative result. Notification of seronegativity has been associated with decreases in interpersonal difficulties (as measured by the Hopkins Symptom Checklist)[37] and reductions in depression, anxiety, fear of getting AIDS, and fear of having infected others.[58] In a cohort of homosexual men, only 43% of seronegative men correctly anticipated their results.[36] Notified seronegative individuals showed significantly lower levels of hopelessness than nonnotified controls at each follow-up. Those authors hypothesize that notification of seronegativity seems to dispel a sense of gloom among those who incorrectly believe they have been infected with HIV.

Among a group heterogeneous for HIV risk behaviors, seronegative individuals experienced relief immediately after notification and at the 2- and 6-week follow-up.[58] Even though most (73%) of this group correctly anticipated that they would be HIV-negative, they experienced significant reductions in distress after notification. In another publication these authors found that psychiatric symptoms were less at 6- and 12-month follow-up than at baseline.[60]

Comparing Counseling and Testing Programs

Public health officials recommend HIV counseling and testing for all persons at increased risk for HIV. Therefore an increasingly important research question concerns the impact of different models of HIV counseling on subsequent risk behaviors and psychological reactions. Representative of such research is a randomized trial currently under development in STD clinics in five cities (Project RESPECT). Clinic attendees who have received pretest counseling and had their blood drawn for HIV testing are assigned to one of three counseling and notification models: clinician-provided counseling; a standard, single session, HIV risk reduction counseling session; or an enhanced counseling intervention consisting of both face-to-face and group counseling sessions. STD/HIV rates and self-reports of sexual risk behaviors are monitored during the follow-up period. Such research should prove helpful in the development of effective counseling and testing programs in different environments and with different at-risk populations.

A second study is representative of methods under development to help buffer notification-induced stress reactions. Antoni et al.[61] studied the use of prenotification interventions with asymptomatic homosexual men. Men were randomly assigned to a cognitive-behavioral stress management group or an assessment-only control group. The intervention groups met twice weekly for 10 weeks and were introduced to stress management techniques, cognitive restructuring techniques, assertiveness skills, behavioral change strategies, and instruction in self-monitoring of environmental stressors. Prenotification psychological scales were administered after 5 weeks of training and 72 hours before notification. Postnotification scales were administered 1 week after notification. A significant increase in depression scores was seen after notification only in seropositive men assigned to the control group. Indeed, the increase in depression score in the control group of seropositive men was twice that among men in the intervention group. Although this extensive intervention may not be feasible in many publicly funded HIV counseling and testing clinics, offering such interventions through mental health clinics or other venues may decrease the number of at-risk persons who refuse HIV counseling and testing.

Expectations for HIV Counseling and Testing

HIV counseling and testing are important public health tools for the identification and medical referral of persons with HIV infection. Although counseling and testing may help facilitate reduction of risk behaviors in some persons, available data suggest that for many the standard pretest and posttest counseling sessions following initial testing may not be sufficient to facilitate behavioral change. It is possible that improvements in the quality of counseling might encourage more behavioral change; improving the quality of counseling should be an important goal of counseling and testing programs. It is also important, however, to assess our expectations of what can be accomplished through pretest and posttest counseling sessions, as currently implemented.

Rugg et al.[2] suggested that there is little theoretical basis for expecting that these brief counseling sessions, usually based on an information dissemination model, can influence behavioral change without other conditions being present. Similarly, Perry and Markowitz[14] suggested that expectations that counseling and testing can reduce or eliminate behaviors that are chronic, reinforced with pleasure, and linked only distantly to negative consequences are unrealistic. Perhaps more realistic goals are (1) providing accurate, up-to-date information that can help reduce the fear of being infected with HIV; (2) helping individuals personalize risk and develop strategies for changing behavior; (3) providing test results in a supportive environment; (4) helping to reduce immediate stress; (5) providing referral for medical care and risk reduction interventions such as drug treatment; and (6) helping individuals develop initial coping strategies. Inducing sustained behavioral change through HIV counseling and testing is probably unrealistic for most individuals. Sustained behavioral change can most effectively be accomplished if HIV counseling and testing are embedded within a larger intervention program. Indeed, behavioral theory and empiric data from intervention research suggest that the most successful intervention programs are multifaceted, combining elements of individual-level, cognitive-behavioral interventions, community-level interventions to encourage changes in social norms, and community mobilization approaches that utilize mass media, social marketing, empowerment, and policy-targeted components.[62] HIV counseling and testing programs are appropriately viewed as an essential component of this larger agenda to encourage behavioral change in at-risk populations.

References

1. Centers for Disease Control and Prevention. Publicly funded HIV counseling and testing—United States, 1991. MMWR 1992;41:613–617
2. Rugg DL, MacGowan RJ, Stark KA, Swanson NM. Evaluating the CDC program for HIV counseling and testing. Public Health Rep 1991;106:708–713
3. Centers for Disease Control and Prevention. Public Health Service guidelines for counseling and antibody testing to prevent HIV infection and AIDS. MMWR 1987;36:509–515
4. Fehrs LJ, Foster LR, Fox V, et al. Public health trial of anonymous versus confidential human immunodeficiency virus testing. Lancet 1988;2:379–382
5. Paringer L, Phillips KA, Hu T. Who seeks HIV testing? The impact of risk, knowledge, and state regulatory policy on the testing decision. Inquiry 1991; 28:226–235
6. Myers T, Orr KW, Locker D, Jackson EA. Factors affecting gay and bisexual men's decisions and intentions to seek HIV testing. Am J Public Health 1993;83:701–704
7. Phillips KA. HIV counseling and testing in developing countries: a research agenda. Unpublished manuscript, 1993
8. Centers for Disease Control and Prevention. Technical guidance on HIV counseling. MMWR 1993;42:11–17
9. Valdiserri RO, Moore M, Gerber AR, et al. A study of clients returning for counseling after HIV testing: implications for improving rates of return. Public Health Rep 1993;108:12–18

10. Miller D. Counselling and psychosocial intervention. In Adler M (ed): ABC of AIDS (2nd ed). London: British Medical Journal, 1990:39–43
11. Celum CL, Coombs RW, Lafferty W, et al. Indeterminate human immunodeficiency virus type 1 Western blots: seroconversion, risk, specificity of supplemental tests, and an algorithm for evaluation. J Infect Dis 1991;164:656–664
12. Kleinman S. The significance of HIV-1-indeterminate Western blot results in blood donor populations. Arch Pathol Lab Med 1990;114:298–303
13. Fillipo BH, Russin SJ. What to do when results of a Western blot test are indeterminate. Postgrad Med J 1991;89:39–40
14. Perry SW, Markowitz JC. Counseling for HIV testing. Hosp Commun Psychiatry 1988;39:731–739
15. Green J. Counseling in HIV Infection and AIDS. Cambridge: Blackwell Scientific, 1989.
16. Weddington WW, Brown BS. Counseling regarding human immunodeficiency virus—antibody testing: an interactional method of knowledge and risk assessment. J Subst Abuse Treat 1989;6:77–82
17. Bresolin LB, Rinaldi RC. Human immunodeficiency virus blood test counseling for adolescents. Arch Fam Med 1993;2:673–676
18. Janosik EH, Smith C, Hardman MM. Crisis Counseling. [City]: Wadsworth, 1984
19. Grace WC, Genser SG, Coslett RN. Psychotherapeutic principles and AIDS counseling for drug injectors. Focus 1992;7:1–4
20. Silverman D, Perakyla A, Bor R. Discussing safer sex in HIV counseling: assessing three communication formats. AIDS Care 1992;4:69–82
21. Prochaska JO, DiClemente CC, Norcross JC. In search of how people change: applications to addictive behaviors. Am Psychol 1992;47:1102–1114
22. Coates TJ, Lo B. Counseling patients seropositive for human immunodeficiency virus: an approach for medical practice. West J Med 1990;153:629–634
23. Wilson RR (ed). Problem Pregnancy and Abortion Counseling. Saluda, NC: Family Life Publications, 1973
24. Huggins J, Elman N, Baker C, Forrester RG, Lyter D. Affective and behavioral responses of gay and bisexual men to HIV antibody testing. Soc Work 1991;36:61–66
25. Siegel K, Levine MP, Brooks C, Kern R. The motives of gay men for taking or not taking the HIV antibody test. Soc Probl 1989;36:368–383
26. Macro International. Assessment of CDC-Funded Counseling and Testing, Referral, and Partner Notification (CTRPN) Services for Prevention of HIV Transmission. 1992
27. Centers for Disease Control and Prevention. HIV counseling and testing services—United States, 1992. MMWR 1992;41:749–752
28. Valdiserri RO, Holtgrave DR, Brackbill RM. American adults' knowledge of HIV testing availability. Am J Public Health 1993;83:525–528
29. Hardy AM, Dawson DA. HIV antibody testing among adults in the United States: data from 1988 NHIS. Am J Public Health 1990;80:586–589
30. Henry K, Maki M, Willenbring K, Campbell S. The impact of experience with AIDS on HIV testing and counseling practices: a study of U.S. infectious disease teaching hospitals and Minnesota hospitals. AIDS Educ Prev 1991;3:314–321

31. Kegeles SM, Coates TJ, Christopher TA, Lazarus JL. Perceptions of AIDS: the continuing saga of AIDS-related stigma. AIDS 1989;3:253–258
32. Wolfson MA, Wartenberg AA, Stein MD. AIDS and substance abuse. Project ADEPT. Providence, RI: Brown University, 1991
33. Cleary P, Van Devanter N, Rogers TF, et al. Depressive symptoms in donors notified of HIV infection. Am J Public Health 1993;83:534–539
34. Higgins DL, Galavotti C, O'Reilly KR, et al. Evidence for the effects of HIV antibody counseling and testing on risk behaviors. JAMA 1991;266:2419–2429
35. Fox R, Odaka N, Brookmeyer R, Polk BF. Effect of HIV antibody disclosure on subsequent sexual activity in homosexual men. AIDS 1987;1:241–246
36. Moulton JM, Stempel RR, Bacchetti P, Temoshok L, Moss AR. Results of a one year longitudinal study of HIV antibody test notification from the San Francisco General Hospital cohort. J AIDS 1991;4:787–794
37. Ostrow DG, Joseph JG, Kessler R, et al. Disclosure of HIV antibody status: behavioral and mental health correlates. AIDS Educ Prev 1989;1:1–11
38. Adib SM, Joseph JG, Ostrow DG, Tal M, Schwartz SA. Relapse in sexual behavior among homosexual men: a 2-year follow-up from the Chicago MACS/CCS. AIDS 1991;5:757–760
39. Doll LS, O'Malley PM, Pershing AL, et al. High-risk sexual behavior and knowledge of HIV antibody status in the San Francisco City Clinic cohort. Health Psychol 1990;9:253–265
40. Dilley JW, Pies C, Helguist M. Face to Face: A Guide to AIDS Counseling. San Francisco: AIDS Health Project, University of California, 1989
41. Chesney MA. Health psychology in the 21st century: acquired immunodeficiency syndrome as a harbinger of things to come. Health Psychol 1993;12: 259–268
42. Lemp GF, Payne SF, Neal D, Temelso T, Rutherford GW. Survival trends for patients with AIDS. JAMA 1990;263:402–406
43. Coates TJ, Morin SF, McKusick L. Behavioral consequences of AIDS antibody testing among gay men. JAMA 1987;258:1889
44. McCusker J, Stoddard AM, Mayer KH, et al. Effects of HIV antibody test knowledge on subsequent sexual behaviors in a cohort of homosexually active men. Am J Public Health 1988;78:462–467
45. Wiktor SZ, Biggar RJ, Melbye M, et al. Effect of knowledge of human immunodeficiency virus infection status on sexual activity among homosexual men. J AIDS 1990;3:62–68
46. McKusick L, Coates TJ, Morin SF, Pollack L, Hoff C. Longitudinal predictors of reductions in unprotected anal intercourse among gay men in San Francisco: the AIDS behavioral research project. Am J Public Health 1990;80:978–982
47. Casadonte PP, Des Jarlais DC, Friedman SR, Rotrosen JP. Psychological and behavioral impact among intravenous drug users of learning HIV test results. Int J Addict 1990;25:409–426
48. Cleary PD, Van Devanter N, Rogers TF, et al. Behavior changes after notification of HIV infection. Am J Public Health 1991;81:1586–1590
49. Landis SE, Earp JL, Koch GG. Impact of HIV testing and counseling on subsequent sexual behavior. AIDS Educ Prev 1992;4:61–70
50. Futterman D, Hein K, Kipke M, et al. HIV testing: adolescents. In: Abstracts from the XI International Conference on AIDS, San Francisco, 1990

51. Skurnick J, Bromberg J, Cordell J, et al. Change in couples' sexual activity after knowledge of HIV discordance: a report from the heterosexual HIV transmission study. In: Abstracts from the XIII International Conference on AIDS, Amsterdam, 1992
52. Otten MW, Zaidi AA, Wroten JE, Witte JJ, Peterman TA. Changes in sexually transmitted disease rates after HIV testing and post-test counseling; Miami, 1988 to 1989. Am J Public Health 1993;83:529–533
53. Zenilman JM, Erickson B, Fox R, Reichart CA, Hook EW. Effect of HIV post-test counseling on STD incidence. JAMA 1992;267:843–845
54. Sunderland A, Minkoff HL, Handte J, Morose G, Landesman S. The impact of human immunodeficiency virus serostatus on reproductive decisions of women. Obstet Gynecol 1992;79:1027–1031
55. Johnstone FD, Brettle RP, MacCallum LR, et al. Women's knowledge of their HIV antibody state: its effect on their decision whether to continue the pregnancy. BMJ 1990;300:23–24
56. Selwyn PA, Carter RJ, Schoenbaum EE, et al. Knowledge of HIV antibody status and decisions to continue or terminate pregnancy among intravenous drug users. JAMA 1990;261:3567–3571
57. Ostrow DG, Leite MC, Beltran E, Adib SM. Long term effects of HIV serostatus on mental health and sexual behavior. In: Abstracts from the XIII International Conference on AIDS, Amsterdam, 1992
58. Perry SW, Jacobsberg LB, Fishman B, et al. Psychological responses to serological testing for HIV. AIDS 1990;4:145–152
59. Ickovics J, Morill A, Beren S, Walsh U, Rodin J. HIV testing & women: behavioral/psychological consequences. In: Abstracts from the XIII International Conference on AIDS, Amsterdam, 1992
60. Perry S, Jacobson L, Card CAL, et al. Severity of psychiatric symptoms after HIV testing. Am J Psychiatry 1993;150:775–779
61. Antoni MH, Baggett L, Ironson G, et al. Cognitive-behavioral stress management intervention buffers distress responses and immunologic changes following notification of HIV-1 seropositivity. J Consult Clin Psychol 1991;59: 906–915
62. Kelly JA, Murphy DA, Sikemma KJ, Kalichman SC. Psychological intervention to prevent HIV infection are urgently needed: new priorities for behavioral research in the second decade of AIDS. Am Psychol 1993;48:1023–1034

19
Legal Aspects of AIDS: The Chasm Between Public Health Practices and Societal Norms

ANN N. JAMES

The purpose of this chapter is to review the key legal issues when addressing the acquired immunodeficiency syndrome (AIDS) and to examine them in light of the unique social issues that surround the legislative, judicial, and regulatory actions that address this disease. No other disease, except perhaps leprosy from the beginning of the millennium to the 1800s, has brought so much societal pressure on its victims and caused such cataclysmic social consequences over such a long period. Therefore all decisions made in regard to legal issues surrounding AIDS are made in the context of the social expectations and attitudes of the participants. To present the volumes of information on legal issues and AIDS is far beyond either the scope of this chapter or the usefulness to the reader, so the topics are limited to these key areas: testing of health care workers (HCWs) and others, disclosure and confidentiality, and requirements for education and safety in the workplcae.

After discovering that AIDS is blood-borne, the Centers for Disease Control and Prevention (CDC) in 1987 provided guidelines for universal precautions, as described elsewhere.[1] Such guidelines are not binding and can be used only to indicate the expectations of good practice; they do not have the force of law. The fact that exposures still occurred in 1989 with HCWs who *knew* that patients were human immunodeficiency virus (HIV)-positive and yet failed to use universal precautions underscores the difficulty of achieving universal compliance and thereby reducing exposure likelihood.[2] The legal question is whether the law, either by imposed statute or established by case law, can change behavior, induce disclosure, or ensure universal compliance with any standard or if more reflective standards based on the circumstances should be established. We examine both approaches and their results and consequences.

To review the law in regard to AIDS, it is impossible not to consider the continuation of discriminatory actions against those who are known to be HIV-positive, actions that go far beyond the workplace to housing, child custody, and family relationships.[3] For the individual, whether HCW or patient, prisoner or prostitute, victim or criminal, the consequences of exposure, as well as of infection, remain significant. For instance, a grand jury in Austin, Texas did not indict an alleged rapist because his victim asked that he wear a condom. Although an indictment was brought by a subsequent grand jury, the fact that an individual attempted to protect herself from this disease when confronted with an armed individual in her apartment was misinterpreted for consent and was not understood as seeking protection against an even greater harm. This situation indicates how strongly local mores affect legal outcomes.

Testing

The value of testing for HIV, as described in other chapters of this book, is not necessarily helpful because a negative test does not mean that the person does not have the virus, only that the antibodies to the virus are not present at the time of testing. Essentially, only a positive test for HIV is meaningful. The issues on testing are focused on various groups of individuals: employees, HCWs, the patient, and others including perpetrators of crimes, prostitutes, and prisoners.

First, can any employer require HIV testing as a condition of employment? The answer is yes, but in limited circumstances. To require such a test means that a positive HIV test would indicate that the applicant fails to meet the bona fide occupational qualification (BFOQ) of the job in question.[4] The burden on the employer is to show that there is a BFOQ for the specific position. A Florida case established the primary principles of law applicable to HIV and employment. A classroom teacher had tuberculosis but was shown medically not to be in an infectious state. The court found that the plaintiff did not present a clear and present danger to her students.[5] This standard was adopted by the state of Rhode Island as the standard for the BFOQ for any workplace in any setting.[6] Because of the concern that such preemployment testing could be challenged as discriminatory, few employers in the health care field have instituted the requirement as a BFOQ.

The circumstances in which an individual conducts the employment can be considered when establishing a BFOQ. This position was clarified by a court ruling on the permissibility of administering mandatory HIV tests to all foreign service employees seeking to qualify or who were already qualified for service abroad. Despite the argument that such testing is prohibited under the Fourth Amendment, which prohibits unreasonable search and seizure, the court found that because HIV-positive individuals placed in some foreign countries could be at substantial risk based on the location such testing was not an unreasonable search because it was closely related to fitness for duty.[7]

The question of whether HIV status can determine fitness to be an HCW depends on what state law demands and if the guidelines of the CDC are adopted in a given state. Some states, among them Delaware, Florida, and Louisiana, have required HCWs to be tested and to obtain informed consent from patients for the delivery of services when they are HIV-positive. Texas and California adopted the initial CDC guidelines and now find themselves in a bind because it is not clear that these standards supersede the court-enforced law.[8] Thus interpretation demands careful consideration by any entity, whether hospital or restaurant, as to what is a BFOQ and how to set rules that are fair to both the workers and the invitees, whether they be the patient or the public.

In the hospital setting, the hospital has an affirmative duty to provide a safe, healthy environment for its employees and its patients as established by case law.[9] To do so, it must ensure that rules are appropriate for the protection of both. The most important case for hospitals is the Leckelt case from Louisiana[10] in which the issue was whether the employee had to disclose his HIV status to the hospital employer, a condition of employment for all infectious or contagious diseases under terms of the employee handbook and the infection control procedures of the hospital. The court concluded that the fact that Mr. Leckelt refused to submit his test results to his employer, not what the submission would have shown, was at issue, and the basis on which the court upheld the right of the hospital to terminate the employee. Whether the hospital could have terminated the employee solely on the strength of a positive HIV test was not at issue. The court concluded that HIV status was not a barrier to hiring but could be a part of post-hiring physical examinations.[10]

The issue of whether patients can or must be tested has also been raised. The fact is that some 39 HCWs and 81 are possibly are HIV-positive from workplace exposures, but only five individuals have been shown to be infected by an HCW, apparently all by the same Florida dentist.[11] The dilemma lies in the fact that even if health care workers know that an individual is HIV-positive there is neither a legal nor an ethical alternative to the delivery of care for the HCW. Consolidated Omnibus Budget Reconciliation Act (COBRA) of course, has established the legal obligation to care for patients in an emergency condition, regardless of ability to pay, HIV status, or any other factor.[12] In addition, the Americans with Disabilities Act (ADA) prohibits places of "public accommodation" from discrimination against handicapped persons and specifically those with AIDS. Places of public accommodation include professional offices of health care providers.[13] Whether individuals present with an emergency condition demanding care under COBRA or seek access under the ADA, HCWs have an obligation to provide that care.

Ethically, both physicians and nurses are obligated to render care, particularly if failure to do so can possibly harm the patient. This obligation has created much discussion among the ethics committees of the medical community, particularly among physicians, who have always maintained their right to select those to whom they will give care. However, the American Medical Society has taken a strong stance on the obligation to deliver services despite HIV status.[14]

An additional liability arising from testing, whether consensual or based on statutory authority, is the possibility of an action for negligence. A number of such cases have been brought based on the emotional and physical risks resulting from either false-negative or false-positive tests. Courts have been reluctant to find for the plaintiff, providing that the defendant hospital and the HCW can demonstrate adequate safeguards in performance of the tests; nonetheless, the possibility of a claim based on negligence is an added liability of testing.

Many states prohibit testing without consent, but the legal dilemma arises when an HCW is exposed to the patient's blood and the patient does not consent to testing. A similar dilemma arises with police and other rescue workers who may be exposed to blood and seek to obtain the HIV status of the person to whose blood they were exposed. Many states have statutes that provide for disclosure, subject to confidentiality, of the HIV status of the individual. The problem arises as the legislative drafters deal with the issue of what constitutes exposure sufficient to warrant involuntary disclosure. The resolution, if not clearly answered under state law, will likely turn on several facts, such as the nature of the risk, the duration and severity of the exposure, and the probability of infection based on reasonable medical judgment.[15] There is a great likelihood of testing without consent where the answers to these questions and state law conflict, but the resulting knowledge can be used only to relieve the concerns of the exposed individual and provides no public health assistance or disclosure to the patient. Resolution of the conflict between societal attitudes on AIDS and confidentiality of medical records is rarely satisfactory to either side.

Deliberate exposure to bodily fluids in a penal setting has usually resulted in testing, and generally the prisoner's diminished expectation of privacy and freedom from search coupled with the state's interest in preventing the spread of AIDS are sufficiently strong to sustain mandatory testing.[16] Among the deliberate exposure cases, one in Texas resulted in the state bringing attempted murder charges against a prisoner who spat on a guard (Weeks v. State, 834 s.w. 2d, 599 [1992]). The charge was upheld in that case and others where a premeditated action was found.

From a legal perspective, then, the legality of testing depends on several questions.

1. Are the results to be used for a prohibited purpose, such as denial of employment without a BFOQ, or to refuse treatment?
2. Are there protocols is place that require disclosure of HIV status, and do they serve a legitimate purpose for the hospital or the employer?

3. Are reasonable accommodations made for an individual with HIV or AIDS in the workplace, and are they based on a general set of criteria applicable to all individuals in the same or similar circumstances?
4. Does the protocol for testing meet the current state standards? Often states require specific protocols to be established before testing can be undertaken.
5. Was the test itself properly done by a capable laboratory, and were all other safeguards against false negatives or false positives in place?

Disclosure and Confidentiality

The right to privacy and expectation of privacy for the medical record and in one's employment have led to a series of conflicting decisions. When AIDS testing was first available, states quickly moved to prohibit all testing without consent; but the concerns of exposed HCWs and rescue personnel resulted in many amendments to those statutes. An expectation of confidentiality remains, however, and is upheld by the courts except for a need-to-know circumstance.

The strongest position on need-to-know was taken in a New Jersey case[17] in which a court ruled that a physician has an affirmative duty to withdraw from performing invasive procedures that pose any risk to patients. The court considered not only the probability of harm from infection but also the harm resulting from fear of exposure until testing could ensure that the patient had not been infected by exposure. This court did not weigh the likelihood of risk; the fact that there was any risk was judged as sufficient to demand that the physician not perform the procedure.

This legal result is bolstered by the requirement that a hospital has a duty to accommodate a handicapped employee (which includes the HIV-positive individual), and to do so it must know of that status, which was the conclusion of the Leckelt court. Following the Arline case, the institution is able to evaluate the HCW's condition only if it has knowledge of the condition; and if the worker is infected, the institution can then determine if there is a risk to patients or other employees and if there can be reasonable accommodations to allow the HCW to function in some other capacity.

If there is a general obligation for both the HCW and the employer to know the HIV status of the HCW, can such individuals still provide patient care and under what conditions? The answer seems to be that informed consent can cure any problem arising from the fears that were the concern of the New Jersey court. Both California and Texas, having adopted the CDC guidelines requiring all HCWs to "know" their status and to obtain informed consent before performing invasive procedures, have not adopted clear rules nor enforced these requirements so the impact of such statutes is not yet know. Because the CDC was unable to develop the list of such procedures, the matter is left ultimately to the discretion of states and individual facilities, which must navigate between the shoals of discrimination against the provider and risk to the patient. To minimize the risk of striking either shoal, every facility should develop a procedure to make such determinations, and care should be taken to protect the confidentiality of the HCW while allowing the patient to make a choice in these situations.

Of course, the patient may have an obligation to disclose HIV status as well, although states have not taken the strong position on this issue that they have on disclosure by HCWs. In California and New York, HCWs sued patients for failure to disclose their HIV-positive status, of which the patients in both cases were aware. In the California case the patient gave false information on her medical form, and during surgery the plaintiff was exposed to the patient's blood. The jury found, in awarding the HCW $102,500, that the patient had a duty to inform and had in fact misrepresented her health status to endanger the life of the HCW. The award was later reduced in a settlement.[18] In New York the plaintiff was the patient, suing his doctor for release of confidential HIV information. Although the court found that the doctor had breached the statutorily imposed duty of confidentiality, the law did not negate the patient's legal duty to disclose his HIV-positive status to the doctor.[19]

In both these cases the courts seem willing to balance the unequal burden on the HCW to disclose and obtain consent, despite the clear evidence that the greater risk is with the HCW, with an obligation of the patient to disclose his or her HIV status. As AIDS has become better understood clinically, it has also become better understood by the courts; and more statutory protections exist for HIV-positive individuals. Thus it seems likely that this trend of balancing the burden of disclosure will continue, to the benefit of all parties.

Workplace Safeguards and Education

Because of the recognized risk to HCWs, Congress forced the Occupational Safety and Health Administration (OSHA) to promulgate regulations defining the employers' responsibility to protect against exposure to blood-borne pathogens in the workplace.[20] These regulations require an employer to determine the exposure likelihood of its employees and to protect appropriately against such exposure. In addition the employer must develop a written exposure control plan if employees will have occupational exposure to potentially infectious materials. The plan must be designed to eliminate or minimize employee exposure and must include, at a minimum, an exposure determination, a procedure for evaluating circumstances surrounding an exposure, and an availability of hepatitis B virus (HBV) vaccinations, postexposure evaluations, and follow-up. The OSHA regulations further mandate that employers implement engineering and work practice controls and provide employees with personal protective equipment. Such safeguards are applicable to any workplace in which employees may be exposed to blood, which includes safety workers in any area of employment.

States have also mandated employee education programs for all employees and educational programs in school systems.[21] The value of such mandatory education is that the workplace becomes less hostile to the HIV-positive employee, and juries are better educated to understand the issues when discrimination cases arise. Obviously, the value of mandatory educational programs in the school system is the possibility of reducing spread of the virus.

Educational programs have not been without controversy, as the use of the term "condoms" in television public service advertisements has only just begun in 1994. The conflict between teaching strict abstinence and the use of condoms has delayed or prevented the use of AIDS educational materials in some school systems.

Conclusion

The legal issues surrounding AIDS are inextricably intertwined with the social fabric of the community and the understanding of the judiciary and the legislature. Since 1990 case law seems to be moving to an accommodation of recognizing the risk to both the HCW and the patient. However, it is clear that cases such as that of Kimberly Bergalis, a young woman infected by the Florida dentist who gave powerful congressional testimony on the patient's right to know, will induce Congress and state legislatures to extend further protection to the patient and, in so doing, extend less protection and less privacy to the HCW.

References

1. Centers for Disease Control. Recommendations for prevention of HIV transmission in health-care settings. MMWR 1987;(suppl 2d):3S–17
2. Seroconversion after exposure found by study to be minimal. AIDS Policy Law 12 BNA 6-7 (June 28, 1989).
3. Dunlap MC. AIDS and discrimination in the United States: reflections on the nature of prejudice in a virus. 34 Vill. L. Rev. 909, 912 (1989); see also Shilts R. And the Band Played On: Politics, People, and the AIDS Epidemic. 1988
4. Florida Stat. Ann. § 760.50(2)(a), (b) (West 1986 & Supp. 1990). The BFOQ exception in employment discrimination jurisprudence is codified at 42 U.S.C. § 2000e-2 (West 1988)
5. School Board of Nassau County v. Arline, 480 U.S. 273 (1987)
6. R.I. Gen. Laws § 23-6-22 (1989 & Supp. 1990)

7. Local 1812, Am. Federation of Government Employees v. United States Department of State, 662 F. Supp. 50 (D.D.C. 1987)
8. Texas Health & Safety Code Ann. §§ 85.201-85.206 (West 1993)
9. Estate of Behringer v. Medical Center at Princeton, 592 A.2d 1251, 1282 (N.J. Super. Ct. Law Div. 1991) and Leckelt v. Board of Comm'rs of Hosp. Dist. No. 1, 714 F. Supp. 1377, 1379, (E.D. La. 1989), aff'd, 909 F.2d 820 (5th Cir. 1990)
10. Leckelt, 714 F. Supp. at 1379
11. Isaacman SH. The Other Side of the Coin: HIV—Infected Health Care Workers, 9 St. Louis U. Pub. L. Rev. (1990); see also Preliminary Analysis: HIV Sero Survey of Orthopedic Surgeons, 40 May 17, 1991, at 309(4); Morbidity and Mortality Wkly. Rep., Update: Transmission of HIV Infection During an Invasive Dental Procedure—Florida, 40 Morbidity and Mortality Wkly. Rep. June 14, 1991, at 377(5)
12. 42 U.S.C. § 1395DD (Supp. III 1993). Several states, including California and Texas, have also enacted status prohibiting "patient dumping." See. e.g., Cal. Health & Safety Code §§ 1317.1-1317.6 (West 1990 & Supp. 1993); Tex. Health & Safety Code § 241.027-.029 (West 1992)
13. Counsel on Ethical and Judicial Affairs, American Medical Association. Ethical Issues Involved in the Growing AIDS Crisis, (1987 and 1988); see also American Nurses' Association Comm'n on Ethics, American Nurses' Association Statement Regarding Risks Versus Responsibility in Providing Nursing Care, NIN Ethics in Nursing Position Statements and Guidelines (1988)
14. 42 U.S.C. § 12182(a) (Supp. III 1993); see also Towers-Crawley S. ADA Primer: A Concise Guide to the Americans with Disabilities Act of 1990
15. The Methodist Hospital in Houston, Texas announced in October 1987 its precise intention to take this position. Because HIV testing requires some time to perform, it clearly could not affect the emergency patient or the patient for whom surgery should or must be scheduled prior to such time as the test returns would be available. The hospital expressed its intention to use the information only to ensure the safety of its own workers and not to discriminate against patients. No test cases with regard to this situation are known.
16. Dunn v. White, 880 F.2d 1188 (10th Cir. 1989), cert. denied, 110 S. Ct. 871 (1990)
17. Estate of Behringer v. Medical Center of Princeton, 592 A.2d 1251 (N.J. Super. Ct. Law Div. 1991)
18. Patient to pay worker in settlement. Mod Health Care 1993; Nov.15:54
19. Doe v. Roe, 588 N.Y.S.2d 236 (N.Y. Sup. Ct. 1992)
20. 29 C.F.R. § 1910.1030 (1991)
21. Texas Health & Safety Code Ann. § 85.010 (West 1992); Florida Stat. Ann. §§ 455.2226, 455.2228 & 381.0034 (West 193)

20
HIV Education in the Workplace

Fred Kroger and Priscilla Holman

Human immunodeficiency virus (HIV) and acquired immunodeficiency syndrome (AIDS), though unknown prior to 1981, now affect nearly every human institution. Families, schools, religious communities, health agencies, civic organizations, and workplaces have felt the disruptions caused by the disease. The social impact of the epidemic has been as profound as its challenge to human life and health.[1] Failure to control the epidemic threatens the survival of numerous human enterprises, including entire cultures.

It is clear that HIV and AIDS represent a major challenge to economic development. Some estimates are that if not checked the disease could consume up to 1.4% of the world's gross domestic product by the year 2000—an amount equal to the total economy of a nation such as India or Australia.[2]

The economic impact of AIDS will be most severely felt by developing countries. In 1990 these countries accounted for more than 80% of the world's infections; by 2000 this disproportionate burden is expected to reach 95%.[3] Multinational companies were among the first to grasp the significance of these trends. Governmental health and commerce agencies have also awakened to the new realities portended by AIDS. Recognition of the workplace as a key arena in which the AIDS wars can and must be fought has been slow in coming.

Most health agencies have identified professions that are associated with HIV transmission risk and have developed intervention programs to reduce risks. The latter have included programs for sex workers, truck drivers, and health care workers.[3–7] During the first decade of prevention programming, general workplaces were commonly overlooked as a logical and strategic focus for attention.[3,4,8,9]

Even though the workplace can pose occupational risks for some, occupation-associated transmission of HIV has not been a major source of disease morbidity. Fear and ignorance resulting in discrimination or lost productivity can be as disruptive to the flow of commerce as are actual illness-related losses.[10]

The second decade of world experience with HIV and AIDS is seeing an increased appreciation for the role that workplace programs play in community and national control efforts.[3] The value of workplaces for HIV prevention will be realized more from their access to audiences and resources than from occupation-specific interventions.

Occupational Risk

The risk of HIV infection in the workplace as with other blood-borne and sexually transmitted diseases, is low.[9,11] In the United States and other countries, concerns about occupational risk of HIV infection are disproportionately high. Studies of effective risk communications suggest that it may be helpful to acknowledge that there is some level of risk associated with specific occupations and emphasize the risk factors that are of greatest relevance.[12]

Persons who engage in frequent acts of sexual intercourse with numerous persons, as is often true of sex industry workers, are at risk of acquiring HIV. From a public health standpoint, prostitution is significant as a means of rapid disease spread to many people.[3,4,13,14] Occupations that expose workers to blood or blood products, particularly in conjunction with the use of sharp instruments, may present workers with occupational risks of HIV exposure. Such occupations can include surgeons, dentists, public safety and emergency responders, morticians, and phlebotomists.

Education programs that teach negotiation and condom usage skills to prostitutes, especially when accompanied by social marketing programs to also educate their clients, have proved beneficial and highly cost-effective.[3–5,14,15] Training and policy requirements for applying "universal precautions" are believed to provide satisfactory levels of protection to workers exposed to blood and blood products.[16] Technologic solutions for reducing the incidence of accidental needlesticks to health care workers are being explored.[16]

A second level of risk is associated with occupations that increase a person's exposure to risky opportunities. The interaction of sex workers and truck drivers in Africa and India, for example, has played an important role in disease spread.[3,4]

A book by American journalist Randy Shilts, *And the Band Played On*, which was also made into a television movie, dramatizes the role that modern travel and travel industry workers may play in disseminating diseases rapidly over broad geographic regions. The book highlighted the story of Patient Zero, a French Canadian airline steward who was epidemiologically linked to 40 AIDS cases in New York City and Los Angeles in 1984. (Researchers have identified HIV in U.S. blood samples from as early as 1969, so the name patient Zero, though useful for dramatic effect, is a misnomer.[17]) There is also evidence that some entertainers and professional athletes may be acculturated to engage in sexual or drug use behaviors resulting from the opportunities and pressures of their nomadic life styles, performance demands, and the fruits of hero worship.[18,19]

In the United States the Entertainment Industry Council and the National Football League, as two examples, have developed highly targeted educational programs for employees in these industries.[18,19] Travel agents in Stockholm County (Sweden) worked with the local AIDS Prevention Program to reach travelers and other migrant populations with prevention messages.[20]

Employee fears of occupation risk of transmission greatly exceed actual risks. Even in occupations for which some level of occupation-associated risk can be demonstrated, excepting sex workers, the acquisition of disease through nonoccupational means has been the norm. Workplace educational programs tailored to meet the unique concerns of various occupations are called for, but general workplace program strategies can be applied to nearly every work group.

Workplace Opportunities

If the workplace is viewed only through the microscope of disease risk, most work sites would generate little interest among disease interventionists. When held up to the prism of opportunity, workplaces offer a richer vision.

The decade preceding the discovery of AIDS saw the rapid growth of workplace "wellness" programs, which often included health education, health promotion, and fitness activities. Some of these programs promised economic benefits to employers in the form of reduced health care costs and improved employee morale, productivity, and loyalty.[21–23] It has been demonstrated, for example, that employees who participated in just 30 minutes of exercise five times a week through company-sponsored fitness programs saved firms up to $7 for every $1 spent.[24]

Companies with well funded wellness programs and full-time health promotion specialists were able to integrate an HIV educational program into existing programs with relative ease.[25–27] The issues of sexuality, homosexuality, and illegal drug use can cause difficulties in some work cultures, just as tobacco cessation programs prove awkward in cigarette factories.

The experiences of Pacific Bell, a telephone company serving several West Coast epicenters in the United States, have proved helpful to other companies looking to turn challenges into opportunities. During the early 1980s, the negative effects of rumor and misinformation were beginning to interfere with Pacific Bell's ability to get its job done. Coins from public telephones in areas of towns dominated by homosexual residents, went uncollected. Telephone installers were refusing to enter some hospitals and offices known to see AIDS patients or requested head-to-toe protective covering. A lineman refused to use the truck of a fellow employee, who was rumored to have died from AIDS, until it was sterilized.[27]

Guided by Pacific Bell's in-house director of health education, the company began its workplace program with the reaffirmation of its humane health policy. ("People with AIDS are sick. We don't fire sick people.") Employee fears were both acknowledged and addressed. A specific work crew of volunteers was formed that agreed to accept work assignments that any employee rejected on the basis of AIDS-related fear. Educational programs were implemented that stressed known facts about disease transmission. Over time, this traditionally conservative corporation took an uncharacteristically public stance by encouraging other companies to adopt similar programs, by financially supporting education programs on AIDS for the community at large, and by helping to defeat a statewide referendum that was thought by AIDS activists to be repressive. It is interesting that by acknowledging and accommodating employee fears, the volunteer work crew was never called into action. Moreover, the company won plaudits for its bold actions in the public arena.[27] In keeping with its roots in company wellness programs, contemporary efforts emphasize HIV prevention with as much vigor as was evidenced in earlier education efforts to prevent hysteria.

This emphasis on disease prevention within the work culture is epidemiologically consistent with known facts about the disease on at least two levels. On one level, the workplace is an appropriate classroom for teaching individuals how to reduce their personal risk.

1. *Access*. Most individuals infected in the United States and other countries acquire their infections between the ages of 18 and 44. The largest number and productive members of most workforces fall within the same age range (Fig. 20.1). Whereas nearly one-fourth of U.S. citizens can be found on a school campus on any given day (making schools and obvious choice for implementing health-promoting programs), more than one-half of this nation's at-risk population can be found at their place of work on any given day.[28]

2. *Vulnerability*. General population surveys have shown that risk behaviors associated with HIV transmission (e.g., sex with multiple partners, with strangers, and while under the influence of alcohol) are prevalent among the younger adults in this age group.[4,29] Even members of the "core groups," who frequently engage in multiple high risk behaviors, are at most times gainfully employed in mainstream occupations. At-risk or vulnerable populations work, and their social networks, including sexual networks, often include coworkers. Approximately 90% of HIV-infected individuals are employed.[1]

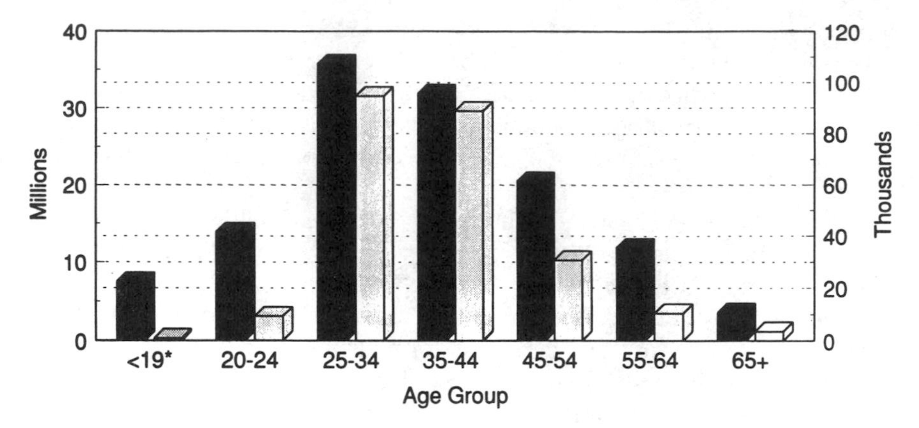

FIGURE 20.1. Age distribution of U.S. workforce and individuals with AIDS. Closed bars = workers (workforce data through 1990); open bars = cases (individuals with AIDS data through June 1992). *Includes only cases in the 13- to 19-year age category; it excludes patients ages 0 to 12.

3. *Readiness*. It is a well known axiom of child education and of adult learning that people learn best when they are ready to learn. Surveys of working adults suggest that they are overwhelmingly (75–80%) desirous of HIV programs in their place of work.[30,31]

4. *Effectiveness*. Employees have reported that they are more likely to heed health advice provided by their employer than advice that comes from their own physicians.[32]

At a second level, employees can be made into family and community AIDS educators through workplace programs. The importance of having health-promoting ideas and behaviors delivered and reinforced through multiple channels and with consistency of message has been well documented.[23]

Educational materials have been developed that can assist adults in their parenting or educational role.[33] The skills needed to talk effectively with young people about sensitive issues can also be taught. Parents desire this knowledge.[31] Putting education materials into their hands increases the likelihood that they will talk with their children about AIDS.[34,35] This type of parental reinforcement of lessons being taught through the schools or in mass media campaigns can contribute to reductions in risk behavior and in the adoption of health-enhancing actions.

One additional opportunity that workplaces offer to disease intervention has not been epidemiologically linked directly to disease prevention but carries the promise of contributing to prevention nonetheless. Businesses and corporations are major contributors of financial and human resources in countries or communities that have a tradition of volunteerism or philanthropy.

Nongovernmental organizations (NGOs) and AIDS service organizations (ASOs) are vehicles for reaching high risk populations with prevention programs and health care services. Ironically, a needs assessment study conducted in a number of developing countries showed that "the full potential of NGOs was not being realized for lack of financial, managerial, and technical support."[3] These areas are among the assets that successful businesses have in abundance. Many corporations have already recognized their capacity for assisting local and national ASOs. As more businesses come to appreciate the heavy toll that AIDS is exacting on manpower availability and productivity, they may be able to support community prevention programs on the basis of "enlightened self-interest" in addition to humanitarian or corporate "good citizen" grounds.

As case numbers escalate, and if medical advances deliver their hoped-for improvements in life expectancy and life quality, the experience of having infected employees remaining productively employed for many years will be commonplace. Maintaining the health of these employees requires employers to focus more attention on the health risks *to* the HIV-compromised person rather than *from* him or her. Corporations must have a large voice in advocating improved health care services and insurance coverage. Uninfected employees must continue to benefit from prevention education that also teaches them to seek reasonable accommodations for infected coworkers.

Perhaps during the second decade of HIV and AIDS, medical epidemiologists will have documented more firmly the broader health benefits of volunteerism, as well as the protective benefits that may derive from the learning gained by participating in disease-specific causes.

Business Responds to AIDS

On World AIDS Day, December 1, 1992, the U.S. Centers for Disease Control and Prevention (CDC) launched a major initiative to increase the involvement of business and labor in HIV prevention efforts. This program serves as a useful model of public, private, and voluntary collaboration to achieve humanitarian, public health, and commerce-benefiting ends.

As the lead governmental agency for HIV prevention programs in the United States, the CDC was able to garner advice and educational products that were available from organizations, agencies, and businesses already addressing HIV workplace issues. From these consultations, and drawing largely on existing material and programmatic resources, the initiative combines modern marketing strategies with nationwide training programs to promote and deliver these services to a wider audience. The CDC's National AIDS Clearinghouse provides an important linkage between businesses and program service by "triaging" requests for material or giving hands-on assistance of a local, national, or multinational nature.

The initiative seeks to engage the country's ten million businesses, most of which are small (fewer than 500 employees), in addressing the challenges of AIDS through comprehensive workplace programs. The recommended program elements are as follows.

1. *Workplace policies.* At a minimum, such policies should address the company's hiring, promotion, transfer, and dismissal policies; the confidentiality protections of employee medical records; benefits available to employees and their families; the manner in which discrimination is handled; and how employee education can be carried out.
2. *Training.* Managers, supervisors, and union leaders should receive training on how the company policies are to be implemented and be given HIV information to answer employee questions and help alleviate unwarranted fears.
3. *Employee education.* All employees should be encouraged to participate in planned educational programs and be offered informational materials that explain company policies and their rationale, provide information on how HIV is and is not transmitted, and explicit information on how to prevent transmission.
4. *Family education.* Families are afforded opportunities to participate in company-supported educational programs, or adult workers are provided materials and instruction to improve and support their skills as family and community educators.
5. *Community involvement.* Employers and employees are encouraged to participate in community planning, financing, and implementation of programs to address the prevention, health care, and social service needs of all citizens and to lend support to school-based comprehensive health programs.

The use of standard marketing strategies and provision of training programs and materials directed toward corporate and labor union leaders, managers, and employee assistance personnel are consistent with theoretic constructs for individual behavioral change. For the Business Responds to AIDS (BRTA) program, the "stages of change" model of Prochaska and DiClementi and Rogers' "diffusion of innovation" provide the theoretic bases for achieving institutional changes.[36,37]

Evaluation of individual workplace programs or national efforts such as BRTA should examine (1) the degree to which theoretically sound programs have been implemented and institutionalized; (2) the degree to which individual employees and their families have participated in the programs and consequently can demonstrate changes in their knowledge, beliefs, health, and health-promoting actions; and (3) ultimately the degree to which workplace, school-based, and community-wide programs have successfully stemmed the tide of an epidemic of portentous impact.

References

1. Jonsen AR, Stryker J (eds). The Social Impact of AIDS in the United States. Washington, DC: National Academy Press, 1993
2. Black RF, Collins S, Boroughs DL. The hidden cost of AIDS. US News and World Report 1992;113(4):48–59
3. World Development Report 1993: Investing in Health. New York: Oxford University Press (for the World Bank), 1993
4. Liskin L, Church CA, Piotrow PT, Harris JA. AIDS Education—A Beginning. Population Reports, 17(8, Series L). Baltimore: Johns Hopkins University, Population Information Program, Center for Communication Programs, 1989
5. AIDSCOM Semiannual Report 6. Washington, DC: Academy for Educational Development (for the United States Agency for International Development), 1990
6. Sepulveda J, Fineberg H, Mann J (eds). AIDS Prevention Through Education: A World View. New York: Oxford University Press, 1992
7. A Curriculum Guide for Public-Safety and Emergency-Response Workers. Atlanta: DHHS (NIOSH), No. 89-108, 1989
8. Information/Education Plan to Prevent and Control AIDS in the United States. Washington, DC: DHHS (PHS), 1987
9. CDC Plan for Preventing Human Immunodeficiency Virus (HIV) Infection—A Blueprint for the 1990s. Atlanta: DHHS (CDC), 1990
10. Adamson-Woods O. An Employer Response Plan for HIV in the Workplace—A Manager's Guidebook. Salem, OR: Oregon Department of Insurance and Finance, 1988
11. Adamson-Woods O. Basic HIV Infection Control for General Work Settings. Salem, OR: Oregon Department of Insurance and Finance, 1989
12. Improving Risk Communication. Washington, DC: National Academy Press, 1989
13. Additional recommendations to reduce sexual and drug abuse-related transmission of human T lymphotropic virus type III/Lymphadenopathy-associated virus. MMWR 1986;35:152–155
14. Hochhauser M, Rothenberger JH. AIDS Education. Dubuque, IA: William C. Brown Publishers, 1992
15. Turner CF, Miller HG, Moses LE. AIDS: Sexual Behavior and Intravenous Drug Use. Washington, DC: National Academy Press, 1989
16. Recommendations for preventing transmission of human immunodeficiency virus and hepatitis B virus to patients during exposure-prone invasive procedures. MMWR 1991;40:(RR-8):1–9
17. Kinsella J. Covering the Plague: AIDS and the American Media. New Brunswick, NJ: Rutgers University Press, 1989
18. Silverstein S. AIDS and the workplace. Los Angeles Times 1992;Nov 7:D1–D2

19. Brown LS, Drotman P. What is the risk of HIV infection in athletic competition? In Abstract Book IXth International Conference on AIDS (PO-21-3102) (Vol. II). Berlin, 1993
20. Country Watch: AIDS Health Promotion Exchange. Geneva: World Health Organization Global Programme on AIDS 1993;1:10–11
21. Blair SN, Piserchia PV, Wilbur CS, Crowder JH. A public health model for work-site health promotion. JAMA 1986;255:921–926
22. Bly JL, Jones RC, Richardson JE. Impact of worksite health promotion on health care costs and utilization. JAMA 1986;256:3235–3240
23. Green LW, Kreuter MW. Health Promotion Planning: An Educational and Environmental Approach. Mountain View, CA: Mayfield, 1991
24. Husted A, Rochell A. Fitness programs pay off. Atlanta Constitution 1993; Sept 10:D4
25. The Workplace Profiles Project. Atlanta: DHHS (CDC), 1992
26. Bunker JF, Eriksen MP, Kinsey J. AIDS in the workplace: the role of EAPs. Almacan 1987;Sept:18–26
27. Kirp DL. Uncommon decency: Pacific Bell responds to AIDS. Harvard Business Rev 1989;May–June:140–151
28. Business Responds to AIDS. Atlanta: DHHS (CDC), 1992
29. Anderson JE, Dahlberg LL. High-risk behavior in the general population—results from a national survey, 1988–1990. Sex Transm Dis 1992;19(6):320–325
30. Employee Attitudes About AIDS: What Working Americans Think. Washington, DC: National Leadership Coalition on AIDS, 1993
31. Kroger CF, Holman P, Sinnock P, et al. Comprehensive workplace response to AIDS. In Abstract Book IXth International Conference on AIDS (PO-D13-3747) (Vol. II). Berlin, 1993
32. Barr JK, Warshaw LJ, Waring JM. AIDS Education in the Workplace: What Employees Think. New York: New York Business Group on Health, 1990
33. AIDS Prevention Guide. Atlanta: DHHS (CDC), 1989
34. Characteristics of parents who discuss AIDS with their children—United States, 1989. MMWR 1991;40:789–791
35. Fisher TD. Parent-child communication about sex and young adolescents' sexual knowledge and attitudes. Adolescence 1986;21:517–527
36. Prochaska JO, Di Clemente CC. Stages and processes of self-change of smoking: toward an integrative model of change. J Consult Clin Psychol 1983;5:390–395
37. Rogers EM. Diffusion of Innovations. New York: Free Press, 1983

21
Standards of Laboratory Practice for HIV Testing

THOMAS L. HEARN

Based on participation in voluntary performance evaluation programs, more than 2000 laboratories across the United States are estimated to be performing human immunodeficiency virus type 1 (HIV-1) antibody testing. Of those that perform HIV-1 testing and are enrolled in the Centers for Disease Control and Prevention (CDC) Model Performance Evaluation Program (MPEP), most are hospital (28%), blood bank including those located in hospitals (27%), independent (16%), and health department (14%) laboratories.[1] Enzyme immunoassays for HIV-1 antibody detection are performed by 654 (97%) MPEP participants who responded to a 1992 survey (n = 674); 224 (33%) performed Western blot testing, 107 (16%) HIV-1 antigen testing, 54 (8%) indirect immunofluorescence tests, 35 (5%) viral culture, 27 (4%) polymerase chain reaction, and 14 (2%) particle agglutination HIV-1 antibody assays.[1] Hence great diversity exists in the types of laboratory performing HIV testing and the scope of HIV-1 testing methods used. In addition to HIV, antibody, and antigen tests, more than 500 laboratories are estimated to be performing $CD4^+$ T lymphocyte enumeration by flow cytometry to assess the status of the immune system of HIV-1-infected persons.

Given the diversity of testing, what is the basis for standards of good laboratory practice in HIV testing, and what are the regulatory requirements that are intended to reassure the public that test results are accurate and reliable? The answer is a combination of voluntary and state and federal regulatory standards.

Voluntary and Regulatory Standards

Advances in the technology for detecting and measuring retrovirus infection and the need to make the technology widely available rapidly stimulated the development of numerous voluntary laboratory practice standards by professional groups and organizations such as the National

Committee for Clinical Laboratory Standards (NCCLS),[2] the Consortium for Retrovirus Serology Standardization (CRSS),[3] and the Association of State and Territorial Public Health Laboratory Directors (ASTPHLD).[4] These voluntary standards are, in general, methodology-specific (e.g., the NCCLS guidelines for flow cytometry[5]) and were devised to assist laboratorians with implementing testing procedures and with maintaining high quality testing practices in laboratories.

Guidance for laboratorians has also been provided through adoption of recommendations that were principally devised for groups of laboratories that perform tests for either a single purpose or for a defined patient population (e.g., the CDC's guidance documents for interpretation and use of the Western blot assay for HIV-1 serodiagnosis[6] and $CD4^+$ T lymphocyte enumeration of HIV-infected persons[7]). Laboratorians are not required to adopt these voluntary practice standards but often do because such standards tend to serve as benchmarks of good laboratory practice at the time they are promulgated.

In addition to the voluntary standards that laboratorians may choose to adopt, all laboratories that perform HIV testing for the purpose of screening, diagnosing, treating, or monitoring patients must comply with applicable standards defined by state and federal laws. (Excluded, for example, are laboratories that perform unlinked testing as part of an epidemiologic investigation.) The standards implementing the Clinical Laboratory Improvement Amendments of 1988 (CLIA '88) are the most comprehensive regulatory standards to date.[8] The net effect of CLIA '88 is to establish a framework for laboratory practice that is uniformly applicable to all facilities that meet the definition of a laboratory stated in the law, regardless of the laboratories' location, clientele, and purposes for testing. In that respect, CLIA '88 is a departure from previous federal laboratory regulations, which were applicable only to laboratories engaged in interstate commerce or those receiving Medicare reimbursement.

CLIA '88 Standards

To be in compliance with CLIA '88 standards, laboratories are required to register with the Health Care Financing Administration (HCFA), apply for a certificate, pay registration and inspection fees, comply with the technical standards, and permit inspections. The extent of the regulations to be applied and met in any laboratory depends on the "complexity" of tests provided by the laboratory. For example, only a Certificate of Waiver is needed for laboratories performing those few tests that have been deemed to have met the criteria for waiver. A Certificate of Waiver exempts laboratories from having to meet the requirements of the technical standards and exempts laboratories from routine inspections. At this

TABLE 21.1. Criteria for categorizing tests as moderate or high complexity.

Knowledge required to perform testing
Training and experience to perform testing
Complexity of reagents or materials preparation
Characteristics of operational steps
Characteristics of calibration, quality control, or proficiency testing materials
Extent or difficulty of troubleshooting and maintenance procedures
Level of interpretation and judgment decisions

time no tests to detect HIV or HIV antibodies (HIV tests) meet the criteria for inclusion as a test qualifying for Certificate of Waiver.

The HIV tests are categorized as either moderate or high complexity tests, with the determination of "moderate" or "high" having been made on the basis of a score derived from an evaluation of seven criteria (Table 21.1). These criteria were applied to essentially every test cleared by the U.S. Food and Drug Administration (FDA) 510k or premarket approval (PMA) processes. A list of more than 12,000 procedures (including 21 HIV-1 antibody tests) has been published with the complexity category noted.[9] Procedures that have not been categorized are all considered high complexity. In fact, with the exception of two commercially available rapid HIV-1 antibody tests—the Cambridge Biotech Recombigen HIV-1 Latex Agglutination Test and the Murex SUDS HIV-1 Test—all HIV tests are high complexity tests. The major differences in the regulatory standards for high and moderate complexity testing are that for moderate complexity tests some of the quality control requirements are to be phased in, and less education and training are required of personnel performing these tests than for performing high complexity testing. Because essentially all HIV tests are in the high complexity group, the standards described are applicable to that category of testing.

CLIA '88 Standards for High Complexity Testing

The CLIA '88 standards fall under the headings of proficiency testing, personnel, quality control, patient test management, and quality assurance (Fig. 21.1). Individually they may be viewed as minimal practice standards that are thought essential for producing safe, reliable test results. Although no one standard alone is likely to be sufficient to give total assurance of high quality results, the assurance comes as a result of their collective synergy. Research has begun and is essential to ascertain and evaluate this concept.

Proficiency Testing

All laboratories are required to participate in DHHS-approved proficiency testing programs for tests or analytes specified in the regulations. For

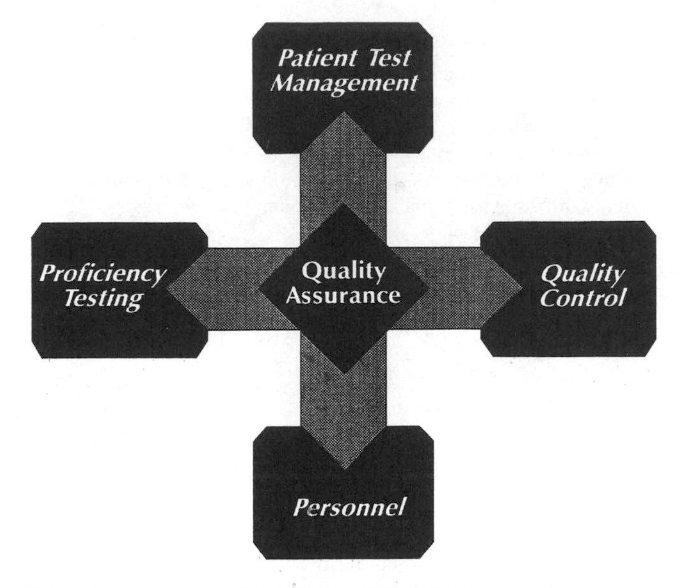

FIGURE 21.1. Components of the CLIA '88 technical regulatory standards.

HIV/acquired immunodeficiency syndrome (AIDS) retrovirus testing, proficiency testing is required only for HIV-1 antibody tests. Proficiency testing providers are approved annually. The providers that offered HIV-1 antibody challenges in their program for the calendar year 1993 were the College of American Pathologists/American Association of Blood Banks, American Association of Bioanalysts, American Proficiency Institute, the states of New York and Wisconsin, and Puerto Rico.

All proficiency testing providers are required to mail five samples three times a year to program enrollees. Laboratories participating in proficiency testing must test the mailed samples in the same way as routine patient specimens, report their results as required by the provider, and take corrective action for unsatisfactory results. A score of less than 80% on a shipment is considered unsatisfactory. Two scores of 80% or less on two consecutive shipments or two of the three most recent shipments is unsuccessful performance and may result in adverse action against the laboratory, including suspension of the laboratory's CLIA certificate to conduct testing.

Personnel

The following personnel functions are described in the technical standards for high complexity testing: laboratory director, clinical consultant, technical supervisor, general supervisor, and testing personnel. Although a minimal level of education and experience has been defined (Table 21.2), it is ultimately the laboratory director's responsibility to be sure that all personnel are capable of and actually do carry out responsibilities described in the regulations.

In some cases a laboratory director may determine that, for its practice, personnel with higher education and more experience than those listed in the regulations are required. In that event the laboratory personnel should meet the laboratory's higher standard. A large percentage of MPEP laboratories (68%), for example, reported that they required their testing personnel to have an academic degree, and most required a bachelor's or master's degree.[1]

A different person is not required for each functional position. So long as an individual meets the educational and experiential requirements, he or she could carry out all of the functions alone. In addition, some tasks related to responsibilities may be delegated to qualified personnel. For example, the laboratory director is responsible for ensuring that the test methodologies selected have the capability of providing the quality of results for patient care; the selecting of test methodology could be delegated to the technical supervisor, although the director ultimately remains responsible.

Patient Test Management

The patient test management portion of the standards emphasizes the importance of the events outside the analytic performance of a laboratory test that profoundly influence the quality or outcome of testing. That is, the preanalytic steps of testing (e.g., specimen collection, transport, and labeling) and the postanalytic steps (e.g., results verification and reporting) are important.

TABLE 21.2. Minimum education and experience required for laboratory personnel performing high complexity testing.

Personnel	Education	Experience
Director	Doctor of Medicine; Doctor of Podiatric Medicine; Doctor of Osteopathy; doctoral degree in a chemical, physical, biologic, or clinical laboratory science plus board certification after September 1, 1994	2 Years (None required for doctoral degree laboratory scientists who have board certification)
Technical supervisor	Bachelor's degree	4 Years
Clinical consultant	Same as Director	None required
General supervisor	Associate degree in a laboratory science or medical technology	2 Years
Testing personnel	High school diploma or equivalent until 1997; thereafter, associate degree in a laboratory science or medical technology	Documentation of training appropriate for testing performed

For HIV testing the issue of specimen labeling is even more critical in settings where anonymous HIV-1 antibody testing is conducted and the usual patient identifiers of name and address are not used. Test request forms must contain a unique patient identifier, the name and address or other suitable identifiers of the authorized person requesting the test, the test to be performed, and the date of the request.

Test reports must indicate the name and location of the laboratory performing the test, the test performed, and the test results. The laboratory is required to keep a record of the results for 2 years from the date issued.

Quality Control

To satisfy quality control requirements, HIV laboratories must establish and follow quality control procedures for monitoring and evaluating the quality of the analytic testing process to assure the accuracy and reliability of patient test results and reports. Critical elements of quality control include having a written procedures manual that describes the entire testing process, verifying the performance characteristics of methods used in the laboratory, conducting calibration and calibration verification at specified intervals and as warranted by testing outcomes, incorporating quality control samples into testing with every run or at least once within every 24 hours that tests are performed, maintaining equipment, taking remedial action when necessary, and documenting records of all quality control activities.

Based on data from MPEP surveys, most HIV testing laboratories are either already complying with quality control standards or can with little additional effort. Compliance with the quality control standards is greatly simplified for most laboratories performing HIV tests, as nearly all of the HIV tests are performed using commercial reagents. When FDA-cleared tests are used, the following items pertain.

1. Laboratories may use all applicable parts of the commercially supplied testing instructions for much of their procedures manual.
2. Laboratories must validate only test performance characteristics established by manufacturers versus establishing them de novo, which would require significantly more effort.
3. Laboratories are generally in compliance if they follow manufacturers' instructions for calibration and quality control. (*Note*: The CLIA '88 standards include provisions for the FDA to review manufacturers' quality control instructions and clear those that comply with the CLIA '88 regulations so the instructions can be included in the manufacturers' labeling for laboratories to follow.)

There are two often asked questions: Can quality control samples included with commercially manufactured tests be used to satisfy the

CLIA '88 quality control requirements? and What must laboratorians do to verify performance characteristics? First, the quality control samples provided with commercially manufactured tests may be used to satisfy the CLIA '88 quality control requirements so long as they exhibit appropriate reactivity (e.g., positive and negative) and mimic, inasmuch as possible, actual patient samples. Recommendations from the ASTPHLD annual conferences on human retrovirus testing, however, have indicated that an independent source of quality control samples may be preferable to the quality control samples included with commercial tests.[10–13]

Second, laboratories are provided great latitude for deciding exactly what constitutes sufficient data to verify performance specifications of commercially manufactured tests. Several runs incorporating patient samples of known reactivity and samples from an appropriate reference population would be a minimal expectation. Guidance on protocols may be found in NCCLS standards.[14] If tests are not commercially manufactured (i.e., if they are in-house tests), a complete method validation is required, including the establishment of accuracy, precision, sensitivity, specificity, reportable ranges of results, and reference ranges.

Quality Assurance

Quality assurance is the linkage among all the components of the regulatory standards (Fig. 21.1). Each laboratory is required to establish and conduct a quality assurance program that assesses the effectiveness of its policies and procedures; identifies and corrects problems; assures accurate, reliable, and prompt reporting of tests results; and assures the adequacy and competence of its staff. Within the context of the CLIA '88 regulations, the quality assurance standards serve as the quality management or quality improvement criteria that require laboratories to evaluate and improve on their practices in an ongoing way.

To be in compliance with the quality assurance standards, each laboratory must take corrective actions and evaluate the effectiveness of those actions. Such actions must be taken when, for example, proficiency testing results are unsatisfactory, problems are detected by quality control, personnel performance problems are identified, or samples are improperly submitted to the laboratory or results improperly reported.

An important element of quality assurance as described in the regulations is the requirement for assuring that results for a given analyte from a laboratory are comparable even if derived by different methodologies or by the same methodologies and instruments used in a different testing location. In those instances and in cases where a test is performed by the laboratory but proficiency testing is not required by regulation, each laboratory is required to verify the accuracy and reliability of its testing at least twice a year. An easy way to satisfy this regulatory requirement is to participate in external performance evaluation programs or to "split" samples with other laboratories. Either way, important independent assessments can be made for each test offered.

Future of Standards

Voluntary and regulatory standards appear to have served the laboratory community and the public well in terms of ensuring high quality HIV testing and, for that matter, testing in general. Both mechanisms are in place. New standards will be developed and old ones deleted or modified as knowledge of the outcomes of testing are better measured and as methodologies improve or change. For example, the CDC guidelines for $CD4^+$ T lymphocyte testing have already been revised. Expectations are that changes in forthcoming CLIA '88 regulations will also occur as implementation brings about new knowledge on how to ensure the quality of testing.

References

1. Centers for Disease Control and Prevention. Results of Survey Mailed to Laboratories August 1992. Report to Participants in the Model Performance Evaluation Program: Testing for Human Immunodeficiency Virus and Human T-Lymphotropic Virus Types I and II. Atlanta: CDC, 1993
2. National Committee for Clinical Laboratory Standards. Nomenclature and Definitions for Use in NRSCL and Other NCCLS Documents (2nd ed). NCCLS Publication NRSCL8-P2. Villanova, PA: NCCLS, 1993
3. Consortium for Retrovirus Serology Standardization. Serological diagnosis of human immunodeficiency virus infection by Western blot testing. JAMA 1988;260:674–679
4. Committee on Human Retrovirus Testing. Report of the Eighth Consensus Conference on Testing for Human Retroviruses. Washington, DC: Association of State and Territorial Public Health Laboratory Directors
5. National Committee for Clinical Laboratory Standards. Clinical Applications of Flow Cytometry. Quality Assurance and Immunophenotyping of Peripheral Blood Lymphocytes. NCCLS Publication H24-T. Villanova, PA: NCCLS, 1992
6. Centers for Disease Control and Prevention. Interpretation and use of the Western blot for serodiagnosis of human immunodeficiency testing virus type 1 infections. MMWR 1989;38:1–7
7. Centers for Disease Control and Prevention. Guidelines for the performance of $CD4^+$ T-cell determinations in persons with human immunodeficiency virus infection. MMWR 1992;41:1–17
8. Department of Health and Human Services. Medicare, Medicaid, and CLIA programs: regulations implementing the clinical laboratory improvement amendments of 1988 (CLIA). Fed Register 1992;57(40):7002–7243
9. Department of Health and Human Services. Compiled list of clinical laboratory test systems, assays, and examinations categorized by complexity: notice. Fed Register 1993;58(141):39860–39973

10. Association of State and Territorial Public Health Laboratory Directors (ASTPHLD). Report of the First State Laboratories Consensus Conference on HTLV III/LAN Serology, Kansas City, MO, May 1986
11. Association of State and Territorial Public Health Laboratory Directors (ASTPHLD). Report of the Second Consensus Conference on HIV Testing, Atlanta, March 1987
12. Association of State and Territorial Public Health Laboratory Directors (ASTPHLD). Report of the Fourth Consensus Conference on Testing for Human Retrovirus, Kansas City, MO, March 1989
13. Association of State and Territorial Public Health Laboratory Directors (ASTPHLD). Report of the Sixth Annual Conference on Human Retrovirus Testing, Kansas City, MO, March 1991
14. National Committee for Clinical Laboratory Standards. User comparison of quantitative clinical laboratory methods using patient samples. NCCLS Publication EP9-P. Villanova, PA: NCCLS, 1986

22
HIV Biosafety: Guidelines and Regulations

JONATHAN Y. RICHMOND

Risk has been defined as hazard plus outrage.[1] When the outrage (perceived concern, outcry, demand for change) concerning an event is high, even when the hazard is low, activities associated with the event are believed to be risky. Alternatively, even when the hazard has been quantified to be real, the risk is construed to be low if the event is perceived to be acceptable.

Some persons handling human blood and body fluids, for example, may not have considered their work particularly risky relative to becoming infected with hepatitis B virus (HBV) because they do not perceive the consequences of infection to be serious. The frequency of HBV infection following a needlestick injury involving a known HBV-contaminated source is about 3 in 10, with sequelae ranging from inapparent infection to cirrhosis of the liver to death.[2,3] About 125 health care workers die every year from HBV infection, and a considerable number become silent carriers.[4] Despite the significant hazard, this perceived low risk has deterred many health care workers from receiving HBV vaccine. Further complicating the acceptance of HBV vaccine is the perceived risk that the human immunodeficiency virus (HIV) could be transmitted in serum-derived HBV vaccines. Even the newer recombinant subunit HBV vaccines, which are not serum-derived, have not been universally accepted by health care workers, even when it is understood that the resulting immunity is excellent.

In contrast to HBV risk, the risk of infection with HIV following a needlestick injury involving blood containing HIV is low, approximately 0.3%,[5] or from a mucous membrane exposure, approximately 0.09%[6]; yet the concern (outrage) is high. Working with materials that may contain HIV is perceived to be hazardous because the consequences of HIV infection are considered more serious than those of HBV infection.

The data continue to indicate that a measurable occupational risk applies to persons working with HIV-infected individuals or infectious samples derived from them, but that the risk is low. We have also accumulated considerable information regarding the activities and manipulations that most often contribute to occupationally acquired HIV infections.

During the dozen or so years health care workers have worked with hundreds of thousands of HIV-infected patients (often not knowing they were infected), there have been 37 documented occupational HIV transmissions.[7] An additional 78 possible occupational transmissions have been reported.[7] The latter health care workers have been investigated and are without identifiable behavioral or transfusion risks. Each worker reported percutaneous or mucocutaneous occupational exposure to blood or body fluids or laboratory solutions containing HIV, but HIV seroconversion specifically resulting from an occupational exposure was not documented. For most of the 78, the facts will not be known. No baseline serum is available, or no exposure route is documented.

Health care workers with patient care responsibilities and those who handle the millions of blood and body fluid samples each year in clinical laboratory settings may also be at risk of HIV infection. As of June 1993, the U.S. Centers for Disease Control and Prevention (CDC)[7] had reports of 14 clinical laboratory workers (mostly phlebotomists) in the United States with documented occupational transmission and 13 reports of possible occupational transmissions. Research laboratorians who manipulate cultures of HIV are also at risk of infection. One person working with large volumes of concentrated HIV became infected, and two others have been infected in the workplace.

The purpose of this discussion is to examine the risks associated with working with HIV in laboratories and to review the biosafety practices and procedures, engineering controls, and managerial activities that have kept the number of occupationally acquired infections remarkably low. This statement is not written to trivialize the fact that some workers have been infected on the job; for them the consequences are severe. Rather, let us understand why so many workers have *not* been infected so we can benefit from that knowledge and strive to reduce the probability of additional occupationally acquired infections.

The three cornerstones that underlie our current standards of practice are discussed below.

1. Beginning during the late 1940s, Sulkin, Pike, and others[8–10] published a series of papers that categorized more than 7000 laboratory-associated infections resulting in about 150 fatalities. Pike concluded in a 1979 review[10] that "the knowledge, the techniques, and the equipment to prevent most laboratory infections are available." At that time the most widely circulated comprehensive guidelines for working safely with infectious microorganisms were found in a publication by the CDC.[11] This booklet introduced the concept of dividing microorganisms into four levels of increasing hazard with general guidelines for each level, and it

provided the basis for the National Institutes of Health (NIH) guidelines for research involving recombinant DNA[12–14] and the 1993 CDC/NIH publication on Biosafety in Microbiological and Biomedical Laboratories (BMBL)[15] and earlier editions.

Each of these sets of guidelines focused on the known human infectiousness of the microorganism being manipulated. They were based on different risks encountered in routine clinical laboratory activities, typical research laboratories, procedures that require infection of laboratory animals, and production facilities where large volumes of infectious microorganisms are grown, manipulated, and concentrated. The guidelines described the facilities and engineering controls required for each level, the recommended practices and procedures, and the specific personal protective equipment appropriate for the work being performed.

The BMBL guidelines can also be used to delineate the practices and procedures and the biosafety level appropriate for working with unknown agents. During the early years of the HIV/acquired immunodeficiency syndrome (AIDS) epidemic, before a virus had been isolated, laboratorians were able to perform their clinical/diagnostic work safely under biosafety level 2 (BSL-2) conditions. This was because epidemiologic evidence suggested that the unknown agent was transmitted in a manner similar to HBV (through sexual contact, blood exposure, or parenteral exposure). Most important, the epidemiologic data demonstrated that this agent was not transmitted through aerosol exposures, and full biosafety level 3 (BSL-3) facilities were not required for research activities, although some BSL-3 practices and procedures were adopted. Laboratorians were thus able to work safely in and around HIV laboratories without having to develop an entirely new set of biosafety guidelines.

2. Early during the HIV epidemic the concern for personal safety (e.g., outrage) was very real in the health care setting and in community activities that involved persons living with HIV/AIDS. Again, epidemiologic evidence weighed against casual transmission of HIV. In any case, it is impossible to conclude that anyone is infected with HIV by looking at them. Except in rare situations, in some states it is not possible to require HIV testing of individuals. Even if serologic testing of all patients were possible, that information would be of limited value because of the relatively long latent period between initial exposure and the appearance of detectable antibody often associated with this infection.

Of real concern is the potential exposure to human blood or to bloody body fluids. During the mid to late 1980s, reports accumulated of occupationally acquired HIV infection following needlestick injuries and exposure of unprotected skin or mucous membranes.[16] The obvious need for standardized guidelines was met in 1987 with universal precautions.[17,18]

Universal precautions (UPs) were developed from the perspective that *all* blood and other body fluids *might* contain HIV, HBV, or other microbial pathogens, and the worker must take certain actions to prevent contact with these materials (e.g., wearing gloves, avoiding sharps, using disinfectants, and washing hands) (Table 22.1). Exquisitely simple and easy to follow, UPs were adapted to a variety of occupations where a potential for blood exposure existed.[19,20]

3. The Occupational Safety and Health Administration (OSHA), drawing on UPs and the BMBL, rigorously expanded the application of these guidelines in 1991.[21] Their document is commonly referred to as the "blood-borne pathogen standard" (BBPS). Most individuals with a reasonably anticipated occupational exposure of the skin, eyes, or mucous membranes or with parenteral contact with blood or other potentially infectious materials were to be covered by this regulation. Under this law, employers classify employees into risk categories, offer HBV vaccine to those at high risk, develop a written exposure control plan, conduct specified annual training, and provide needed engineering controls and personal protective equipment (Table 22.2).

Among the 15 laboratorians with documented occupational HIV transmission discussed earlier, 11 had percutaneous exposures, 3 had mucocutaneous exposures, and 1 had both percutaneous and mucocutaneous exposures. Thirteen were exposed to blood from an HIV-infected person, one to visibly bloody body fluid, and one to concentrated live HIV. The procedure most often associated with transmission was phlebotomy. Several of these workers have developed AIDS.[22]

The remainder of this chapter focuses on those biosafety issues gleaned from UPs, the BMBL, the BBPS, and other[23,24] guidelines and regulations

TABLE 22.1. Essential elements of universal precautions.

Barrier precautions are used at all times.
Gloves are worn when handling BBFs; they are optional at volunteer donor sites.
Phlebotomists wear gloves with an uncooperative patient, with nonintact skin, when performing skin puncture, and when in training.
Gloves are changed between patients.
Face shields/guards are worn for protection from splattering.
Gowns or aprons are worn to protect street clothing.
Hands are washed if contaminated.
Hands are washed after removing gloves.
Care is taken to minimize injuries.
Appropriate needle discard containers are used.
Needles are not removed, sheared, or bent.
All sharps are handled carefully.
Care is taken to avoid spills and splatters.
All surfaces and devices are decontaminated after use.

BBF = blood or body fluids.

TABLE 22.2. Major components of the blood-borne pathogen standard.[a]

Written exposure control plan
Work exposure determinations
Schedule and implementation of the plan
Procedure for evaluating exposure incidents
Exposure determinations detailing job classifications where all or some employees are at risk of exposure and a list of tasks that put them at risk
Methods of compliance
Following universal precautions
Following a hierarchy of controls
Engineering
Work practice
Personal protective devices
Maintaining a clean and sanitary work site
Meeting requirements for regulated waste
Safely handling contaminated laundry
Providing appropriate training
Offering free HBV vaccinations to at-risk employees
Providing postexposure evaluation and follow-up
Maintaining specified medical records
Providing employees copies of requested medical records
Communicating through use of biohazard signs and labels

[a] OSHA's rule applies to all persons occupationally exposed to human blood or other potentially infectious materials, such as human semen, vaginal secretions, cerebrospinal fluid, saliva during dental procedures, any body fluid visibly contaminated with blood, and all body fluids in situations where it is difficult to differentiate between body fluids; unfixed human tissues or organs; HIV- or HBV-containing cells or tissue cultures, organ cultures, and culture medium or other solutions as well as blood, organs, or other tissues from experimental animals infected with HIV or HBV.

that apply specifically to HIV. In addition to HIV, HBV and other primary and opportunistic pathogens may be present in body fluids and tissues. Laboratory and other workers who follow these biosafety practices can ensure maximum protection against inadvertent exposure to any pathogens that may be present in clinical specimens.

Potential Exposures to HIV in Laboratories

The HIV has been isolated from blood, semen, saliva, tears, cerebrospinal fluid, amniotic fluid, breast milk, cervical secretions, and tissue of infected persons and experimentally infected nonhuman primates.[25] Since 1987 the CDC has consistently recommended the use of UPs when handling blood or any blood-contaminated body fluids; this practice precludes the need to identify clinical specimens obtained from patients known or suspected to be infected with a blood-borne pathogen.[17,18]

Identifying Persons at Risk

It is generally easy to determine which job classifications include persons at occupational risk of exposure to blood or body fluids: Phlebotomists, health care professionals who work directly with patients, clinical laboratory workers, and laboratorians working with human retroviruses are examples. Identifying people by job classification alone is not necessarily an accurate determination, however. For example, some house-cleaning personnel work in operating rooms, emergency rooms, laboratories, or morgues, whereas others work only in the cafeteria, administrative offices, and similar nonpatient areas. It is therefore necessary to identify both the position and the tasks or procedures associated with the position, as the tasks define the risk of exposure to blood-borne pathogens. These requirements are part of OSHA's exposure control plan.[21]

Tracking at-risk personnel over time is difficult, particularly in large institutions where career advancement or other job changes may affect the exposure risk of individual employees. Tracking is necessary, however, so new at-risk employees can be offered HBV vaccination, for example, or be provided appropriate training and subsequent annual updates. Personnel officials must ensure that job announcements and position descriptions properly identify and address particular positions as having potential occupational risk of exposure to blood-borne pathogens. Supervisors also play a critical role by ensuring that new employees receive proper orientation to their jobs and are included in the necessary programs.

Engineering and Work Practice Controls

Wherever possible, engineering and work practice controls are used to eliminate or minimize worker exposure. Personal protective devices are also used to minimize remaining routes of occupational exposure to HIV (from blood and body fluids). Because these routes include needlestick or other sharps injuries, mucous membrane exposures (eyes, nose, mouth), or contact of open wounds (e.g., cuts) or broken skin (e.g., dermatitis), several approaches apply to each: Use a safer device, develop a safer procedure, or place a barrier between the worker and the infectious materials.

Extensive training has heightened awareness to the risks associated with manipulating hollow-bore needles for injection or sample collection purposes, as well as solid needles used for suturing. In addition, devices have been developed, such as self-sheathing needles, needleless intravenous systems, and numerous other new products engineered specifically to eliminate or reduce the risks. When needles must be used, they should not be recapped, bent, clipped, or removed by hand. If recapping or removal is necessary, however, a protective device that allows one-handed manipulation is available. Collection boxes for used needles should be puncture-resistant and leak-proof.

Within the context of the BBPS, responsibility is placed on the employer to ensure that safety-related equipment is provided and particular work practices are followed. For example, sinks for hand-washing must be available at or near the work station. When the occupational exposure potential is at a remote site (e.g., blood collections in nonmedical facilities or rooms, accident or crime scenes), the employer is required to provide either an appropriate antiseptic hand cleanser and clean towels or antiseptic towelettes. Employees still should wash their hands with soap and water as soon as is practical. Both the employer and employee share responsibility for *rigorously* following all prudent practices and procedures.

When blood or body fluids are manipulated, particular attention must be given to minimizing the possibility of splatter, splashing, spraying, or creation of droplets. Most blood and body fluid manipulations are done in laboratories, and special practices are described below. When blood and body fluid samples must be moved (e.g., from the collection site to the laboratory), they are placed in a secondary container. If the sample container breaks, the sample does not contaminate public space (e.g., elevators, hallways, cars). The secondary container must be rigid and leak-proof. This principle of "double packaging" is applied when blood and body fluids are be shipped in the mail or by courier service.[26] The clinical specimen must be packaged so it does not break; the container should be surrounded by absorbent material sufficient for the volume of the liquid. If the primary container breaks, the contents must be absorbed so the secondary container does not leak.

This situation introduces another biosafety principle, particularly for laboratorians: Make sure the work is safe for others. Spills on work surfaces, equipment, or anything else that may have been splattered with blood and body fluids must be decontaminated promptly. Furthermore, potentially infectious materials remaining after or generated by the research activities must be discarded properly. Chemical disinfection, steam autoclaving, incineration, and other recognized treatments, or double packaging for off-site treatment or burial, are appropriate. Waste materials are handled in such a way that the next person who handles them does not sustain an occupational exposure to the blood and body fluids.[27]

Personal Protective Equipment

The BBPS recognizes as "appropriate" only that personal protective equipment (PPE) that offers significant barrier protection. Barriers must

TABLE 22.3. Examples of personal protective equipment.

Gloves
Gowns
Laboratory coats
Face shields
Masks
Eye protection
Mouthpieces
Resuscitation bags
Pocket masks
Other ventilation devices

be impervious to blood and other infectious fluids, protecting an employee's clothing, skin, eyes, mouth, or other mucous membranes under normal conditions of use and for the duration of the time the PPE is in use. Examples of PPE listed in the BBPS are enumerated in Table 22.3. Employers must supply, clean, replace, and repair PPE free of charge to the employees.

All PPE must be removed and left at the work site, including laboratory coats, scrub suits, and gowns. Laboratory coats are worn to protect clothing from splatters or spills of various hazardous materials in the laboratory (see discussion on biosafety cabinets below). Laboratory clothing must not be worn to the office, classroom, library, or cafeteria. Gloves, masks, and eye protection also must be removed and left at the work site.

Wearing gloves during activities that may involve blood or body fluid (BBF) exposure is one example of using an appropriate barrier. The type of glove selected should be matched to the task. Latex gloves may be necessary for laboratory or surgical activities, but a more durable household glove may offer more protection to a person responding to an accident scene (where the gloved hands may encounter broken glass, for example) or for someone handling potentially infectious wastes. Household gloves have the added advantage that (if intact) they can be surface-decontaminated and washed for reuse; latex surgical gloves cannot be reused.

To date no gloves have been developed that are puncture-proof and still provide the tactile sensitivity and dexterity needed for many delicate procedures. Protective gloves are available that guard against slicing accidents and are useful in some activities, such as autopsy or necropsy.

Surgical masks may be worn for two fundamental reasons: to protect the mucous membranes of the nose and mouth when the potential for splatter of BBF is high, and to keep the person's fingers out of their nose and mouth. HIV is not transmitted by the aerosol route, so a respirator is not required.

Eye protection should also be worn when the potential for BBF splatter is high. Many persons prefer a face shield to goggles for comfort reasons.

It is a legitimate reason for selecting one type of PPE over another, as uncomfortable or overly restrictive ones may be discarded by the user.

Work in Research Laboratories

The BBPS does not include specific recommendations for workers in clinical or diagnostic laboratories engaged solely in the analysis of blood, tissues, or organs. The specific recommendations for HIV and HBV research laboratories and production facilities provided in the BBPS are found in the BMBL; the following discussion is therefore based on the BMBL guidelines,[15] which recommends that human serum from any source that is used as a control or reagent in a test procedure should be handled at BSL-2.

In addition to the information presented in the BBPS, the CDC/NIH recommends the following precautions.[15]

1. BSL-2 standard and special practices, containment equipment, and facilities are recommended for activities involving all blood-contaminated clinical specimens, body fluids, and tissues from all humans or from HIV- or simian immunodeficiency virus (SIV)-infected or inoculated laboratory animals.
2. Activities such as producing research laboratory-scale quantities of HIV or SIV, manipulating concentrated virus preparations, and conducting procedures that may produce droplets or aerosols are performed in a BSL-2 facility but using the additional practices and containment equipment recommended for BSL-3.
3. Activities involving industrial-scale volumes (often produced in multiliter fermenters) or preparations of concentrated HIV or SIV are conducted in a BSL-3 facility using BSL-3 practices and containment equipment.
4. Nonhuman primates or other animals infected with HIV or SIV are housed in animal biosafety level 2 (ABSL-2) facilities using ABSL-2 special practices and containment equipment.

Table 22.4 lists the key elements of the BSL-2 guidelines; and the reader should refer to the BMBL[15] for complete details. All laboratory personnel working at BSL-2 must have specific training in handling blood and body fluids and HIV cultures. Furthermore, they must be directed by competent scientists. When working in the laboratory with infectious materials, access to the laboratory is restricted. Extreme caution applies while handling sharp items. A biological safety cabinet or other physical containment equipment is used whenever there is a potential of creating infectious aerosols or splashes.

The essential difference between BSL-2 and BSL-3 is that BSL-3 facilities are designed for work with infectious disease agents that can be

TABLE 22.4. Major components of biosafety level 2.

Standard microbiologic practices
Restricted access to laboratory
Wash hands after handling HIV materials, after removing gloves, before leaving the laboratory
Eating, drinking, smoking, handling contact lenses, applying cosmetics prohibited
Mouth pipetting prohibited
Minimize creation of aerosols and splatters
Decontaminate work surfaces daily and after an HIV spill
Decontaminate HIV contaminated materials before disposal
Special practices
Immunocompromised persons restricted from the laboratory
Laboratory director advises of potential risks in laboratory
Biohazard warning sign on door listing requirements for entry and emergency contact persons
HBV immunization given, as appropriate
Baseline serum sample obtained; laboratorians participating in an appropriate medical surveillance program
Biosafety manual prepared; read by laboratorians
HIV-specific biosafety training provided
Special precautions taken with sharps
Leak-proof containers used for transport or storage
Appropriate disinfectants used on work surfaces, equipment
HIV and other pathogen spills and accidents reported to supervisor; medical evaluation and surveillance initiated; confidential written records maintained
Safety equipment (primary barriers)
Use of certified biological safety cabinets (BSCs)
Use of appropriate PPE when manipulating HIV outside the BSC
Two pairs of gloves worn when handling HIV cultures or other materials containing HIV
Laboratory facilities (secondary barriers)
Hand-washing sink and eyewash facility present
Laboratory easily cleanable; no rugs
Bench tops impervious to water, resistant to acid
Method for decontamination of infectious waste available (autoclave, chemical disinfection system, incinerator, other approved decontamination system)

transmitted by the aerosol route. Certain manipulations of potentially lethal agents in BSL-2 facilities are carried out using some of the practices and procedures specified for BSL-3 (refer to the BMBL for complete details).

The BSL-3 laboratories must be set apart from other facilities, which is accomplished by having a double-door entry into the laboratory. It may be a "laboratory within a laboratory" arrangement, creation of an anteroom (which would provide a place for gowning, handwashing, and so on), or, where several BSL-3 laboratories exist, installing a door in a dead-end hallway to create this barrier space.

Directed single-pass air movement is an engineering control for containing agents potentially transmitted by the aerosol route. Air moves from an area of lesser contamination (generally the hallway) to the area of greater contamination (the laboratory) in BSL-3 facilities, where it is then exhausted out of the building. Another engineering control results from sealing *all* penetrations through the walls and by installing floor, wall, and ceiling coverings that are monolithic (e.g., no seams or other sources of potential leakage). Should the need arise to decontaminate the laboratory totally, the air supply and exhaust systems can be blocked and a gaseous decontaminant (e.g., formaldehyde gas) can be released. A sealed laboratory is critical to preventing escape of the decontaminating gas.

Practices and Procedures

Laboratorians who work with cultures of HIV require special training and must be supervised by scientists experienced in working with retroviruses. A performance-based set of guidelines (such as those of the BMBL) should be adopted by the laboratory and augmented to contain site-specific detailed biosafety protocols to be followed by the laboratory workers. Biosafety manuals based on the BMBL but created by the individuals doing the HIV work are much more likely to reflect the nuances of the work location and to be followed.

Particular attention should be given to three areas: rigorously following a sharps injury prevention program, manipulating HIV in containment devices, and having an active counseling/medical surveillance program.[28] In addition to the safe handling of needles already discussed, attention must be focused on scalpel blades, glass Pasteur pipettes, coverslips, and any broken glass. Whenever possible, glass should be avoided. Virtually all sizes of culture flasks, transfer devices, and so on are available in plastic or metal. Should there be a need to pick up a dropped or broken sharp object, care should be exercised to use a forceps, hemostat, brush and pan, or other manipulating device instead of the gloved hand alone.

Biological Safety Cabinets

Biological safety cabinets (BSCs) are the most widely used primary containment devices for activities involving manipulations of viable HIV; suggestions for working safely with body fluids in BSCs have been published.[29] The BMBL[15] contains diagrams and simple explanations of the airflow patterns found in each of the classes and types of BSC; all cabinets must be certified by qualified technicians (at the time of installation, after being moved, and annually) to ensure that these airflow patterns and the integrity of the filters are maintained.

Both supply and exhaust air passes through high efficiency particulate air (HEPA) filters designed to remove 99.97% of particles 0.3 μm in size; particles larger and smaller are trapped with even greater efficiency. The consequence of passing through the HEPA filter therefore is that the air flowing onto the work surface or out of the cabinet is virtually sterile. By observing good microbiologic techniques, including surface decontamination with 70% alcohol of items placed in the BSC, it is possible to maintain sterility during standard tissue culture techniques without the use of an open flame.

All required items for a particular procedure should be placed in the BSC (which has been turned on) before work commences. The laboratorian should be able to complete the task without removing his or her arms from the cabinet, as such motion would create turbulence at the face of the cabinet and disrupt the delicate airflow balance. Should it be necessary to remove one's arms and subsequently replace them in the BSC, the arm motion should be slow and perpendicular to the plane of the sash. The chair should be adjusted to a height such that the laboratorian's armpits are at the same height as the bottom of the sash. This positioning keeps the breathing zone of the worker above the area where there is greatest potential for exposure to contaminated air; it is also the best position to view the work surface through the sash.

While manipulating materials containing HIV, including tissue cultures, the laboratorian's arms should not rest on the edge of the cabinet. Such a position disrupts the inward airflow and can create an opportunity for potentially contaminated cabinet air to flow along the laboratorian's arms and into his or her lap. (Personal clothing can be protected from possible contamination by wearing a buttoned or solid-front laboratory coat when working in a BSC.)

The downward-flowing HEPA-filtered air supply splits when it reaches the work surface at a line approximately one-third the distance from the front air grill; the air then flows forward or backward, to be entrained in the cabinet's internal recirculation system. Manipulations of materials containing HIV should be done in the back two-thirds of the work surface; splatters or small droplets are then drawn away from the laboratorian's arms.

Plastic-backed absorbent laboratory pads should be placed on the work surface; they should not be allowed to interfere with the front or back air return grills. Should drops of HIV-containing liquids fall to the work surface, they are absorbed and so do not splatter as they would if they fell on bare stainless steel. These pads should be discarded as infectious waste when cleaning up after the procedure is finished. All materials should be removed from the BSC and the work surface of the cabinet decontaminated with 70% ethanol or other suitable disinfectant. It is not necessary to turn the cabinet off at the end of the work day.

Medical Surveillance

As indicated in Table 22.4, laboratorians working with HIV (or materials suspected of containing HIV) must be enrolled in an appropriate medical surveillance and counseling program.[28] Before work begins, a baseline serum sample should be obtained and stored for future reference. A comprehensive program would include additional serum collections and HIV antibody testing every 3 months or at the time of an overt exposure to blood and body fluids or HIV-containing material.

Another significant part of the medical surveillance program is the educational component. Training should be provided before work begins and at least annually thereafter.[28] Included in the program is a discussion of the risk factors associated with working with HIV, the mechanism for reporting spills of infectious materials and any injuries resulting in (potential) exposures, the availability and possible efficacy of prophylactic zidovudine (AZT),[30] and the follow-up surveillance that is made available to any exposed workers. Details of these programs are made available in writing to all laboratorians, with periodic reminders of the procedures to be followed in the event of an exposure.

Conclusion

The risk of acquiring HBV or HIV infection in occupational settings is small but real. Wherever possible, engineering controls are used to reduce the risk. Specific work practices and procedures have also proved effective for safely handling blood and body fluids or manipulating materials containing these viruses. In some instances specific personal protective safety devices are also recommended. Supervisors are required by law to train employees and provide safety-related materials. Individual employees are ultimately responsible for rigorously following all biosafety guidelines, practices, and procedures that have been implemented to minimize the risk of acquiring either HBV or HIV in the workplace.

References

1. Hance BJ, Caron C, Sandman PM. Industry Risk Communication Manual. Boca Raton, FL: CRC Press/Lewis Publishers, 1990
2. Grady GF, et al. Hepatitis B immune globulin for accidental exposures among medical personnel. J Infect Dis 1978;138:625–638
3. Seef LB, et al. Type B hepatitis after needlestick exposure: prevention with hepatitis B immune globulin. Ann Intern Med 1978;88:285–293
4. CDC. Unpublished data, 1993
5. Bell DM. Human immunodeficiency virus transmission in health care settings: risk and risk reduction. Am J Med 1991;91(suppl 3B):294S–300S

6. Ippolito G, Puro V, De Carli G. The risk of occupational human immunodeficiency virus infection in health care workers; Italian Multicenter Study, the Italian study group on occupational risk of HIV infection. Arch Intern Med 1993;153:1451–1458
7. Centers for Disease Control. HIV/AIDS Surveillance Report. 1993;5(2):13
8. Sulkin SE, Pike RM. Viral infections contracted in the laboratory. N Engl J Med 1949;241:205–213
9. Pike RM, Sulkin SE, Schulze ML. Continuing importance of laboratory-acquired infections. Am J Public Health 1965;55:0–9
10. Pike RM. Laboratory associated infections: incidence, fatalities, causes and prevention. Annu Rev Microbiol 1979;33:41–66
11. Centers for Disease Control. Classification of Etiologic Agents on the Basis of Hazard (4th ed). Atlanta: US Department of Health, Education and Welfare, Public Health Service, 1974
12. Recombinant DNA guidelines. Fed Register 1976;41:27902–27943
13. Guidelines for research involving recombinant DNA molecules. Fed Register 1986;51:16958–16968
14. Office of Research Safety, National Cancer Institute and the Special Committee of Safety and Health Experts. Laboratory Safety Monograph: A Supplement to the NIH Guidelines for Recombinant DNA Research. Bethesda: National Institutes of Health, 1978
15. Centers for Disease Control and Prevention and the National Institutes of Health. Biosafety in Microbiological and Biomedical Laboratories (3rd ed). HHS Publ. No. (CDC) 93-8395. Atlanta: US Department of Health and Human Services, Public Health Service, 1993
16. Marcus R, Kay K, Mann JM. Transmission of human immunodeficiency virus (HIV) in healthcare settings worldwide. Bull WHO 89;67:577–582
17. Centers for Disease Control. Recommendations for prevention of HIV transmission in health-care settings. MMWR 1987;36(suppl 2):3S–18S
18. Centers for Disease Control. Update: universal precautions for prevention of transmission of human immunodeficiency virus, hepatitis B virus and other bloodborne pathogens in healthcare settings. MMWR 1988;37:377–382, 387, 388
19. Centers for Disease Control. Guidelines for prevention of transmission of human immunodeficiency virus and hepatitis B virus to health-care and public-safety workers. MMWR 1989;38:5–6
20. Centers for Disease Control and Prevention. Recommended infection control practices for dentistry. MMWR 1993;42:RR-8
21. US Department of Labor, Occupational Safety and Health Administration. Occupational exposure to bloodborne pathogens, final rule. Fed Register 91;56:64175–64182
22. Metler R, Ciesielski C, Ward J, Marcus R. HIV seroconversions in clinical laboratory workers following occupational exposure, United States [abstract PoC 4147]. In: Poster Abstracts, VIII International Conference on AIDS, Amsterdam, 1992:C269
23. Kennedy ME. Biosafety: principles and practices in the human immunodeficiency laboratory. In Schochetman G, George JR (eds): AIDS Testing: Methodology and Management Issues. New York: Springer-Verlag, 1992: 181–186

24. National Committee for Clinical Laboratory Standards. Protection of Laboratory Workers from Infections Transmitted by Blood, Body Fluids and Tissues. NCCLS Document M29-T2. 1992:11(14)
25. Schochetman G, George JR. AIDS Testing: Methodology and Management Issues. New York: Springer-Verlag, 1991
26. Interstate shipment of etiologic agents. Fed Register 1942;CFR part 72
27. Richmond JY. Developing a response to medical waste issues. ASM News 1990;56:420–423
28. Castro KG, Polder JA. Management of occupational exposure to the human immunodeficiency virus. In Schochetman G, George JR (eds): AIDS Testing: Methodology and Management Issues. New York: Springer-Verlag, 1992: 204–207
29. Richmond JY. Safe practices and procedures for working with human specimens in biomedical research laboratories. J Clin Immunoassay 88;11(3): 115–121
30. Centers for Disease Control. Public health statement on management of occupational exposure to human immunodeficiency virus, including considerations regarding zidovudine postexposure use. MMWR 1990;39(No. RR-17)

23 Management of Occupational Exposure to HIV

DENISE M. CARDO, KENNETH G. CASTRO, JACQUELYN A. POLDER, and DAVID M. BELL

Many health care workers (HCWs) are potentially at risk for occupationally acquired human immunodeficiency virus (HIV) infection.[1,2] Although improved methods of exposure prevention remain the mainstay of further risk reduction efforts, the likelihood that exposures will nevertheless continue to occur has made it necessary for institutions to develop and continually update procedures for postexposure management.[3–7] This chapter focuses on the management of occupational exposures and discusses issues regarding attempted postexposure prophylaxis with antiviral agents such as zidovudine. Many such incidents also require evaluation and management of possible exposure to other blood-borne pathogens (e.g., hepatitis B and C viruses) and tetanus prophylaxis, as discussed elsewhere.[8,9]

Postexposure Management

To provide clear guidance, each institution should adopt an appropriate postexposure management plan. This plan should include an explanation of what constitutes an occupational exposure that may place a worker at risk of HIV infection, procedures for promptly reporting and evaluating such exposures, and recommended follow-up management of the exposed worker, including the possible use of antiviral agents.

The U.S. Public Health Service (PHS) has defined an exposure that may place a worker at risk of HIV infection as[4]:

> . . . a percutaneous injury (e.g., a needlestick or cut with a sharp object), contact of mucous membranes, or contact of skin (especially when the exposed skin is chapped, abraded, or afflicted with dermatitis or the contact is prolonged or involving an extensive area) with blood, tissues, or other body fluids to which universal precautions apply including: (a) semen, vaginal secretions, or other body fluids contaminated with visible blood, because these substances have been

implicated in the transmission of HIV infection; (b) cerebrospinal fluid, synovial fluid, and amniotic fluid, because the risk of transmission of HIV from these fluids has not yet been determined; and (c) laboratory specimens that contain HIV (e.g., suspensions of concentrated virus), during the performance of job duties.

Reporting Exposures

The prompt reporting of exposures is essential, not only for management of the exposed worker but also for the identification of continuing hazards and the evaluation of preventive measures. Reporting systems should include prompt access to expert consultants and safeguarding the confidentiality of the exposed worker. Some institutions have found it useful to implement hotlines staffed by expert clinicians 24 hours a day to coordinate reporting and initial management of exposed workers.[6] Regional hotlines, similar to regional poison control systems, have also been proposed as a potentially efficient way to provide coverage for multiple institutions, workers in clinics and private offices, and first responders. Implementation of regional systems depends on the availability of sufficient resources and coordination of reporting and follow-up procedures among participating employers.

Prompt, complete reporting depends on appropriate education of workers and a supportive and nonpunitive response by employers. Educational programs for workers, including orientation and in-service activities, should familiarize HCWs with their personal risk of occupational exposure to HIV, measures to prevent these exposures, and the principles of postexposure management. HCWs must understand the importance of reporting exposures immediately after they occur, because certain interventions that may be indicated must be initiated promptly to be effective.[4–6]

On reporting an exposure, the worker should be promptly evaluated and counseled regarding the risk of HIV infection, the procedures for postexposure management including the availability of potentially useful antiviral agents such as zidovudine, and precautions to prevent HIV transmission to others.[4]

Evaluating the Exposure

Evaluation of the exposure to assess its potential for HIV transmission is important for counseling and follow-up management, including assessment of the suitability of postexposure zidovudine use (see below).

Prospective studies of HCWs who have had an occupational exposure have estimated that the risk of HIV infection after percutaneous exposure to HIV-infected blood is approximately 0.3% [upper limit of 95% confidence interval (CI) = 0.6%],[1,2,10–14] and after a mucous membrane exposure 0.09% (upper limit of 95% CI = 0.5%).[15] Transmission of HIV infection after skin exposure has also been documented; the risk of transmission after exposures to skin, whether intact or nonintact, has not been precisely quantified.[10] In one study the upper limit of 95% CI for HIV transmission after skin exposure to HIV-infected blood was 0.04%.[16] The risk of transmission after occupational exposure to potentially infectious tissues of fluids other than blood (i.e., fluids for which universal precautions are recommended) is unknown.

Risk evaluation should also include an assessment of factors that may increase or decrease the probability of HIV transmission after an individual occupational exposure. Risk factors for seroconversion after a percutaneous exposure to HIV-infected blood have not been well defined or quantified epidemiologically. The titer of HIV in the specimen, which may vary by several orders of magnitude, depending on the stage of illness of the source patient and whether the source patient was taking antiviral agents to which the virus was sensitive, is likely to be the most important factor.[17] In vitro studies have indicated that less blood is transferred across membranes by a needle that passes through a glove or is of smaller gauge or solid rather than hollow bore.[18–21]

Emergency Management

Each incident of occupational exposure to blood or body fluids requiring universal precautions should be considered a medical emergency. After an occupational exposure, first aid should be administered as necessary. Puncture wounds and other cutaneous injuries should be washed with soap and water, and exposed oral and nasal mucosa should be decontaminated by vigorous flushing with water. Eyes should be irrigated with clean water, saline, or sterile irrigants designed for this purpose.[4,6] No scientific evidence indicates that the use of antiseptics for wound care offers additional benefit in reducing the risk of transmission of HIV, but their use is not contraindicated. The use of a caustic agent (e.g., bleach), which causes tissue trauma and is of no proved efficacy for this purpose, is not recommended.

After an exposure, efforts should be made to identify and evaluate clinically and epidemiologically the source individual for evidence of HIV infection. The source patient should be informed of the incident and tested for serologic evidence of HIV infection after consent is obtained. If consent cannot be obtained from the source individual (e.g., because the individual is unconscious), policies should be developed for testing source individuals in compliance with applicable state and local laws.[4]

The circumstances of the exposures are recorded in a confidential medical record. Data to be collected include demographic information about the exposed worker, the date and time of exposure, the job duty being performed at time of exposure, a detailed account of the exposure (e.g., type of exposure, amount and type of fluid or material, type of device, severity of exposure), a description of infection-control precautions (e.g., gloves, eye protection, masks), conditions of the source person (e.g., stage of disease, use of zidovudine), and details about counseling, postexposure management, and follow-up plans.[4,6]

Because of the possibility of an incubating HIV infection, to prevent further transmission to his or her contacts the HCW is advised to refrain during the follow-up period from donating blood, semen, or organs and to refrain from breast-feeding when safe and effective alternatives to breast-feeding are available. To prevent HIV transmission to sexual partners, all exposed HCWs, including pregnant women, should abstain from, or use latex condoms during, sexual intercourse throughout the follow-up period.[4]

Follow-up Management

If the source individual has acquired immunodeficiency syndrome (AIDS), is known to be HIV-seropositive, or refuses testing, the HCW should be evaluated clinically and serologically for evidence of HIV infection as soon as possible after the exposure (baseline) and, if seronegative, retested periodically for at least 6 months (e.g., at 6 weeks, 3 months, and 6 months) to determine if HIV infection has occurred.[4]

Beause of reports of delayed seroconversion in some studies of homosexual men, several investigators have used the polymerase chain reaction (PCR) to examine specimens from HCWs after exposure to HIV-infected blood.[22–24] Data from these examinations combined with data on several hundred HCWs who have remained seronegative when tested 2 years or more after exposure suggest that seroconversion beyond 6 months after an occupational exposure, if it occurs, is uncommon.[12,22,25]

The HCW is advised to report and seek medical evaluation for any acute illness that occurs during the follow-up period. Such illness, particularly if characterized by fever, rash, myalgia, fatigue, malaise, or lymphadenopathy, may be indicative of acute HIV infection, drug reaction, or another medical condition. In 50% to 70% of patients with primary HIV infection, an acute retroviral syndrome develops approximately 3 to 6 weeks after the exposure.[26]

If the source individual is HIV-seronegative and has no clinical manifestations of AIDS or HIV infection, no further HIV follow-up of the exposed HCW is necessary unless there is epidemiologic evidence to suggest that the source individual may have recently been exposed to HIV. If the source patient cannot be identified, decisions regarding appropriate follow-up must be individualized based on factors such as whether the potential sources are likely to include an individual at increased risk of HIV infection.[4]

Information on the psychological impact of an occupational exposure is limited, but experts have found that supportive counseling during the follow-up period is an important part of management.[6] During all phases of follow-up, it is vital that the confidentiality of the HCW and the source patient be protected.[4,6]

Use of Antiviral Agents as Postexposure Prophylaxis

Zidovudine

Zidovudine is a thymidine analogue in which the 3′-hydroxy (-OH) group is replaced by an azido ($-NH_3$) group. In vitro, zidovudine inhibits replication of some retroviruses, including HIV, by interfering with the action of viral ribonucleic acid (RNA) -dependent deoxyribonucleic acid (DNA) polymerase (reverse transcriptase) and possibly also by other mechanisms. Although it inhibits reverse transcriptase, it has no intrinsic antiviral activity and does not prevent the virus from entering cells. It is rapidly absorbed from the gastrointestinal tract after oral dosing, and a peak serum concentration occurs within 0.5 to 1.5 hours.[4,27,28]

In a double-blind, placebo-controlled trial, zidovudine prolonged survival, decreased the severity of opportunistic infections, and improved the quality of life in persons with AIDS and in those with advanced HIV infection.[29] Later studies have indicated that zidovudine can delay disease progression in patients with HIV infection and an absolute CD4 count of less than 500/mm, and also appears to reduce HIV transmission from mother to infant.[3,30,31,31b] However, although the use of zidovudine clearly benefited patients with AIDS and AIDS-symptomatic HIV infection, the indications for use of this drug in persons with asymptomatic HIV infection are controversial.[32]

Animal Studies

Animal studies using murine and feline retroviruses are of limited value because these viruses have pathogenic mechanisms different from those of HIV. Ruprecht et al. demonstrated that zidovudine administered 4 hours after inoculation with Rauscher murine leukemia virus suppressed the viremia.[33] Tavares et al. showed that zidovudine prevented feline leukemia virus infection when administered to cats immediately after the virus exposure.[34] McCune et al., studying immunodeficient mice that had received transplants of human hematolymphoid organs (SCID-hu mice), observed that HIV infection was suppressed during the use of zidovudine, but infection was detected in all animals after discontinuation of zidovudine.[35] Because the mice were exposed to HIV by intrathymic injection of a sizable virus inoculum, the relevance of this experiment to percutaneous exposures in HCWs is unclear. In a subsequent study, the SCID-hu mice were treated with zidovudine at different times after intravenous infection with a standard dose of HIV. When zidovudine was given within 2 hours of injection, virus replication was suppressed in all animals 2 weeks later, but the investigators did not determine if infection was present after zidovudine was discontinued.[36]

Some studies with simian immunodeficiency virus (SIV) showed that zidovudine did not prevent SIV infection in monkeys injected with moderate or high inocula or with a rapidly lethal variant of SIV, even when zidovudine was combined with interferon and administered prior to virus inoculation.[37–39] When lower challenge doses of SIV were used and zidovudine was injected subcutaneously 24 hours prior to moderate inocula of SIV and continued for 28 days, two of two monkeys in one study and one of six monkeys in the highest dose treatment group in another study did not become infected.[40,41]

Data involving studies of laboratory animals must be interpreted with caution, as they have most often been derived using nonhuman retroviruses having pathogenic mechanisms different from the pathogenesis of HIV infection in humans. Also, many variables, including the size of the viral inoculum, the dose of zidovudine, the route of administration, and the timing of drug administration, may affect the apparent effectiveness of the treatment under study.[5]

Human Studies

Little information exists with which to assess the efficacy of postexposure zidovudine. Because of the relatively low rate of seroconversion after an occupational exposure to HIV-infected blood, a large sample size is necessary for a trial to have the statistical power to assess the efficacy of postexposure prophylaxis. The Burroughs-Wellcome Company sponsored a double-blind, placebo-controlled study to evaluate 6 weeks of zidovudine prophylaxis involving HCWs who had experienced occupational percutaneous, mucous membrane, or nonintact skin exposures to HIV-infected blood.[42] Because the study failed to enroll enough participants, it was discontinued a year later.

Failures

Failure of zidovudine to prevent HIV infection after exposure has been reported in at least 13 instances.[1,4,43–50] Tokars et al. summarized data for eight HCWs who became infected with HIV despite zidovudine treatment after percutaneous exposure to HIV-infected blood.[1] In these cases, zidovudine was begun 30 minutes to 12 hours (median 1.75 hours) after exposure and was used in doses of 800 to 1200 mg/day (median 1000 mg/day) for 8 to 54 days (median 21 days).

An additional five instances of failure of zidovudine have been reported after exposures in which the quantity of HIV-infected blood injected was larger than would be expected from a needlestick. They include two instances of accidental intravenous inoculation of HIV-infected blood or other body fluid during nuclear medicine procedures,[47,48] one blood transfusion,[4] one suicidal self-inoculation,[49] and one assault on a prison guard with a needle and syringe.[50] These case reports indicate that if zidovudine is protective any protection afforded is not absolute.

Toxicity

Information about the toxicity of orally administered zidovudine comes primarily from animal studies and from studies of HIV-infected patients who were treated with zidovudine.[51] Prolonged treatment is associated with hematologic toxicity (anemia, granulocytopenia, thrombocytopenia), myopathy, nausea, fatigue, insomnia, and headaches.[51]

Serious toxicity associated with short-course zidovudine therapy in healthy persons appears to be rare. In several prospective studies, the most commonly reported symptoms and signs of acute toxicity included nausea, vomiting, headache, malaise, fatigue, anemia, and granulocytopenia.[1,42,52,53]

In the Italian Multicentre Study on occupational risk of HIV infection, 56% of the 211 HCWs who took prophylactic zidovudine developed side effects. The most frequent reactions reported were nausea (40%), asthenia (17%), vomiting (15%), headache (9%), or gastric pain (8%); most of the adverse effects began within the first 10 days of prophylaxis. In 29 (14%) cases toxicity led to prophylaxis interruption; in 9 cases side effects resolved after reduction from the initial dose. Major hematologic side effects were rare: anemia in 5 cases (2.5%) and neutropenia in 1 case (0.5%). All hematologic side effects were observed within 10 days of beginning prophylaxis, which was continued at 1000 mg per day for 3 to 4 weeks without further decrease in hematologic values. A transient increase in serum alanine aminotransferase levels was observed in 2 (1.6%) of 124 HCWs for whom data on renal and hepatic function were available.[52]

In a CDC surveillance project of prospective evaluation of HCWs exposed to HIV-infected blood, 176 (75%) HCWs who took zidovudine after exposure reported one or more symptoms during 6 weeks after beginning prophylaxis; the most frequent symptoms were nausea (50%), malaise or fatigue (33%), headache (25%), vomiting (11%), and myalgia or arthralgia (10%). Seventy four (31%) HCWs did not complete their planned regimen of zidovudine because of adverse symptoms.[1]

In studies at San Francisco General Hospital and the National Institutes of Health, side effects were reported by 33% and 24% of the HCWs, and the drug was discontinued in 17% and 50% of HCWs, respectively.[13] In a placebo-controlled trial (Burroughs-Wellcome), nausea, vomiting, and arthralgia were reported more commonly among HCWs taking zidovudine than among those taking placebos.[42] In a prospective open-label trial to evaluate postexposure zidovudine toxicity, including 19 U.S. centers, HCWs were evaluated at 2, 4, 6, 12, 26, and 52 weeks. Among 130

HCWs exposed to HIV-infected blood or body fluids, 44 (34%) discontinued the drug before 28 days; 30 (68%) had subjective symptoms, including fatigue (69%), nausea (58%), headaches (40%), and insomnia (25%).[53]

For healthy persons not infected with HIV, the risk of long-term toxicity, including teratogenic and carcinogenic effects, is not known. Vaginal tumors, including carcinomas, have been observed in mice receiv- doses of zidovudine that the U.S. Food and Drug Administration (FDA) has determined resulted in plasma levels in mice approximately equal to those in humans receiving a dose of 200 mg every 4 hours.[4]

Zidovudine has a direct effect on the developing mouse embryo,[54] but it is not known if zidovudine causes fetal harm when administered to a pregnant woman or if it affects reproductive capacity. In limited studies of the outcomes of newborns of HIV-infected mothers treated with zidovudine during pregnancy, adverse outcomes specifically attributable to zidovudine were not observed.[55] The PHS recommends that pregnancy be avoided throughout the time zidovudine is taken.[4]

Current Practice

Among 200 investigators from participating hospitals who enrolled HCWs in the CDC's ongoing surveillance of workers exposed to HIV-infected blood from October 1988 through June 1992, 110 (55%) reported that at least one HCW at their institution used zidovudine after exposure.[1] The proportion of enrolled HCWs using zidovudine increased from 5% during the fourth quarter of 1988 to 50% during the third quarter of 1990 and has been stable subsequently, averaging 43% during 1992. Physicians, dentists, and medical students were more likely to use zidovudine than were other health care professionals, as were HCWs with percutaneous injury compared with those exposed through contact of mucous membranes or skin.[1] A significant increase in the proportion of HCWs consenting to be treated with zidovudine was also observed during 1988–1990 among those enrolled at the 30 centers participating in the Italian Multicentre Study; from April 1990 through June 1991 in this study, 20.8% of workers took zidovudine after exposure.[52]

Because of the absence of conclusive data regarding efficacy, adequate dose, dose intervals, and duration and method of administration, there are diverse opinions among physicians about the use of postexposure zidovudine. Specific recommendations for use of zidovudine after an occupational exposure have varied, depending on the institution.

In the CDC surveillance project, zidovudine was prescribed by collaborating investigators in doses ranging from 200 to 1800 mg/day (median 1000 mg/day) and for periods of 1 to 180 days (median 42 days). The interval from exposure to first dose of zidovudine ranged from less than 5 minutes to 17 days (median 4 hours).[1]

At San Francisco General Hospital, HCWs are eligible to be offered zidovudine when they have had an occupational exposure (percutaneous, mucous membrane, or nonintact skin) with HIV-infected blood or other body fluid that is a potential source for transmission of HIV, unless the HCW is pregnant, lactating, or immunosuppressed. Employees who elect to receive zidovudine after exposure are given 200 mg by mouth every 4 hours (six times daily) for 3 days and then 200 mg by mouth every 4 hours (five times daily; no dose is given at 4:00 a.m.) for 25 days. HCWs are monitored periodically with clinical and laboratory evaluation during the period they take the drug and at 3, 6, and 12 months after the exposure. This schedule is also followed at the National Institutes of Health.[6,53]

In Italy the national protocol of the Ministry of Health recommends that zidovudine be offered to HCWs after all percutaneous, mucous membrane, and nonintact skin exposures to HIV-infected blood or other body fluid to which the CDC's universal precautions apply. The recommended dose is 1000 to 1200 mg per day for 4 to 6 weeks, with the first dose given within 4 hours of the exposure. To monitor zidovudine toxicity, all HCWs who decide to take zidovudine are followed periodically for at least 2 weeks after interruption of the drug.[52]

PHS Statement Regarding Zidovudine

As noted previously, the PHS recommends that the employer make available a system for managing occupational exposure that includes reporting, evaluation, counseling, and follow-up. Because of limitations of current knowledge, the PHS does not recommend for or against the use of zidovudine after exposure.

The PHS also recommends that HCWs who may be at risk for occupational exposure be aware of the considerations that pertain to the use of zidovudine after exposure. Ideally, HCWs should be familiarized with these considerations prior to exposure to facilitate prompt and rational decision-making after exposure. They include (1) the postexposure risk of HIV infection and the factors that have been postulated to influence this risk; (2) the limitations of current knowledge of the efficacy of zidovudine as postexposure prophylaxis; (3) the apparent need to begin prophylaxis promptly if prophylaxis is given (i.e., within minutes of exposure); (4) the relatively frequent short-term toxicity; (5) the lack of knowledge of potential long-term toxicity; and (6) the need for postexposure follow-up, regardless of whether zidovudine is taken.[4]

Of the eight HCWs with documented zidovudine failures summarized by Tokars et al.,[1] all seroconverted to being HIV antibody-positive within 6 months of exposure. However, because of theoretic concerns regarding the potential for delayed seroconversion in HCWs who have taken postexposure zidovudine a longer follow-up interval (e.g., 1 year) should be considered.[14]

The FDA has approved the use of zidovudine for certain patients with symptomatic and asymptomatic HIV infections but not for postexposure prophylaxis. Nonetheless, because an approved drug may be prescribed for an unlabeled indication, a physician could prescribe zidovudine for postexposure prophylaxis and would not be required to obtain written informed consent from the patient before doing so. However, because no data exist to demonstrate the efficacy of zidovudine as postexposure prophylaxis and little information is available on zidovudine toxicity, it is important to provide adequate information to the exposed HCWs so they can make informed choices about whether to receive the drug.[56,57] The PHS recommends that if a physician offers zidovudine as prophylaxis after an occupational exposure and the exposed HCW elects to take the drug, the physician or other appropriate health care provider should obtain written informed consent from the HCW for this use of this drug. The consent document should reflect the information about use of zidovudine as postexposure prophylaxis, emphasizing the need for follow-up medical evaluations and for precautions to prevent transmission of HIV infection during the follow-up period.[4]

Resistance

The emergence of zidovudine-resistant strains of HIV from AIDS patients previously treated with zidovudine has been reported by several authors.[58–62] In a study in which the investigators examined the zidovudine susceptibility of 372 HIV isolates obtained before, during, and after treatment with the drug, most patients developed resistant strains 6 to 9 months after the beginning of treatment.[59]

Zidovudine-resistant HIV can be transmitted and can cause primary infections.[46,63–66] Exposure to a strain of HIV already resistant to zidovudine could potentially explain some failures of attempted prophylaxis with zidovudine.[46] It is unlikely to be the only explanation for such failures, however. Among eight documented failures of postexposure zidovudine, three involved source patients who had no known history of zidovudine use. For two of the eight cases, laboratory studies were conducted to assess possible zidovudine resistance of the infecting strain. In one case the source patient was not known to have taken zidovudine before the exposure. Because the infecting strain could not be isolated from the HCW's blood, the CDC performed direct sequencing of the HIV reverse transcriptase gene (amplified by PCR) from the HCW's peripheral blood mononuclear cells. No mutations were found at positions associated with zidovudine resistance (positions 41, 67, 70, 215, and 219), indicating that the strain was apparently sensitive to zidovudine.[1] In the other case, the source patient had been receiving zidovudine for 18 months at the time of the exposure. The HIV isolated from the HCW showed substantially decreased sensitivity to zidovudine.[46]

Other Antiretroviral Agents

Chemoprophylactic regimens with a variety of other agents have been proposed, including the use of other nucleoside analogue agents, such as 2′,3′-dideoxycytidine and 2′,3′-dideoxyinosine. These drugs are effective in the treatment of patients with established HIV infection. Their toxicity is different from that of zidovudine, consisting primarily of neurotoxicity and pancreatitis.[28] Use of these agents individually or in combination with zidovudine has been considered for HCWs occupationally exposed to blood from patients who have been receiving zidovudine for extended periods and who may be infected with HIV isolates that are relatively resistant to zidovudine.[67] Aside from a few anecdotal reports, there are no data to evaluate this practice.

To collect information on postexposure chemoprophylaxis with zidovudine and other antiretroviral agents, the CDC has an ongoing surveillance project of HCWs with occupational exposures to HIV. The CDC does not receive the HCW names or other personal identification. All institutions are encouraged to enroll HCWs with occupational exposures to HIV-infected blood in this surveillance project. Additional information and enrollment materials can be obtained from the Hospital Infections Program, National Center for Infectious Diseases, Centers for Disease Control and Prevention, 1600 Clifton Road NE, Atlanta, Georgia 30333, Mail stop A-07; telephone 404-639-1547.

References

1. Tokars JI, Marcus RA, Culver DH, et al. Surveillance of human immunodeficiency virus (HIV) infection and zidovudine use among health care workers with occupational exposure to HIV-infected blood. Ann Intern Med 1993;118:913–919
2. Marcus R, CDC Cooperative Needlestick Study Group. Surveillance of health-care workers exposed to blood from patients infected with the human immunodeficiency virus. N Engl J Med 1988;319:1118–1123
3. Centers for Disease Control and Prevention (CDC). Recommendations for prevention of HIV transmission in health-care settings. MMWR 1987;36(no. 2S)
4. Centers for Disease Control and Prevention (CDC). Public health service statement on management of occupational exposure to human immunodeficiency virus, including considerations regarding zidovudine postexposure use. MMWR 1991;39(RR-1)
5. Henderson DK, Gerberding JL. Prophylactic zidovudine after occupational exposure to the human immunodeficiency virus: an interim analysis. J Infect Dis 1989;160:321–327
6. Gerberding JL, Henderson DK. Management of occupational exposures to bloodborne pathogens: hepatitis B virus, hepatitis C virus, and human immunodeficiency virus. Clin Infect Dis 1992;14:1179–1185

7. Occupational Safety and Health Administration (OSHA). Occupational exposure to bloodborne pathogens; final rule. Fed Register Dec 6 1991;56: 64175–64182, 29 CFR Part 1910.1030
8. Centers for Disease Control and Prevention (CDC). Protection against viral hepatitis: recommendations of the immunization practices advisory committee (ACIP). MMWR 1990;39(RR-2)
9. Centers for Disease Control and Prevention (CDC), Immunization Practices Advisory Committee. Diphtheria, tetanus, and pertussis: recommendations for vaccine and other preventive measures. MMWR 1991;40(RR-10)
10. Centers for Disease Control and Prevention (CDC). Update: human immunodeficiency virus infections in health care workers exposed to blood of infected patients. MMWR 1987;36:285–289
11. Gerberding JL, Bryant-LeBlanc CE, Nelson K, et al. Risk of transmitting the human immunodeficiency virus, cytomegalovirus, and hepatitis B virus to health care workers exposed to patients with AIDS and AIDS-related conditions. J Infect Dis 1987;156:1–7
12. Henderson DK, Fahey BJ, Willy M, et al. Risk for occupational transmission of human immunodeficiency virus type 1 (HIV-1) associated with clinical exposures: a prospective evaluation. Ann Intern Med 1990;113:740–746
13. Henderson DK. Postexposure chemoprophylaxis for occupational exposure to human immunodeficiency virus type 1: current status and prospects for the future. Am J Med 1991;91:3125–3195
14. Bell DM. Human immunodeficiency virus transmission in health care settings: risk and risk reduction. Am J Med 1991;91(suppl 3B):294S–300S
15. Ippolito G, Puro V, DeCarli G, et al. The risk of occupational human immunodeficiency virus in health care workers. Arch Intern Med 1993;153: 1451–1458
16. Fahey BJ, Koziol DE, Banks SM, et al. Frequency of nonparenteral occupational exposures to blood and body fluids before and after universal precautions training. Am J Med 1991;90:145–153
17. Ho DD, Mougil T, Alam M. Quantitation of HIV type 1 in the blood of infected persons. N Engl J Med 1989;321:1622–1625
18. Shirazian D, Herzlich BC, Mokhtarian F, et al. Needlestick injury: blood, mononuclear cells, and acquired immunodeficiency syndrome. Am J Infect Control 1992;20:133–137
19. Gaughwin MD, Gowans E, Ali R, et al. Blood needles: the volumes of blood transferred in simulations of needlestick injuries and shared use of syringes for injection of intravenous drugs. AIDS 1991;5:1025–1027
20. Napoli VM, McGowan E. How much blood is in a needlestick? J Infect Dis 1987;155:828
21. Woolwine J, Mast S, Gerberding J. Factors influencing needlestick infectivity and decontamination efficacy: an ex vivo model [abstract 1188]. In: Program and Abstracts, 32nd Interscience Conference on Antimicrobial Agents and Chemotherapy (Anaheim). Washington, DC: American Society for Microbiology, 1992:309
22. Horsburgh CR, Ou CY, Jason J, et al. Duration of human immunodeficiency virus infection before detection of antibody. Lancet 1989;2:637–640
23. Wormser GP, Joline C, Bittker S, et al. Polymerase chain reaction for seronegative health care workers with parenteral exposure to HIV-infected patients. N Engl J Med 1989;321:1681–1682

24. Henry K, Campbell S, Jackson B, et al. Long-term follow-up of health care workers with work-site exposure to human immunodeficiency virus [letter]. JAMA 1990;263:1765
25. Gerberding JL, Littell C, Brown A, et al. Cumulative risk of HIV and hepatitis B (HBV) among health care workers (HCW): long-term serologic follow-up & gene amplification for latent HIV infection [abstract 959]. In: Program and Abstracts, 30th Interscience Conference on Antimicrobial Agents and Chemotherapy (Atlanta). Washington, DC: American Society for Microbiology, 1990:246
26. Pontaleo G, Graziosi C, Fauci A. The immunopathogenesis of human immunodeficiency virus infection. N Engl J Med 1993;5:327–335
27. Yarchoan R, Mitsuya H, Myers C, Broder S. Clinical pharmacology of 3′-azido-2′,3′-dideoxythimidine (zidovudine) and related dideoxynucleosides. N Engl J Med 1989;321:726–738
28. Kamali F. Clinical pharmacology of zidovudine and other 2′,3′-dideoxynucleoside analogues. Clin Invest 1993;71:392–405
29. Fischl MA, Richman DD, Grieco MH, et al. The efficacy of azidothymidine (AZT) in the treatment of patients with AIDS and AIDS-related complex: a double-blind, placebo-controlled trial. N Engl J Med 1987;317:185–191
30. Volberding PA, Lagakos SW, Koch MA, et al. Zidovudine in asymptomatic human immunodeficiency virus infection. N Engl J Med 1991;322:941–949
31. Fischl MA, Richman DD, Hansen M, et al. The safety and efficiency of zidovudine (AZT) in the treatment of subjects with mildly symptomatic human immunodeficiency virus type-1 (HIV) infection. Ann Intern Med 1990;112:727–737
31b. Centers for Disease Control and Prevention (CDC). Zidovudine for the prevention of HIV transmission from mother to infant. MMWR 1994;43: 285–287
32. Sande MA, Carpenter CCJ, Cobbs CG, et al. Antiretroviral therapy for adult HIV-infected patients: recommendations from a state-of-the-art conference. JAMA 1993;270:2583–2589
33. Ruprecht RM, O'Brien LG, Rossoni LD, Nusinoff-Lehrman S. Suppression of mouse viraemia and retroviral disease by 3′-azido-3′-deoxythymidine. Nature 1986;323:467–469
34. Tavares L, Roneker C, Johnston D, et al. 3′-Azido-3′-deoxythymidine in feline leukemia virus-infected cats: a model for therapy and prophylaxis of AIDS. Cancer Res 1987;47:3190–3194
35. McCune JM, Namikawa R, Shih CC, et al. Suppression of HIV infections in AZT-treated SCID-hu mice. Science 1990;247:564–566
36. Shih C-C, Kaneshima H, Rabin L, et al. Postexposure prophylaxis with zidovudine suppresses human immunodeficiency virus type 1 infection in SCID-hu mice in a time-dependent manner. J Infect Dis 1991;163:625–627
37. McClure HM, Anderson DC, Fultz P, et al. Prophylactic effects of AZT following exposure of macaques to an acutely lethal variant of SIV (SIV/SMM/PBj-14) [abstract TCO42]. In Proceedings of the Vth International Conference on AIDS, Montreal, 1989:522
38. Fazely F, Haseltine WA, Rodger RF, Ruprecht RM. Postexposure chemoprophylaxis with ZDV or ZDV combined with interferon: failure after

inoculating rhesus monkeys with a high dose of SIV. J Acquir Immune Defic Syndr 1991;4:1093–1097
39. Lundgren B, Bottiger D, Ljungdahl-Stahle E, et al. Antiviral effects of 3′-fluorothymidine and 3′-azidothymidine in cynomolgus monkeys infected with simian immunodeficiency virus. J Acquir Immune Defic Syndr 1991;4:489–498
40. Tsai CC, Follis KE, Grant R, et al. Effect of dosing frequency on zidovudine prophylaxis against simian immunodeficiency virus in Macaca facicularis [abstract 58]. In: Program and Abstracts, 32nd Interscience Conference on Antimicrobial Agents and Chemotherapy (Anaheim). Washington, DC: American Society for Microbiology, 1992:120
41. Van Pampay KK, Marthas ML, Ramos RA, et al. Simian immunodeficiency virus (SIV) infection of infant rhesus macaques as a model to test antiretroviral drug prophylaxis and therapy: oral 3′-azido-3′-deoxythymidine prevents SIV infection. Antimicrob Agents Chemother 1992;36:2381–2386
42. LaFon SW, Mooney BD, McMullen JP, et al. A double-bind, placebo-controlled study of the safety and efficacy of retrovir (zidovudine, ZDV) as a chemoprophylactic agent in health care workers exposed to HIV [abstract 489]. In: Program and Abstracts, 30th Interscience Conference on Antimicrobial Agents and Chemotherapy (Atlanta). Washington, DC: American Society for Microbiology, 1990:167
43. Tait DR, Pudifin DJ, Gathiram V, et al. HIV seroconversions in health-care workers, Natal, South Africa [abstract PoC 4141]. In: Proceedings of the VIIth International Conference on AIDS, Amsterdam, 1992:268
44. Lot F, Abiteboul D. Infections professionnelles par le V.I.H. en France: le point au 31 mars 1992. Bull Epidemiol Hebd 1992;26:117–119
45. Looke DF, Grove DI. Failed prophylactic zidovudine after needlestick injury [letter]. Lancet 1990;335:1280
46. Anonymous. HIV seroconversion after occupational exposure despite early prophylactic zidovudine therapy [letter]. Lancet 1993;341:1077–1078
47. Centers for Disease Control and Prevention (CDC). Patient exposures to HIV during nuclear medicine procedures. MMWR 1992;41:575–578
48. Lange JMA, Boucher CAB, Hollak CEM, et al. Failure of prophylactic zidovudine after accidental exposure to HIV-1. N Engl J Med 1990;322:1375–1377
49. Durand E, LeJenne C, Hugues FC. Failure of prophylactic zidovudine after suicidal self-inoculation of HIV-infected blood [letter]. N Engl J Med 1991; 324:1062
50. Jones PD. HIV transmission by stabbing despite zidovudine prophylaxis [letter]. Lancet 1992;338:884
51. Richman DD, Fischl MA, Grieco MH, et al. The toxicity of AZT in the treatment of patients with AIDS and AIDS-related complex: a double-blind, placebo-controlled trial. N Engl J Med 1987;317:192–197
52. Puro P, Ippolito G, Guzzanti E, et al. Zidovudine prophylaxis after accidental exposure to HIV: the Italian experience. AIDS 1992;6:963–969
53. Beekman S, Fahrner R, Koziol DE, et al. Safety of zidovudine administered as post-exposure chemoprophylaxis to health care workers sustaining occupational exposures to HIV [abstract 1121]. In: Proceedings of the 33rd Inter-

science Conference on Antimicrobial Agents and Chemotherapy (New Orleans). Washington, DC: American Society for Microbiology, 1993:324
54. Toltzis P, Marx CM, Kleinman N, et al. Zidovudine-associated embryonic toxicity in mice. J Infect Dis 1991;163:1212–1218
55. Sperling RS, Shatton P, O'Sullivan M, et al. A survey of zidovudine use in pregnant women with human immunodeficiency virus infection. N Engl J Med 1992;326:857–861
56. Gerberding JL. Is antiretroviral treatment after percutaneous HIV exposure justified? [editorial]. Ann Intern Med 1993;118:979–980
57. Polder JA, Bell DM, Martone WJ, et al. Zidovudine use after occupational exposure to the human immunodeficiency virus [reply to the editor]. N Engl J Med 1991;324:266–267
58. Smith MS, Koerber KL, Pagano JS. Zidovudine-resistant human immunodeficiency virus type 1 genomes detected in plasma distinct from viral genomes in peripheral blood mononuclear cells. J Infect Dis 1993;167:445–448
59. Land S, McGavin C, Lucas R, et al. Incidence of zidovudine-resistant human immunodeficiency virus isolated from patients before, during, and after therapy. J Infect Dis 1992;166:1139–1142
60. Larder BA, Kemp SD. Multiple mutations in HIV-1 reverse transcriptase confer high-level resistance to zidovudine (AZT). Science 1989;246:1155–1158
61. Larder BA, Kellam P, Kemp SD. Zidovudine resistance predicted by direct detection of mutations in DNA from HIV-infected lymphocytes. AIDS 1991; 5:137–144
62. Kellan P, Boucher CA, Larder BA. Fifth mutation in human immunodeficiency virus type 1 reverse transcriptase contributes to the development of high-level resistance to zidovudine. Proc Natl Acad Sci USA 1992;89: 1934–1938
63. Erice A, Mayers DL, Strike DG, et al. Primary infection with zidovudine resistant human immunodeficiency virus type 1. N Engl J Med 1993;328:1163 –1165
64. Hermans P, Sprecher S, Clumeck N. Primary infection with zidovudine resistant HIV [letter]. N Engl J Med 1993;329:1123
65. Masquelier B, Lemoigne E, Pellegrin I, et al. Primary infection with zidovudine resistant HIV [letter]. N Engl J Med 1993;329:1123–1124
66. Fitzgibbon JE, Gaur S, Frenkel LD, et al. Transmission from one child to another of human immunodeficiency virus type 1 with a zidovudine-resistance mutation. N Engl J Med 1993;329:1835–1841
67. Malcolm JA, Dobson PM, Sutherland DC. Combination chemoprophylaxis after needlestick injury. Lancet 1993;341:112–113

24
Preventing HIV Transmission in Health Care Settings

CAROL A. CIESIELSKI and DAVID M. BELL

Among the most controversial purposes for which HIV testing has been proposed is the provision of information to assist in preventing human immunodeficiency virus (HIV) transmission in health care settings. Initial debate focused on the possible role of testing patients in order to protect health care workers; more recent controversies have involved proposals for testing health care workers to prevent HIV transmission to patients. In this chapter, we summarize available data on the risk of HIV transmission in health care settings, including data on whether knowledge of a patient's HIV status affects a health care worker's likelihood of exposure to the patient's blood. We also summarize preventive strategies recommended by the Centers for Disease Control and Prevention (CDC).

HIV Transmission from Patient to Health Care Worker

Information on the occupational risk of HIV infection in health care workers is derived from several sources, including surveillance data and risk assessment studies.

Surveillance Data

Cases of acquired immunodeficiency syndrome (AIDS) are reported to the CDC from state and local health departments; case reports include information on whether the patient had been employed in health care settings. Reports are also received through a separate surveillance system of occupationally acquired HIV infection, regardless of whether the worker has developed AIDS. Through June 30, 1993 the CDC had received reports of 37 health care workers with documented seroconversion after occupational exposures and 78 health care workers with possible occupationally acquired HIV infection (Table 24.1). Of the seroconverters, 32 had percutaneous exposures, 4 had mucocutaneous

TABLE 24.1. Health care workers[a] with documented and possible occupationally acquired AIDS/HIV infection, by occupation, reported through June 1993, United States.

Occupation	Documented occupational transmission[b] (no.)	Possible occupational transmission[c] (no.)
Dental worker, including dentist	—	7
Embalmer/morgue technician	—	3
Emergency medical technician/paramedic	—	7
Health aide/attendant	1	8
Housekeeper/maintenance worker	1	6
Laboratory technician, clinical	14	13
Laboratory technician, nonclinical	1	1
Nurse	13	15
Physician, nonsurgical	4	8
Physician, surgical	—	2
Respiratory therapist	1	1
Technician, dialysis	1	1
Technician, surgical	1	1
Technician/therapist, other than those listed above	—	3
Other health-care occupations	—	2
Total	37	78

[a] Health care workers are defined as those persons, including students and trainees, who have worked in a health care, clinical, or HIV laboratory setting at any time since 1978. See *MMWR* 1992;41:823–825.

[b] Health care workers who had documented HIV seroconversion after occupational exposure: 32 had percutaneous exposure, 4 had mucocutaneous exposure, 1 had both percutaneous and mucocutaneous exposures. Thirty-four exposures were to blood from an HIV-infected person, 1 to visibly blood fluid, 1 to an unspecified fluid, and 1 to concentrated virus in a laboratory. Eight of these health care workers have developed AIDS.

[c] These health care workers have been investigated and are without identifiable behavioral or transfusion risks; each reported percutaneous or mucocutaneous occupational exposures to blood or body fluids, or laboratory solutions containing HIV; but HIV seroconversion specifically resulting from an occupational exposure was not documented.

exposures, and 1 had both a percutaneous and a mucocutaneous exposure.[1] There were 34 exposures to HIV-infected blood, 1 to a visibly bloody fluid, 1 to an unspecified fluid, and 1 to concentrated virus in a laboratory. These figures provide minimum estimates on the number of health care workers infected through occupational transmission because not all occupational exposures are reported, and not all infected health care workers are reported to health departments.

Risk Assessment Data

HIV Seroprevalence Among Health Care Workers

Risk assessment data are available from HIV seroprevalence surveys conducted among health care workers and from prospective studies of workers exposed to HIV-infected blood. Important determinants of risk include the HIV seroprevalence among patients, the risk of HIV transmission after a single blood contact, and the nature, frequency, and preventability of blood contacts.

Among 3420 orthopedic surgeons in the United States and Canada who participated in an anonymous serosurvey, 2 (0.06%) were HIV-seropositive.[2] Both of these surgeons reported nonoccupational risks for HIV infection. Among 770 surgeons surveyed from moderate to high AIDS incidence areas in the United States, 1 (0.14%) was HIV-seropositive (a general surgeon who did not report nonoccupational risks on a questionnaire).[3] Among physicians, surgeons, and dentists in the U.S. Army Reserve, the seroprevalence is 0.09%.[4] Among 356 hemodialysis staff in five cities, none was seropositive.[5–9] In six regions in the United States with a high incidence of AIDS, more than 9000 health care workers were evaluated after donating blood between March 1990 and May 1991; 3 (0.03%) were HIV-seropositive. Two of the three had nonoccupational risk factors for HIV infection, and the third was lost to follow-up.[10] Studies of dental workers in the United States and Denmark have also shown low seroprevalence rates.[11–14] An important limitation of all of these studies is that for most of the workers tested the extent of their occupational exposure to HIV is not known. Also, some of these rates may be underestimates if workers who knew or suspected that they might be positive declined to be tested. Nevertheless, these surveys indicate low rates of previously undetected HIV infection among the health care workers studied.

HIV Seroprevalence Among Patients

In the United States the HIV seroprevalence of patients varies widely by geographic area and by the patient's age, sex, race/ethnicity, presenting clinical condition, and other factors. In one study conducted in 20 sentinel hospitals in 15 cities, the HIV seroprevalence in patients with diagnoses not associated with HIV infection ranged from 0.2% to 14.2%.[15] It is especially noteworthy that several studies have found unrecognized HIV infection among many patients. For example, a CDC study conducted in New York, Chicago, and Baltimore found that the percentage of patients whose HIV infection was unknown to hospital emergency department staff was 66% to 70% in three inner-city emergency departments and 40% to 91% in three suburban emergency departments.[16] Studies of consecutive admissions to the Department of Veterans Affairs Hospital in Washington, DC, and of emergency department visits at the Johns Hopkins Hospital in Baltimore have also found that the infection status of approximately two-thirds of HIV-infected patients was unknown to the health care providers at the time of presentation.[17,18] The results of these studies, showing variable rates of HIV prevalence with many unrecognized infections, emphasize the need for strict adherence to universal precautions (discussed below).

Risk of HIV Transmission After Occupational Exposure

The risk of HIV infection has been estimated to be approximately 0.3% after a single percutaneous exposure to HIV-infected blood, based on prospective evaluation of more than 3000 exposed health care workers.[19–21] The risk after a mucous membrane exposure to HIV-infected blood is less well defined. In one study a health care worker became infected after extensive splash to the hands, eyes, and mouth with blood from an asymptomatic HIV-infected patient; aggregated with other studies, the risk after mucous membrane exposure was estimated at 0.09%.[21] The risk after a skin exposure to HIV-infected blood is believed to be substantially less but has not been precisely quantitated because no health care workers enrolled in prospective cohort studies became infected after only a skin exposure.

It is likely that the risk of HIV transmission after a needlestick is influenced by the titer of HIV in the source patient's blood and the quantity of blood injected. Studies using laboratory models suggest that the quantity of blood injected is influenced by various factors, including whether the needle was solid or hollow bore, the diameter of the needle, the severity of the injury, and whether the injured health care worker was wearing gloves.[22,23] It is also possible that host factors are important. Unfortunately, there are currently insufficient epidemiologic data to quantify the possible effect of these and other variables on the risk of HIV transmission after a needlestick.

Prospective studies have made substantial progress in describing the epidemiology of occupational blood contact, defining risk factors for blood contact, and directing prevention efforts. Observation studies and questionnaire surveys conducted by the CDC and others suggest that in the United States the number of percutaneous injuries sustained per year may be about 12 for a surgeon, 5 for a dentist, 4 for an obstetrician, 1.8 for a physician on a medical ward, 1 for a scrub assistant, 0.98 for a nurse on a medical ward (H. Smith and L. Aiken, personal communication), and 0.4 for a hospital emergency department worker.[24,25] These estimates are clearly not applicable to every worker in each occupational group, and not all injuries may be comparable.

Exposure Prevention

Risk reduction can be accomplished most effectively by reducing the frequency of blood exposures among health care workers. Exposure prevention requires a combination of engineering controls (e.g., self-sheathing needles), safe work practices and techniques, personal protective equipment (e.g., gloves, gowns, and face shields), and training in and enforcing compliance with their proper use. Infection-control programs should incorporate principles of universal precautions, a system of infection control under which specimens of blood and certain other body fluids of *all* patients are assumed to be infectious. Universal precautions include the appropriate use of hand-washing, protective barriers (e.g., gloves), and care in the use and disposal of needles and other sharp instruments in all health care settings. Also, instruments and other reusable equipment used when performing invasive procedures should be appropriately disinfected or sterilized.[26,27]

Although currently available barrier precautions such as gloves, gowns, and masks can often prevent skin and mucous membrane blood contacts, they cannot prevent percutaneous injuries. Prevention of these injuries in surgical and obstetric settings requires changes in technique and personal protective equipment, such as development of thimbles or gloves that resist needle puncture while preserving tactile sensation. Changes in the design of surgical instruments may also be important.

Further progress in preventing injuries related to common procedures such as venipuncture is likely to be accomplished primarily by changes in the design of needles and other medical devices. Data from studies of needlestick injuries suggest that at least half of such injuries outside the operating room might be preventable with safer needle devices.[28,29] All such changes in technique, protective equipment, and devices must be carefully evaluated to document that they enhance worker safety without compromising patient care.

Routine Testing of Hospitalized Patients for HIV Antibody

Protection of Health Care Workers

Personnel in some hospitals have advocated testing patients to protect health care workers in settings where frequent exposure of health care workers to patients' blood may be anticipated. Specific patients for whom testing has been advocated include those undergoing major surgical procedures or treatment in critical care units, especially if they have conditions that involve uncontrolled bleeding. For surgical patients who test HIV antibody-positive, additional precautions have been advocated, including the use of stapling instruments rather than hand suturing to perform tissue approximation, use of electrocautery devices and scissors rather than scalpels as cutting instruments, use of gowns that totally prevent seepage of blood onto the skin of operative team members, use of "space suits" or respirators with an air filter or self-contained air supply to prevent inhalation of aerosols, and exclusion of inexperienced personnel from the operating team. The efficacy of these and other measures in protecting workers and their effect on patient care is unknown.

Studies have not shown that testing patients, in addition to implementing universal precautions, is helpful for protecting health care workers.[30] In a prospective study of 1307 consecutive operations at San Francisco General Hospital, Gerberding et al. found that neither knowledge of diagnosed HIV infection nor awareness of a patient's high risk status for HIV infection influenced the rate of blood exposure among surgical personnel.[31] The authors concluded that there was no evidence to suggest that preoperative testing for HIV infection would reduce the frequency of accidental exposures to blood.

Tokars et al. observed a sample of 1382 operations at four hospitals in the New York City and Chicago areas, including two inner city and two suburban hospitals.[32] A statistically significant reduction in exposure frequency could not be demonstrated when the patient was known or suspected to be HIV-positive (Table 24.2). Studies of surgical and obstetric procedures at a large public hospital in Atlanta also suggested no benefit from preoperative HIV screening of patients.[33,34]

These data suggest that when compliance with universal precautions is high (which might be easiest to achieve in hospitals with high HIV seroprevalence among patients), knowledge of a patient's HIV risk or serostatus confers no additional benefit in preventing blood exposures during surgery. In hospitals with lower HIV seroprevalence among patients, the measured exposure rates were lower in the small number of patients who were known or suspected to have HIV infection. Compliance with universal precautions may have been more likely for these patients

TABLE 24.2. Relation of HIV status of the patient to percutaneous injuries sustained during surgical procedures.

Patient HIV status[a]	No. of observed procedures	No. (%) of procedures with ≥1 injury	Relative risk (95% confidence interval)
Negative or unknown	1342	93 (7%)	—
Known negative	28	0	—
Suspected negative	756	52 (7%)	—
Unknown	558	41 (7%)	—
Positive	32	1 (3%)	0.4 (0.1–3.1)[b]
Known positive	5	0	—
Suspected positive	27	1 (4%)	—

Source: Tokars et al.,[32] by permission.
[a] As perceived by the primary surgeon.
[b] All positives were compared with all negatives and unknowns.

but not demonstrable statistically because of the relatively small number of such patients.

Evaluation of the practical aspects of a voluntary HIV admission screening program in a large private hospital concluded that the program was of greater benefit to the patient than the health care worker[35]; another program was discontinued, in part because of logistical difficulties.[36]

Thus the limited studies available do not demonstrate that routine preoperative HIV testing of patients would protect surgical personnel. If universal precautions are routinely followed, knowledge of a patient's HIV serostatus should not reduce the likelihood of blood exposure. Moreover, health care workers whose infection control practices are based on patients' HIV test results risk a false sense of security, as patients with negative HIV tests may nevertheless be infected with HIV prior to seroconversion or with other blood-borne pathogens, such as hepatitis B and hepatitis C viruses, human T lymphotrophic virus type I (HTLV-I), and doubtless other pathogens yet undiscovered. To best protect themselves from occupational infection with a blood-borne infection, health care workers should receive appropriate vaccines (e.g., hepatitis B vaccine) and observe infection control measures that incorporate the principles of universal precautions.

Testing for Benefit of Patients

The U.S. Public Health Service (PHS) has recommended that routine voluntary testing of patients be considered in health care settings with a high rate of HIV infection.[37] This voluntary testing has been recommended as a means to diagnose persons with HIV infection earlier in the course of their infection and should not be used as a substitute for universal precautions or other infection-control techniques.

Decisions regarding the need to establish routine, voluntary testing programs for patients should be made by individual institutions in consultation with public health authorities and based, in part, on the HIV seroprevalence among patients in their institutions. Patient testing should include provisions for (1) obtaining informed consent, (2) providing appropriate counseling, (3) ensuring confidentiality of results, and (4) providing optimal care, regardless of test results. Testing programs should be evaluated to see if they reduce the frequency of adverse exposures and to determine their effect on patient care.[26,37]

HIV Transmission from Infected Health Care Workers to Patients

Florida Dental Investigation

The only known instance of HIV transmission to patients occurred in one cluster of six patients of a Florida dentist. None of these six patients had other known exposures to HIV, and their viruses were shown to be genetically similar to the virus infecting the dentist.[38–40] Each of these six patients received dental care from the dentist after he was diagnosed with AIDS and had evidence of severe immunosuppression. Low $CD4^+$ T lymphocyte counts are associated with high virus titers in the blood and therefore could be associated with an increased likelihood of transmission if an injury to the dentist occurred.[41] The dentist performed many invasive procedures after he was diagnosed with AIDS, during which there were many opportunities for injury. In addition, he frequently experienced fatigue, which may have increased the likelihood of injury. Although he routinely wore gloves, such protection does not prevent most injuries caused by sharp instruments.

The specific incident(s) that led to HIV transmission in this office could not be determined. Breaches in infection control and other office practices explaining these transmissions could not be identified. All six patients received multiple injections of local anesthetic, and a percutaneous injury to the dentist during anesthetic administration could have resulted in contamination of the syringe apparatus with the dentist's blood, after which additional anesthetic may have been injected into the same patient. Such an injury could also result in direct contact of the dentist's blood with the patient's inflamed or nonintact oral tissues during the procedures.[38]

There was no evidence to indicate that a contaminated instrument or equipment was the principal factor in these transmissions.[42] Although some visit-days between patients were shared, they probably occurred by chance alone.[38] Also, the procedures performed on these shared visit days probably did not involve the use of the same instruments, including the high-speed dental handpiece, on more than one of the infected patients.[42]

Interviews with family, staff, health care providers, patients, and other who knew the dentist have not provided support for the hypothesis that the infections were intentionally transmitted. The dentist initially cooperated with the investigation and provided a blood specimen for genetic sequencing. Also, most of the procedures done by the dentist were routinely observed by staff, all patients were awake during the procedures, and no unusual behavior was noted or suspected by either patients or staff.[38]

The epidemiologic investigation concluded that the available evidence supported direct dentist-to-patient, rather than patient-to-patient, transmission.[38,42]

Risk Assessment Data

Because hepatitis B virus (HBV), another blood-borne virus, has been transmitted from health care workers to patients during invasive procedures, the report that HIV could be transmitted similarly was not surprising to many hospital infection-control experts. The critical questions concern the magnitude of the risk, the factors that increase or decrease the risk, and how the risk should be managed.

Available data indicate that the risk of HIV transmission from an infected health care worker to a patient during an invasive procedure is small, certainly much smaller than the occupational risk of HIV infection faced by health care workers.[43]

"Lookback Studies" of Patients of HIV-Infected Health Care Workers

Supportive evidence comes from retrospective investigations of patients of other HIV-infected health care workers. As of September 1993 the CDC was aware of test results of more than 22,000 patients who had been treated by 63 HIV-infected health care workers, including 33 dentists (excluding the Florida dentist discussed previously), 13 surgeons and obstetricians, 13 physicians, and 4 other health care workers. No additional cases of HIV transmission to patients have been documented.[44] Efforts are under way to aggregate procedure-specific data from multiple investigations for patients whose procedures and test results are known.

Although these results are reassuring, they have certain limitations.[43] For example, it is often unknown what percent of the patients in the health care workers' practices were tested and how many of the patients tested actually underwent invasive procedures, particularly procedures that have been associated with HBV transmission from health care workers to patients or with elevated percutaneous injury rates for health care workers. Little is known about the skill, technique, and infection control practices of these health care workers or about their medical status. Factors such as whether the health care workers had progressed to AIDS when the procedure was performed, was taking antiretroviral medication (shown to decrease p24 antigenemia and possibly associated with decreased transmissibility[45]), or had conditions that might predispose him or her to bleeding or injury (e.g., thrombocytopenia, peripheral neuropathy, encephalopathy, or skin lesions) are unknown but may affect the risk of HIV transmission to patients.

Current data from the retrospective studies are consistent with estimates derived using modeling techniques, which indicate that the average risk of HIV transmission from a surgeon to a patient due to percutaneous injury during an invasive procedure is 2.4 to 24.0 per million.[46] For the reasons noted above, these data and estimates represent population averages and do not necessarily apply to a particular health care worker performing a particular invasive procedure under a particular set of circumstances.[43]

HBV Transmission from Infected Health Care Workers to Patients

Since the discovery of AIDS, transmission patterns of HBV have been helpful as a model for understanding HIV transmission. Both viruses are transmitted via sexual, perinatal, and blood-borne routes. Both viruses present an occupational risk to health care workers and can potentially be transmitted from health care worker to patient during invasive procedures. There are notable differences, however. HBV is much more transmissible than HIV. The risk of infection after a percutaneous exposure is 30% if the exposure is to hepatitis B e antigen-positive blood (a marker correlated with high virus titers in blood and with increased transmissibility of HBV) compared to 0.3% for HIV-infected blood.[24] Also, HBV is more environmentally stable than HIV.

As of 1992, more than 350 instances of HBV transmission to patients have been identified in association with 34 HBV-infected health care workers, including 21 surgeons (7 obstetrician/gynecologists, 7 cardiac surgeons, 6 general surgeons, and 1 orthopedic surgeon), 9 dentists (of whom 5 were oral surgeons), 1 general practitioner, 1 respiratory therapist, and 2 cardiopulmonary bypass pump technicians.[43]

All of the implicated health care workers who were tested for hepatitis B e antigen were positive. In several investigations, the risk of HBV transmission was correlated with the invasiveness of the procedure. For example, transmission was more likely during dental extractions than with dental prophylaxis (cleaning) and during hysterectomy than with dilatation and curettage.[43]

Percutaneous blood exposure was believed to be the most likely mechanism of these transmissions; but as in the Florida dental case, the actual incidents that resulted in transmission were not identified. Each of these reports described situations in which the contamination of surgical wounds or traumatized tissue was possible, either from deficient infection-control practices, such as not wearing gloves, or from injury to the infected health care worker with a sharp object, which may have recontacted the patient (e.g., needlesticks incurred while manipulating needles without being able to see them). In several instances, surgeons who resumed practice after reported modification of techniques subsequently transmitted HBV to additional patients.[43]

Observational Studies of Injuries During Surgery

Observation studies of injuries during surgery provide additional evidence that certain procedures are more likely than others to involve injury to a surgeon and possible exposure of the patient to the surgeon's blood. In a prospective CDC study conducted in four hospitals, at least one percutaneous injury occurred among surgical personnel during 96 (6.9%) of 1382 operative procedures observed; this rate was comparable to rates of 1.3% to 15.4% reported in other studies.[32] Injury rates were highest on the gynecology service, especially during hysterectomy. At least one member of the surgical team was injured during 21.3% of vaginal and 10.3% of abdominal hysterectomies. In this study, a sharp object was observed to recontact the patient's wound after injuring the health care worker in 28 (2.0%) of 1382 procedures. This "recontact rate" was also highest on the gynecology service (4.2%), and within that service it was highest for vaginal hysterectomy (8.5%), a procedure that often involves manipulation of poorly visualized needles. In a logistic regression model, risk factors for injury included the type of procedure performed, the duration of the procedure, and holding the tissue being sutured with fingers, rather than with an instrument. Although the risk of infection transmission due to the observed injuries and recontacts is uncertain, these observations are highly consistent with data on HBV transmission from surgeons to patients, showing elevated surgeon injury rates and patient recontact rates in the specialties and procedures most often implicated in HBV transmission to patients. The results also suggest that variations in surgical technique may play an important role in preventing injuries.

Fewer data are available on injuries sustained during dental procedures. In retrospective questionnaire surveys, dentists reported an average rate of approximately 5 percutaneous injuries per year in 1991, compared with 12 per year in 1987.[25] The CDC is currently sponsoring prospective studies to better characterize the nature, frequency, and preventability of blood exposures in dentistry.

Preventing HIV Transmission from Health Care Workers to Patients During Invasive Procedures

Both health care workers and patients are protected best by compliance with infection control precautions and the development of new instruments, techniques, and protective equipment that reduce the likelihood of sharp injuries to the health care worker without adversely affecting patient care.

Available data indicate that the risk of transmission of a blood-borne pathogen from a health care worker to a patient during an invasive procedure is likely to depend on the procedure performed and on the technique and medical condition of the health care worker. The CDC has recommended that evaluation of the appropriate patient care duties of a health care worker with HIV infection should consider these three factors.[43] Mandatory testing of health care workers is not justified based on the small risk to patients.[47]

References

1. Centers for Disease Control and Prevention. HIV/AIDS Surveillance Report. Atlanta: CDC, July 1993
2. Tokars JI, Chamberland ME, Schlable CA, et al. A survey of occupational blood contact and HIV infection among orthopedic surgeons. JAMA 1992; 268:489–494
3. Panlilio A, Chamberland M, Shapiro C, Schable C, Srivastava P. Human immunodeficiency virus (HIV), hepatitis B virus (HBV), and hepatitis C virus (HCV) serosurvey among hospital-based surgeons [abstract PO-C18-3024]. Presented at the IXth International Conference on AIDS, Berlin, 1993
4. Cowan DN, Brundage JF, Pomerantz RS, Miller RN, Burke DS. HIV infection among members of the U.S. Army Reserve components with medical and health occupations. JAMA 1991;265:2826–2830
5. Chirgwin K, Rao TKS, Landesman SH. HIV infection in a high prevalence dialysis unit. AIDS 1989;3:731–735
6. Assogba U, Ancelle Park RA, Rey MA, et al. Prospective study of HIV-1 seropositive patients in hemodialysis centers. Clin Nephrol 1988;29:312–314
7. Peterman TA, Lang GR, Mikos NJ, et al. HTLV-III/LAV infection in hemodialysis patients. JAMA 1986;255:2324–2326
8. Goldman M, Liesnard C, Vanherweghem JL, et al. Markers of HTLV-III in patients with end stage renal failure treated by hemodialysis. BMJ 1986; 293:161–162
9. Comodo N, Martinelli F, DeMajo E, et al. Risk of HIV infection on patients and staff of two dialysis centers: seroepidemiological findings and prevention trends. Eur J Epidemiol 1988;4:171–174
10. Chamberland ME, Peterson L, Munn V, et al. Low rate of HIV-1 infection among health care workers who donate blood [abstract M.D. 62]. Presented at the VIIth International Conference on AIDS, Florence, 1991
11. Klein RS, Phelan JA, Freeman K, et al. Low occupational risk of human immunodeficiency virus among dental professionals. N Engl J Med 1988;318: 86–90
12. Ebbesen P, Melbye M, Scheutz F, et al. Lack of antibodies to HTLV-III/LAV in Danish dentists. JAMA 1986;256:2199
13. Flynn NM, Pollet SM, Van Horne JR, et al. Absence of HIV antibody among dental professionals exposed to infected patients. West J Med 1987;146: 439–432
14. Siew C, Gruninger SE, Hojvat S. Screening dentists for HIV and hepatitis B. N Engl J Med 1988;318:1400–1401
15. Janssen RS, St. Louis ME, Satten GA, et al. HIV infection among patients in U.S. acute care hospitals: strategies for the counseling and testing of hospital patients. N Engl J Med 1992;327:445–452
16. Marcus R, Culver DH, Bell DM, et al. Risk of human immunodeficiency virus infection among emergency department workers. Am J Med 1993;94: 363–370
17. Gordin FM, Gibert C, Haeley HP, Willoughby A. Prevelence of human immunodeficiency virus and hepatitis B virus in unselected hospital admissions: implications for mandatory testing and universal precautions. J Infect Dis 1990;161:14–17

18. Kelen GD, DiGiovanna T, Bisson L, et al. Human immunodeficiency virus infection in emergency department patients: epidemiology, clinical presentations, and risk to health care workers: the Johns Hopkins experience. JAMA 1989;262:516–522
19. Tokars JI, Marcus R, Culver DH, et al. Surveillance of human immunodeficiency virus (HIV) infection and zidovudine use among health care workers with occupational exposure to HIV-infected blood. Ann Intern Med 1993;118:913–919
20. Henderson DK, Fahey BJ, Willy M, et al. Risk for occupational transmission of human immunodeficiency virus type 1 (HIV-1) associated with clinical exposures: a prospective evaluation. Ann Intern Med 1990;113:740–746
21. Ippolito G, Puro V, De Carli G, et al. The risk of occupational human immunodeficiency virus infection in health care workers: Italian multicenter study. Arch Intern Med 1993;153:1451–1458
22. Woolwine J, Mast S, Gerberding JL. Factors influencing needlestick infectivity and decontamination efficacy: an ex vivo model [abstract 1188]. Presented at the 32nd Interscience Conference on Antimicrobial Agents and Chemotherapy, Anaheim, 1992
23. Shirazian D, Herzlich BC, Mokhtarian F, Spatoliatore G, Grob D. Needlestick injury: blood cells and acquired immunodeficiency syndrome. Am J Infect Control 1992;20:133–137
24. Bell DM. Human immunodeficiency virus transmission in health care settings: risk and risk reduction. Am J Med 1991;91(suppl 3B):294S–300S
25. Gruninger SE, Siew C, Chang SB, et al. Human immunodeficiency virus type I infection among dentists. J Am Dent Assoc 1992;123:57–64
26. Centers for Disease Control. Recommendations for prevention of HIV transmission in health care settings. MMWR 1987;36(suppl 2S):3S–18S
27. Centers for Disease Control. Update: universal precautions for prevention of transmission of human immunodeficiency virus, hepatitis B virus, and other bloodborne pathogens in health-care settings. MMWR 1988;37:377–388
28. Jagger J, Hunt EH, Brand-Elnaggar J, Pearson RD. Rates of needlestick injury caused by various devices in a university hospital. N Engl J Med 1988;319:284
29. Marcus R, CDC Cooperative Needlestick Surveillance Group. Surveillance of health care workers exposed to blood from patients infected with the human immunodeficiency virus. N Engl J Med 1988;319:1118–1123
30. Gerberding JL. Does knowledge of HIV infection decrease the frequency of occupational exposures to blood? Am J Med 1991;91(suppl 3B):308S
31. Gerberding JL, Littell C, Tarkington A, Brown A, Schecter W. Risk of exposure of surgical personnel to patients' blood during surgery at San Francisco General Hospital. N Engl J Med 1990;322:1788–1793
32. Tokars JI, Bell DM, Culver DH, et al. Percutaneous injuries during surgical procedures. JAMA 1992;267:2899–2904
33. Panlilio AL, Foy DR, Edwards JR, Gerberding JL, Schecter WP. Blood contacts during surgical procedures. JAMA 1991;265:1533–1537
34. Panlilio AL, Welch BA, Bell DM, et al. Blood and amniotic fluid contact sustained by obstetrical personnel during deliveries. Am J Obstet Gynecol 1992;167:703–708

35. Harris RL, Boisaubin EV, Salver PD, Semands DF. Evaluation of a hospital admission HIV antibody voluntary screening program. Infect Control Hosp Epidemiol 1990;11:628–634
36. Hobratsch LV, Hurley DL. Preoperative HIV screening in a low prevalence population [abstract 287]. Presented at the 29th Interscience Conference on Antimicrobial Agents and Chemotherapy. Houston, September 1989
37. Centers for Disease Control and Prevention. Recommendations for HIV testing services for inpatients and outpatients in acute-care hospital settings. MMWR 1993;42(RR-2):1–6
38. Ciesielski C, Marianos D, Ou CY, et al. Transmission of human immunodeficiency virus in a dental practice. Ann Intern Med 1992;116:798–805
39. Ou CY, Ciesielski CA, Myers G, et al. Molecular epidemiology of HIV transmission in a dental practice. Science 1992;256:1165–1171
40. Centers for Disease Control and Prevention. Update: investigations of persons treated by HIV-infected health care workers—United States. MMWR 1993; 42:329–331, 337
41. Ho DD, Moudgil T, Alam M. Quantitation of human immunodeficiency virus type 1 in the blood of infected persons. N Engl J Med 1989;321:1621–1625
42. Gooch B, Marianos D, Ciesielski C, et al. Lack of evidence for patient-to-patient transmission of HIV in a dental practice. J Am Dent Assoc 1993; 124:38–44
43. Bell DM, Shapiro CN, Gooch BF. Preventing HIV transmission to patients during invasive procedures. J Public Health Dent 1993;53(3):170–173
44. Robert L, Chamberland M, Marcus R, et al. Update: epidemiologic and laboratory investigations to evaluate the risk of HIV transmission for HIV-infected health-care workers (HCWs) [abstract 57]. Presented at the 33rd Interscience Conference on Antimicrobial Agents and Chemotherapy. New Orleans, October 1993
45. Chaisson RE, Allain JP, Leuther M, Volberding PA. Significant changes in HIV antigen level in the serum of patients treated with azidothymidine. N Engl J Med 1986;315:1610–1611
46. Bell DM, Shapiro CN, Culver DH, et al. Risk of hepatitis B and human immunodeficiency virus transmission from an infected surgeon due to percutaneous injury during an invasive procedure: estimates based on a model. Infect Agents Dis 1992;1:263–269
47. Centers for Disease Control. Recommendations for preventing transmission of human immunodeficiency virus and hepatitis B virus to patients during exposure-prone invasive procedures. MMWR 1991;40(RR-8):1–9

Glossary

Acquired immunodeficiency syndrome (AIDS) Set of serious clinical ailments (including numerous opportunistic infections and neoplasms) resulting from severe immune dysfunction due to infection with the human immunodeficiency virus (HIV). In addition, a CD4 T-helper lymphocyte count below 200/mm^3 in the presence of HIV infection constitutes an AIDS diagnosis.

Adult T cell leukemia/lymphoma (ATLL) Malignant proliferation of mature T cells associated with infiltrative lesions of the skin and viscera, lytic bone lesions, and hypercalcemia.

AIDS-related complex (ARC) Term not officially defined or recognized by the Centers for Disease Control and Prevention (CDC). ARC is meant to describe an assortment of chronic symptoms and physical findings that are found in persons infected with HIV but do not meet the CDC definition of AIDS. Included in the symptoms are lymphadenopathy, chronic diarrhea, unintentional weight loss, lethargy, recurrent fevers, oral thrush, and certain changes in the patient's immune system. ARC may or may not develop into AIDS.

Antibody Complex set of proteins (immunoglobulins) found in the blood and produced by B cells in response to exposure to specific foreign molecules or antigens. Antibodies have the ability to combine with the specific antigen that stimulated antibody production. At present, five classes of antibody are distinguishable: immunoglobulins G (IgG), M (IgM), A (IgA), D (IgD), and E (IgE). Antibodies can neutralize toxins and microorganisms, and interact with various elements of the immune system to eliminate infectious organisms from the body.

Antibody-dependent cell-mediated cytoxicity (ADCC) Immune system response in which antibodies attach or bind to target cells, identifying them for attack by other components of the immune system.

Antigen Substance that when introduced into the body of a human or other animal stimulates the production of specific antibodies or T cell responses to that antigen.

Antigenemia Presence of antigen in the blood.

Asymptomatic Without symptoms.

Autoradiography Production of an image on an x-ray film by a radioactively labeled substance.

B lymphocyte (B cell) Type of lymphocyte (white blood cell) that, in response to stimulation initiated by an antigen or pathogen, is transformed into a plasma cell, which produces antibodies against the specific antigen.

BSL-4 Level of containment required for safe handling of the most contagious pathogenic microorganisms.

CD4 Protein found on the cell surface of helper-T lymphocytes and to a lesser degree on the surface of monocyte/macrophage cells, Langerhans cells, and dendritic cells. HIV infects these cells by first attaching to the CD4 protein (also known as the CD4 receptor).

CD4 lymphocyte (T4 cell) T lymphocyte that expresses the cell-surface marker molecule CD4. These cells are believed to consist mainly of helper/inducer lymphocytes, which secrete many soluble molecules and play a significant role in the regulation of the human immune system. These cells are the main target for infection with HIV.

CD8 Protein found on the surface of suppressor T lymphocytes.

CD8 lymphocyte (CD8 T cell) T lymphocyte that expresses the cell-surface marker molecule CD8. These cells are believed mainly to be suppressor/cytotoxic lymphocytes, which play significant functional and regulatory roles in the human immune system.

Cell-mediated immunity Defense mechanism of the immune system involving the coordinated activity of at least two groups of T lymphocytes, namely, the helper T cells and killer T cells. The helper cell produces various substances that regulate and stimulate other cellular components of the immune system. The killer T lymphocyte's role is to destroy cells in the body bearing foreign antigens (e.g., cells infected with viruses).

Chromosome Rod-like structure found in the cell nucleus that contains the genes of that cell. Chromosomes are composed of DNA and proteins. They can be seen by light microscopy during certain stages of cell division.

Cohort In epidemiology, a group of persons with some characteristics in common.

Core antigens Protein antigens that constitute the internal structure or core of the virus.

Cytomegalovirus (CMV) Member of the herpes group of viruses that rarely causes disease in healthy adults. CMV is known to cause a severe congenital infection of infants and life-threatening infections in patients who require immune suppression. In AIDS patients, however, CMV may result in pneumonia or inflammation of the retina, liver, colon, and kidney.

Cytopathic Ability to induce pathologic changes to cells.

Cytotoxic T lymphocyte (CTL) Lymphocyte capable of killing foreign cells that have been marked for destruction by the cellular immune system.

DNA (deoxyribonucleic acid) Substance of heredity. This large nucleic acid, found mainly in the chromosomes of the nucleus of living cells, is responsible for transmitting hereditary characteristics of an organism. DNA is composed of the sugar deoxyribose, phosphate, and the bases adenine, thymine, guanine, and cytosine.

DNA denaturation Separation of DNA into its two strands of nucleotides (e.g., by exposing it to near-boiling temperatures).

DNA probe Specific sequence of single-stranded DNA used to seek out a complementary sequence in other single strands of DNA or RNA. The probe is usually tagged with a radioactive molecule or various nonradioactive molecules so it can be detected.

DNA sequencing Process of determining the nucleotide sequence of DNA.

EIA or ELISA Acronyms for enzyme immunoassay or enzyme-linked immunosorbent assay. The terms are used interchangeably and represent a test used to detect antibodies against HIV in blood samples. An assay based on antigen–antibody interactions, it uses enzymes attached to an antibody to measure the reaction.

Electrophoresis Method of separating molecules, such as DNA fragments or proteins, by using an electrical field to make them move through a medium (e.g., agarose or polyacrylamide gel) at rates that correspond to their electrical charge and size.

Encephalitis Inflammation of the brain.
Encephalopathy Any of a variety of degenerative diseases of the brain.
Endemic Regarding diseases associated with particular locales or population groups.
env Represents the gene for, or the structural proteins of, the envelope of retroviruses.
Envelope antigens Proteins that comprise the envelope or surface of a virus.
Epidemic Circumstance where a disease spreads rapidly through a community in which that disease is normally not present or is present at low prevalence.
Epstein-Barr virus Member of the herpes family of viruses. It is believed to be the etiologic agent of mononucleosis in young adults and has been implicated in the development of Burkitt's lymphoma in Africa.
Etiologic agent Organism that causes a disease.
Etiology Cause or origin of a disease.

False negative Negative test result for a sample from a person who is truly positive for the condition. The patient is incorrectly diagnosed as not having a particular disease or characteristic.
False positive Positive test result for a sample from a person who is truly negative for the condition. The person is incorrectly diagnosed as having a particular disease or characteristic.

gag Literally means group-specific antigen. Represents the gene for the structural proteins of the core of retroviruses.
Gene Unit of heredity; a segment of the DNA molecule containing the code for a specific function.
Gene expression Manifestation of the genetic material of an organism as specific traits. Specific gene products are expressed as proteins.
Genome Full genetic or gene complement of an organism.
Gold standard In biomedical testing it is the independent test that unequivocally verifies the presence or absence of the condition for which the test is run.

Helper/suppressor ratio Ratio of CD4 T-helper lymphocytes to CD8 T-suppressor lymphocytes.
Hemophilia Hereditary bleeding disorder of males, inherited through the mother, and caused by a deficiency in the ability to make one or more blood-clotting factors.
Herpes simplex Acute diseases caused by the herpes simplex viruses types 1 and 2. Painful blisters form on the skin and mucous membranes, especially on the borders of the lip (type 1) or the mucous surface of the genitals (type 2).
Herpesvirus group Group of large DNA-containing viruses that include the herpes simplex viruses, varicellazoster virus (etiologic agent of chickenpox and shingles), cytomegalovirus, and Epstein-Barr virus.
Histocompatibility testing Matching of self antigens (HLA) on the tissues of a transplant donor with those of a recipient.
HIV (human immunodeficiency virus) Name selected by the International Committee on the Taxonomy of Viruses for the etiologic agent of AIDS (originally termed HTLV-III, LAV, or ARV). HIV is a member of the retrovirus family of viruses.
Human T lymphotropic virus type I (HTLV-I) First known human retrovirus, it is unrelated to HIV. It infects T lymphocytes and is associated etiologically with adult T cell leukemia/lymphoma (ATLL) and tropical spastic paraparesis (TSP), also known as HTV-I-associated myelopathy (HAM).
Human T lymphotropic virus type II (HTLV-II) Second human retrovirus to be

identified. HTLV-II is similar to HTLV-I, and there is extensive serologic cross-reactivity between proteins of the two viruses. HTLV-II is currently a virus in search of a disease.

Human leukocyte antigen (HLA) Human protein markers of self used for histocompatibility testing.

Humoral immunity Part of the human immune defense mechanism that involves the production of antibodies and associated components present mainly in body fluids such as serum and lymph.

Hybridization Coming together of single strands of nucleic acids so they adhere (owing to hydrogen bonding) and form a double strand. The technique of hybridization is used in conjunction with probes to detect the presence or absence of specific complementary sequences of nucleic acids.

Immune system System of defense mechanisms of the body in which specialized cells and proteins (including antibodies) in the blood and other body fluids work together to eliminate disease-producing microorganisms and other foreign substances.

Immunoglobulin Group of serum proteins with antibody activity.

Incidence In epidemiology, the number of new cases of a disease that occur in a defined population within a specified time period (i.e., the rate of occurrence).

IND (Investigational New Drug) Prior to human testing, a drug or diagnostic test sponsor must file an IND application with the Food and Drug Administration (FDA) to evaluate the consistency of product manufacturing, safety and efficacy based on preclinical data, suitable credentials of the investigators, adequacy of the clinical study design, and approval of the study by the Institutional Review Board (IRB). The IND becomes effective if the FDA does not disapprove the application in 30 days.

Indirect immunofluorescence assay (IFA) Assay based on the detection of antigen–antibody interactions (the antigen is expressed on an infected cell) using a second antibody tagged with a fluorescent compound (e.g., fluorescein) to detect the first antibody.

Interferons Class of proteins involved in immune function and containing inhibitory capability for certain viral infections.

Interleukin-2 (IL-2) Substance produced by T lymphocytes capable of stimulating activated T lymphocytes and possibly selected B lymphocytes to proliferate.

Interstitial pneumonia Localized acute inflammation of the lung.

In vitro Refers to those experiments conducted in an artificial environment, such as in tissue culture.

In vivo Refers to those experiments conducted in living animals or humans.

Kaposi sarcoma Tumor or cancer of the blood and/or lymphatic vessel walls. It generally appears as blue-violet to brownish skin blotches or bumps. AIDS-associated Kaposi sarcoma is much more aggressive than the earlier, rarer form of the disease found in the United States and Europe, where it occurs primarily in men over age 50 to 60 and usually of Mediterranean origin.

Lentiviruses Subfamily of retroviruses that includes HIV. These viruses produce diverse chronic diseases in their host. They produce an acute cytocidal infection followed by a slowly developing multisystem disease including encephalitis. Associated with this process is a persistent viral infection and the development of viral latency following the initial viremic phase.

Lymph nodes Small organs of the immune system that are distributed throughout the body. They filter lymph fluid, which contains all types of lymphocytes that can temporarily reside in the nodes. Foreign antigens in the lymph fluid or blood are filtered out by the lymph nodes for attack by the immune system.

Lymphadenopathy Condition in which the lymph glands or nodes are swollen.

Lymphocyte White blood cell that is part of the immune system.

Macrophage Form of white blood cell that has the ability to ingest or phagocytize foreign particulate matter, such as bacteria. Macrophages have the further job of presenting foreign antigen to the appropriate lymphocyte subgroup, thereby activating it.

Messenger RNA (mRNA) RNA copy of the genetic information contained in DNA and used to direct the synthesis of specific proteins outside the nucleus.

Mitogen Substance that induces cell division.

Monoclonal antibodies (mAbs) Antibodies derived from a single cell clone of antibody cells. mAbs are a homogeneous population of antibodies that recognize only one type of antigen.

Monocyte Phagocytic white blood cell that develops into a macrophage and engulfs and destroys bacteria and other disease-causing microorganisms.

Mycobacterium avium-intracellulare Bacterium related to the organism that causes tuberculosis in humans but that rarely caused disease in humans prior to AIDS.

Nosocomial infection Hospital-acquired infection; an infection not present or incubating in a patient prior to admittance to the hospital.

Nucleic acids DNA and RNA, the molecules that carry genetic information.

Nucleotides Building block of DNA or RNA. It is composed of one base, one phosphate, and one sugar molecule (deoxyribose in DNA, ribose in RNA).

Oligonucleotide Short string of nucleotides.

Oligonucleotide probe Short DNA sequence synthesized from a known gene sequence.

Opportunistic infection Infection caused by an organism that ordinarily does not cause disease but under circumstances such as impaired immunity becomes pathogenic.

Pandemic Epidemic that occurs worldwide.

Pathogen Microorganism that causes disease.

Pathogenic Capable of causing disease.

Peripheral blood mononuclear cells (PBMCs) Group of white blood cells found in the peripheral blood compartment consisting of monocytes, large granular lymphocytes, B cells, and T cells.

Persistent generalized lymphadenopathy (PGL) Condition characterized by persistent, generalized swollen lymph glands.

Phagocyte Blood or tissue cell that binds to, engulfs, and destroys microorganisms, damaged cells, and foreign particles.

Plasma Liquid portion of the blood in which the particulate components are suspended.

***Pneumocystits carinii* pneumonia** Pneumonia caused by the parasite *Pneumocystis carinii*; it is the most common life-threatening opportunistic infection diagnosed in AIDS patients.

pol Polymerase gene of retroviruses.

Polymerase chain reaction (PCR) Technique originally developed for the selective in vitro enzymatic amplification of targeted DNA sequences by 10^6 times or more.

Polymorphism Single gene trait that exists in two or more forms (e.g., HLA class II antigen).

Predictive value Proportion of individuals with positive test results who have the condition for which the test is being run.

Prevalence In epidemiology, the total number of cases of a disease in existence at a specific time and within a well defined area; the percentage of a population affected by a particular disease at a given time.

Provirus DNA copy of the genetic information of a retrovirus that can be integrated into the DNA of the infected cell. Copies of the provirus are passed on to each of the infected cell's daughter cells.

Radioimmunoprecipitation assay (RIPA) Assay based on precipitation of an antigen–antibody complex. One component of the complex, usually the antigen, is radioactive so as to be able to measure the precipitated complex.

Restriction enzyme (restriction endonuclease) Enzyme that recognizes a specific nucleotide sequence (usually four to six nucleotides) in double-stranded DNA. The enzyme cuts both strands of the DNA molecule at every place where this sequence appears.

Retrovirus Family of viruses that contain their genetic information in the form of RNA and that have the ability to copy the RNA into DNA (using the viral enzyme reverse transcriptase) inside an infected cell. The resulting DNA is incorporated into the genetic structure of the cell in the form of a provirus.

Reverse transcriptase Enzyme possessed by all retroviruses that allows them to produce a DNA copy (provirus) of their RNA genetic information. This action is an early step in the virus's natural life cycle.

RNA (ribonucleic acid) Ribose-containing nucleic acid associated with the control of chemical activities inside a cell. It can also be the form in which various viruses contain their genetic information.

SDS-PAGE (sodium dodecyl sulfate-polyacrylamide gel electrophoresis) Technique for the separation of molecules (principally proteins) by their differential migration through a gel (composed of a polymer of acrylamide termed polyacrylamide) according to the size of the molecules in an electrical field.

Sensitivity In serologic testing, the percentage of individuals who test positive in a particular test and who in fact do have the condition for which the test is run.

Seroconversion Initial development of detectable antibodies specific to a particular antigen; the change of a serologic test result from negative to positive as a result of antibodies induced by the introduction antigens or microorganisms into the host.

Seronegative Condition in which antibodies to a specific antigen are not found in the blood, such as antibodies to HIV.

Seropositive Condition in which antibodies to a specific antigen are found in the blood, such as antibodies to HIV.

Serum Clear liquid that separates from the blood when it is allowed to clot completely. It is blood plasma from which fibrinogen has been removed during the process of clotting.

SIV (simian immunodeficiency virus) Virus of subhuman primates (e.g., sooty mangabey monkeys) that can cause immunodeficiency and AIDS and that is closely related to the human immunodeficiency virus type 2 (HIV-2).

Specificity In serologic testing, the percentage of people who test negative in a particular test and who in fact do not have the condition for which the test is run.

T lymphocyte (T cell) Cell that matures in the thymus gland. These cells are found primarily in the blood, lymph, and lymphoid organs. Subsets of T cells have a wide variety of specialized functions within the immune system.

T4 lymphocyte (T4 cell) Synonym for CD4 T-helper/inducer cell.

T4/T8 cell ratios Ratio of T4 cells (helper cells) to T8 cells (suppressor cells). Persons with AIDS have a deficiency of T4 cells and a low T4/T8 ratio.

T8 lymphocyte (T8 cell) Synonym for CD8 T-suppressor cell.

Tropical spastic paraparesis (TSP) Degenerative neurologic disease characterized by weakness in the lower extremities, sensory disturbances, and urinary incontinence.

Viremia Presence of virus in circulating blood; it implies active viral replication.

Virion Complete virus particle.

Western blot assay Test that involves identification of an individual's antibodies against specific proteins of the virus. Western blot test is at least as specific as the EIA and is the most widely used confirmatory test on samples found to be repeatedly reactive on EIAs.

Zidovudine (AZT) HIV reverse transcriptase inhibitor 3′-azido-3′-deoxythymidine. It is licensed by the U.S. Food and Drug Administration for the treatment of HIV infection and AIDS. The drug is known to inhibit the multiplication of HIV.

Index